The Ethics of Research Biobanking

Jan Helge Solbakk • Søren Holm • Bjørn Hofmann
Editors

The Ethics of Research Biobanking

Editors

Dr. Jan Helge Solbakk
University of Oslo
Fac. Medicine
Section for Medical Ethics
0318 Oslo
Box 1130 Blindern
Norway
j.h.solbakk@medisin.uio.no

and

University of Bergen
Centre for International Health
5009 Bergen
Box 7804
Norway

Dr. Bjørn Hofmann
University College of Gjøvik
Department of Health, Care and Nursing
Section for Radiography and Health
Technology
bjoern.hofmann@hig.no

and

University of Oslo
Fac. Medicine Section for Medical Ethics
0318 Oslo
Box 1130 Blindern
Norway
b.m.hofmann@medisin.uio.no

Dr. Søren Holm
Cardiff University
Cardiff Law School
Museum Avenue
Cardiff
United Kingdom CF10 3AX
holms@cardiff.ac.uk

ISBN 978-0-387-93871-4 e-ISBN 978-0-387-93872-1
DOI 10.1007/978-0-387-93872-1
Springer Dordrecht Heidelberg London New York

Library of Congress Control Number: 2009927504

Printed on acid-free paper

Springer is part of Springer Science + Business Media (www.springer.com)

Preface

This study is the result of an interdisciplinary research collaboration and co-production between scholars from eight different countries (Denmark, Iceland, France, Norway, Portugal, Sweden, the UK and USA). This collaboration, which has lasted since 2004, has been made possible through generous funding from the Norwegian Research Council and the Norwegian Institute of Public Health. Besides, several of the contributors have received additional forms of funding and support from other agencies, institutions and individuals. All these are duly acknowledged in the relevant chapters.

The aim of this study has been to investigate some of the ethical, legal and social challenges raised by research biobanking in its different modern forms and formats. The ambition has been to communicate the results of this endeavour in such a form that it may reach relevant academic and professional audiences (e.g. biobank curators, biobank researchers, ethicists, gene-epidemiologists, health law experts, philosophers, social scientists and advanced and graduate students in the relevant disciplines) as well as health and research regulators, ministries, politicians and the general public.

The editing of this book has been greatly facilitated by the excellent working conditions offered to the editors by the Brocher Foundation in Geneva. Finally, the editors would like to express their gratitude to three particular individuals: senior advisor Helge Rynning of the Norwegian Research Council for his gentle, open-minded and continuous support, research assistent Isabelle Budin Ljøsne for making it technically and logistically possible for the authors behind this book to regularly meet and collaborate and associate professor Anastasia Maravela-Solbakk for her willingness to proofread the whole manuscript.

Oslo, Norway — Jan Helge Solbakk
Cardiff, UK — Søren Holm
Oslo, Norway — Bjørn Hofmann

Contents

Contributors

Vilhjálmur Árnason is professor of philosophy and chair of the Centre for Ethics at the University of Iceland. He works mainly in the fields of moral theory, bioethics and political philosophy. He is a co-editor of *The Ethics and Governance of Human Genetic Databases. European Perspectives*, Cambridge University Press, 2007.

Anne Cambon-Thomsen, MD is director of research in CNRS (French National Centre for Scientific Research). She is an immunogeneticist who presently leads a multidisciplinary team on genomics and public health in the context of research in epidemiology and public health at INSERM (National Institute for Health and Medical Research) at the Faculty of Medicine of University Paul Sabatier, Toulouse, France. She is involved in several European projects in transplantation medicine, genomics sciences, public health and biobanks. She is a member of several international scientific advisory boards as well as of the European Group on ethics in science and new technologies (EGE).

Arthur Caplan is the Emmanuel and Robert Hart Professor of Bioethics and the director of the Center for Bioethics at the University of Pennsylvania in Philadelphia. His most recent book is the *Penn Guide to Bioethics* (Springer, 2009) which he co-edited with Autumn Fiester and Vardit Ravitsky.

Erik Christensen is a Ph.D. research fellow in bioethics at the Department of Philosophy and Bioethics Research Group, The Norwegian University of Science and Technology.

Paula Lobato de Faria, Ph.D. is associate professor of Health Law and BioLaw at the National School of Public Health of the Universidade Nova de Lisboa (New University of Lisbon). She has authored a large number of papers published both in Portuguese and in international scientific journals and she collaborates in several international projects in the fields of biobanks and genetics.

Lynn G. Dressler, Dr.P.H. is assistant professor in the Eshelman School of Pharmacy at the University of North Carolina (UNC) at Chapel Hill. She serves as associate director for ethics and policy for the UNC Institute for Pharmacogenomics and Individualized Therapy and is co-director of the Research Ethics Consultation and Education Service of the UNC Center for Genomics and Society. Dressler's background spans translational research in breast cancer, health policy and bioethics.

Her research focuses on research ethics in genetics/genomics and the process of translating genetic/genomic findings into clinical practice.

Pascal Ducournau is a sociologist and senior lecturer at the J.F. Champollion University and University of Toulouse. He is a researcher at the Unit 558 of the INSERM – National Institute of Health and Medical Research (team "Genomics and Public Health: Interdisciplinary Approach"), and at the LISST (a team of the CNRS – French National Center for Scientific Research). He is a specialist on issues related to the ethics of public health, biomedical research and biomedical innovation.

Anne Hambro Alnæs is a medical anthropologist, educated at Oslo University. In 2001 she defended her doctoral thesis "Minding Matter. Medical Modernity's Difficult Decisions" in which she analysed the cultural and ethical quandaries of necro organ donation. She is currently involved in several EU-sponsored research projects focusing on various aspects of transplantation medicine and was awarded a 3-year post-doctorate research fellowship in 2008. Anne Hambro Alnæs is affiliated to the Section for Medical Ethics, Faculty of Medicine, University of Oslo.

Klaus Hoeyer (M.A. Anthropology, Ph.D. Medical Ethics) is associate professor of organizational analysis in the Department of Public Health, University of Copenhagen. His research interests primarily revolve around the social and ethical implications of biobanking for research and therapeutic purposes, and he draws on a wide set of methodologies from the social sciences and the humanities.

Bjørn Hofmann is professor at the University College of Gjøvik and adjunct professor at Section for Medical Ethics, Faculty of Medicine, University of Oslo. He is trained both in the natural sciences (electronics, physics and biomedical engineering) and in the humanities (philosophy and history of ideas). He holds a Ph.D. in philosophy of medicine from the University of Oslo (2002). His fields of research are philosophy of science, medical ethics, science and technology studies and health technology assessment.

Søren Holm is a professorial fellow in bioethics and director of the Cardiff Centre for Ethics, Law and Society at the Cardiff University, UK, as well as an adjunct professor of medical ethics at Section for Medical Ethics, Faculty of Medicine, University of Oslo. He holds degrees in medicine, philosophy and health care ethics, and two doctoral degrees in medical ethics. He is the editor in chief of the *Journal of Medical Ethics* and has written extensively on issues in research ethics.

Jan Reinert Karlsen is a research fellow at Section for Medical Ethics, Faculty of Medicine, University of Oslo. Karlsen studied philosophy at the University of Bergen and holds the degree of Cand. Philol. He is currently completing a Ph.D. thesis on the biopolitics of human research biobanks, while working as a researcher at the EU-funded project, GeneBanC.

Elisabeth Rynning, LL.D. is professor of Medical Law at the Department of Law, Uppsala University (www.jur.uu.se). She is the leader of the Nordic Network for Research in Biomedical Law. She has been a longstanding member of the Swedish National Council on Medical Ethics and she has served on a number of law commissions in the area of biomedicine. Her research interests include autonomy and privacy in healthcare and biomedical research, as well as various other medico-legal issues.

Anne Maria Skrikerud is a philosopher with the degree of Cand. Philol. from the University of Oslo. She completed her Cand. Philol. thesis in 2003 on the ethics of embryonic stem cell research. Currently, she is completing a Ph.D. thesis on information, privacy and risk in biobank research at Section for Medical Ethics, Faculty of Medicine, University of Oslo.

Jan Helge Solbakk is trained as a physician and a theologian and he holds a Ph.D. in ancient Greek philosophy as well. He is professor of medical ethics at Section for Medical Ethics, Faculty of Medicine, University of Oslo. He is also adjunct professor of medical ethics and philosophy of medicine at Centre for International Health, Faculty of Medicine, University of Bergen. In the period of February 2007 to August 2008 Solbakk served as the chief of bioethics at the UNESCO Headquarters in Paris. He is/has been a member of several national and international expert committees and working parties dealing with research ethics, biotechnology, human genetics and stem cell research and he has written extensively on these issues.

Berge Solberg is associate professor at the Department of Social Work and Health Science at The Norwegian University of Science and Technology (NTNU). He is trained as a philosopher. From 2004, Solberg coordinated an ethics project on biobanks entitled "Consentenced to contribute to the common good?". Currently, he is in charge of the project "In genes we trust? Biobanks, commercialization and everyday life". Solberg is a member of the Norwegian Biotechnology Advisory Board.

Roger Strand, dr. scient. in biochemistry, is professor and director at Centre for the Study of the Sciences and the Humanities, University of Bergen, Norway. He is a member of the Norwegian National Committee for Research Ethics in Science and Technology. His main research interests relate to the significance of scientific uncertainty and complexity in the science–policy interface.

Lars Ursin is a post-doctoral fellow in philosophy at the Philosophy Department and Bioethics Research Group, the Norwegian University of Technology and Science, Norway.

Laurie Zoloth is professor of medical humanities, bioethics, Jewish studies and religion. She is also the director of the Northwestern University Center for Bioethics, Science and Society, and director of the Brady Program for Ethics and Civic Life, in the Weinberg College of Arts and Sciences. She is the former president of the American Society for Bioethics and Humanities and she is currently the chair of the Howard Hughes Bioethics Advisory Board. She has also served on the NASA National Advisory Council. She received an M.A. from San Francisco State University and a second M.A. and Ph.D. from the Graduate Theological Union.

Introduction

Jan Helge Solbakk, Søren Holm, and Bjørn Hofmann

Why is it that we talk about collections of biological materials or samples as "biobanks"? What is the rationale behind using the word *bank* to name these institutions? And to what extent is it justifiable to frame these institutions within a vocabulary of hard currency, i.e., of economical values? These were some of the queries that started off this project. The etymology behind the word "bank" may give us some initial clues to address these questions: The word originates from the ancient Greek word *trapeza*, which, literally speaking, means a "four-footed table," from *tra* (akin to *tettares* four) + *peza* foot. In modern Greek, this is still the word used for bank. Similarly, the Italian word "banco" or "banca" and the French word "banque" refer to the money-changers and lender's exchange table or counter. In Renaissance Italy, these were benches and/or counters located in *public* places, from which the money-changers and lenders used to operate and on which they would display their material in terms of different currencies (Rochet 2002, Liddell et al. 1940).

We will probably never get to know exactly *why* the analogy of table came into use in relation to the activity of money-changers and lenders. What made this analogy attractive may, however, be somewhat easier to grasp. Our suggestion is that it functioned as a simple analogy to characterize the dynamics of change and exchange – over a table – that took place between different stakeholders, i.e., the money-changers and lenders and their different customers.

We believe that this suggestion may also be of help in explaining why the analogy of bank has come into such abundant use in relation to collections of biological material or samples. It captures in a succinct way the ethical core activities of these institutions, i.e., the different activities of exchange and change that take place between donors of biological samples and guardians or curators of such samples (Editorial 2007) as well as between bio-guardians/curators and users of biological samples, i.e., health professionals and researchers. Because of the last decades' advances within genetic and gene-epidemiological as well as genomic research, stem cell research and regenerative medicine, the amount of such activities of exchange and change have increased enormously. Besides, a whole range of new forms of exchange and change of biological samples have seen the day, while at the same time old collections of diagnostic samples stored for decades in dusty hospital cellars have gained new life and attention due to the interest from different research communities, in particular genetics, gene-epidemiology and genomics research communities. At the same time in many countries, new and huge – public as well as private – collections of biological samples and health information have been – or are in the process of being – established to advance and encourage further research within these areas. Such forms of research may not only be important to develop new diagnostic tests and tools for gene-related diseases or unveil the possible causes and origins of gene-related forms of disease in a sample population; allegedly such forms of research may also give rise to better and more efficient methods of treatment, including drugs aimed at prevention as well as therapy. Interesting to observe is also that stakeholders outside the research communities, i.e., private and public investors, have become heavily involved in this enterprise. The reason for this is evident: the great expectations raised by the economic potential of research biobanking and biobank research. This observation gives perhaps some further insight into the question why the bank analogy has gained so much conceptual and persuasive power and space within this field of research and development.

All these different activities and initiatives in relation to setting up new research biobanks, converting dusty collections of pathological samples into biobanks for research, developing new research programs and investing into the set up of such enterprises and programs have also generated a lot of attention and interest among ethicists, policymakers, and health law experts, in particular in relation to questions about the best and safest way of regulating such enterprises and activities. This book is the result of a research collaboration between ethicists, health law experts, social scientists, and gene-epidemiologists from eight different countries (Denmark, Iceland, France, Norway, Portugal, Sweden, UK, and USA). The aim of this collaboration has been to investigate some of the ethical, legal, and social challenges raised by research biobanking in its different modern forms and formats.

The ethical, legal, and social issues raised by research biobanking can be divided into four main clusters of issues:

- Issues concerning how biological materials are entered into the bank
- Issues concerning research biobanks as institutions
- Issues concerning under what conditions researchers can access materials in the bank, problems concerning ownership of biological materials and of intellectual property arising from such materials
- Issues related to the information collected and stored, e.g., access-rights, disclosure, confidentiality, data security, and data protection

The first cluster of issues has been much discussed. Relevant problems are, for instance what kind of consent should be given by persons who give material to a research biobank, under what conditions can material in diagnostic or therapeutic biobanks be "converted" into research materials, and under what conditions can materials obtained without consent or against the will of the "donor" be "converted" into research materials? Other problems in this cluster of issues concern exactly what rights the donor gives to the bank and what rights the donor retains, questions about incentives for giving to, grounds for withdrawal from the bank as well as renewed consent from children with stored tissues when they reach the age of legal maturity.

The second cluster of issues is concerned with the biobank as an institution. What kind of institution is it? Under what conditions can it be sold, merged with other biobanks, exported, divided, or destroyed? These issues are much less discussed in the literature, but may be of importance for the two other clusters of issues (is the distinction between public and private biobanks for instance important when regulating consent procedures?)

The third cluster of issues raises questions concerning research ethics governance of the use of stored biological materials as well as questions concerning how a biobank should set priorities among a number of competing research projects. This cluster is also concerned with ownership and intellectual property issues, including various modes and levels of profit sharing, if any. Thereby it also touches upon the basic question what biological material *is*.

The fourth cluster of issues concerns the long-term relations between researchers and users of the biobanks on one side and the sampled population on the other. It includes access to results on individual or global level, ways of dissemination of information about biobank use and data protection and confidentiality issues. There is a considerable interplay between the ethical and legal issues in each of the described clusters. If, for instance, relatively liberal rules are implemented concerning the entry of materials into biobanks, stricter rules concerning the use of these materials are likely to be needed and vice versa.

The book is organized in two separate parts. The first part represents an attempt to gain new knowledge about the different regulative issues implicated in the establishment of biobanks for research, conversion of old collections of pathological samples or more recent collections of therapeutic samples (blood, bone marrow, umbilical cord blood, sperm, oocytes, fertilized eggs, embryos, aborted fetuses, etc.) into such enterprises, and the development and conduct of research based on samples and health information stored in such banks. In this part of the book, we undertake an investigation along traditional pathways, i.e., we pursue the different regulatory options possible to envisage within a normative terrain dictated by different conceptions and interpretations of the informed consent doctrine.

As already alluded to, the traditional approach to the ethical and legal issues raised by research biobanking has been to extend the informed consent and other research ethics procedures that are already in place and to supplement them by measures directly transferred from the area of data protection. Informed consent was originally developed in the context of the doctor–patient relationship, and later extended to the researcher–research participant relationship but still mainly in the clinical setting. In this setting, it is normally possible to inform the potential research participant about the exact nature and purpose of the research project in considerable detail. However, in the context of research biobanking, this level of specification is hardly achievable, because the research performed on banked materials is, by nature, open ended. We cannot know how the materials stored today will be used in 20 years' time, because we have no idea what will be possible in 20 years' time. Furthermore, it is practically impossible to obtain actual informed consent for each new use of the stored materials. This problem of specification might be an indication that current consent procedures are insufficient to provide the donors of biomaterials with adequate protection of their rights. In the biobank setting, consent is required not only for a specific research procedure, but also for a transfer of some or all of the rights of control over the actual material and its use. If this is conceptualized as a "transfer of ownership," informed consent suddenly looks like a very odd procedure for such a transfer, since ownership is usually transferred by means of contracts, and based on the advice of lawyers, not on information given by medical doctors. If, on the contrary, one considers this not as ownership, but as "right of control" over its use without proprietary rights, then the direct consent or the consent to transfer this right over use to a body or another person looks more appropriate. In any case, the clarification of this issue seems to be crucial, so even if the traditional approach is problematic, it is, nevertheless, important to continue analyses along these lines, since they are the currently acknowledged legitimate basis for biobanking. In the book ten different chapters are dedicated to researching these questions within the different possible normative options offered by the traditional approach.

Although the primary objective of this book is to critically assess the traditional regulatory approach to research biobanking and develop alternative ways of conceiving of and regulating research biobanks, our main hypothesis is that besides attempting to directly extend the analysis of the traditional approach described above, it will be equally fruitful to investigate the conceptual potential of analogies from a range of areas outside medical research, i.e., analogies other than that of a table or counter where people change and exchange valuables between themselves and a common institution. The second part of this book represents an attempt to pursue this goal. The analogies we propose are not intended as templates or models to be followed, but as tools for analysis. Examples of such analogies are ordinary commercial banking, voluntary associations, clubs (e.g., book clubs) or unions, libraries, conscription, taxation, and management of art pieces. By developing these analogies, analysing their implications, and identifying their limitations we believe that it will be possible to:

- Achieve a deeper understanding of the structural arrangement of research biobanking
- Critically assess the vocabulary prevailing in the field of biobanking, in particular the labels employed and the roles, rights and duties ascribed to the different parties affected by or involved in biobanking (donor, "biobanker," researcher, research ethics committees, the sampled population)
- Make recommendations regarding different ways in which a biobank could be structured as a social institution

This part of the book, which consists of 11 chapters, aims more specifically at pursuing the following objectives: To explore a range of analogous social contexts where people transfer something to a common institution and to draw out possible implications for biobanking. This will (among many other aspects) include the following:

- Investigating to what extent different forms of incentives employed by associations, clubs, and unions to recruit new members might be of help in resolving the question about appropriate forms and uses of incentives in relation to donors of biological materials for research purposes
- Investigating the possibilities provided by the notion of currency exchange in addressing the problem of converting materials stored in already existing diagnostic banks and therapeutic banks (e.g., blood banks, sperm banks, and IVF-banks) and health registries into "bio-currency" for research purposes
- Studying whether commercial banking notions such as savings account, savings, saver or depositor, small saver, interest rates, loan, etc. might be of help in clarifying the relation between the right of ownership and the right of disposal of materials stored in biobanks and health registries for research purposes
- Addressing the question whether notions from commercial banking might be of help in clarifying the issue of commercial utilisation of research biobanks, as for example the issue whether individual donors of bio-currency for research purposes should be attributed with economical rights analogous to the rights of small savers with savings accounts in ordinary banks
- Analysing whether the notions of conscription and taxation might be of help in resolving the question whether there could be a duty to donate biological materials for research under certain circumstances
- Investigating the possibilities provided by the notion of "loan" employed in the running of libraries to address the question of researchers' rights and duties in relation to materials "borrowed" from research-biobanks
- Further exploring alternatives to individual informed consent procedures, identified through the use of analogies. In particular, critically analysing the ethical and legal justifications of employing collective information and consent-procedures when converting materials stored in diagnostic banks, therapeutic banks, and health registries into materials for research purposes
- Analysing the issue of valid consent procedures in future research projects not yet conceived and/or not sufficiently specified
- Studying the institutional context of biobanking and its implications for the legal and ethical regulation of biobanks
- Outlining different possibilities for resolution of the problems in the four clusters of issues defined above on the basis of the analysis performed

Finally, in a last chapter entitled, "In the Ruins of Babel: Should Biobank Regulations Be Harmonized?," we return to the recurrent bioethical and biopolitical question about harmonisation of biobank practices and regulations. The question is critically addressed in terms of the social and ethical robustness of such regulations. Our conclusion is that current proposals for integrating and harmonizing research biobanking in an increasingly international and industrial context are not socially and ethically robust. Rather, they amplify the inherent problems of such a challenge.

References

Liddell HG, Scott R, Jones HS. ([9]1940) *A Greek-English Lexicon.* Oxford: Clarendon Press. http://www.perseus.tufts.edu/cgi-bin/ptext?doc = Perseus%3Atext%3A1999.04.0057%3Aentry%3D%23104539)

Rochet JC (2002) 'Why are there so many Banking Crises?', Lecture at London School of Economics, http://www.lse.ac.uk/collections/LSEPublicLecturesAndEvents/pdf/20020709t0946z001.pdf)

Editorial (2007) What is the Human Variome Project? Nature Genetics 39: 423

Part I
Research Biobanking: The Traditional Approach

Consent to Biobank Research: One Size Fits All?

Bjørn Hofmann, Jan Helge Solbakk, and Søren Holm

Abstract Express informed consent has become a standard requirement in research related to human beings, so also in biobank research. However, it has been argued extensively that this approach is inappropriate for biobank research and that it seriously hampers beneficial research. This chapter analyses biobank research to see whether it has particular features that require exceptional regulation. The conclusion drawn is that biobank exceptionalism is not defensible. Nevertheless, it is acknowledged that certain types of biobank research challenge so many of the traditional approaches in research ethics that alternative approaches need to be pursued. Four alternatives to informed consent are explored: broad consent, the confidentiality/privacy approach, submission to the researcher, and conditioned authorization. Pros and cons related to all of them indicate that a contextual approach has to be taken; one size does not fit all. The question in biobank research is not "to consent or not to consent", but how to protect and promote the interests of individuals contributing to research at the same time as benefiting society and future patients.

Informed Consent, the Ideal Solution to Biobank Research?

Express informed consent has become the standard procedure for including subjects in health-care research. The reasons for this are manifold: to protect research persons[1] from abuse (curb on research risk), to assure that the person is in control

[1] In this chapter we use the term "research person" instead of "research participant" or "research subject" because, in our opinion, it is more neutral. It has no connotations of active participation or passive submission.

B. Hofmann (✉)
Section for Radiography and Health Technology, Department of Health, Care and Nursing, University College of Gjøvik, e-mail: e-mail:bjoern.hofmann@hig.no
and
Section for Medical Ethics, Faculty of Medicine, University of Oslo, Oslo, Norway
e-mail: b.m.hofmann@medisin.uio.no

J.H. Solbakk, et al. (eds.), *The Ethics of Research Biobanking*,
DOI: 10.1007/978-0-387-93872-1_1,

of issues concerning his or her health (autonomy), because making one's own decisions promotes one's well-being (beneficence), to promote rational decisions, and to maintain the public's trust in research.[2]

Consent to research on biological material appears to have a long tradition (Rosoff 1981: 254). Although the laws vary considerably from country to country, informed consent is used extensively in biobank research (Maschke 2005). One of the reasons for this is that many of the obstacles to informed consent in clinical research do not present themselves in biobank research. In clinical medicine, it can be argued, patients are suffering, are in pain and despair, and thus are not able to make autonomous decisions (Friedson 1970). In biobank research such impediments to informed consent are absent, and research persons are better able to understand the scope and consequences of participating, and they normally are free from undue influence from family members in shock or grief. In other words, informed consent appears to be particularly, not to say exemplarily, suitable in biobank research.

On the other hand, because the ordinary research person in biobank research is a healthy person in command of herself/himself, he or she is less vulnerable and needs less protection. Express informed consent would thus appear unnecessary (Helgeson et al. 2007). Moreover, empirical studies show that research persons do often not need or want to make decisions themselves (Schneider 1998; Ring and Kettis-Lindblad 2003: 204; Kettis-Lindblad et al., 2006, 2007). Hence, biobank research seems not to need informed consent, at least not in the same degree or manner as is the case with clinical research.

This leads to a paradox: in biobank research the preconditions for the protective and affirmative function of informed consent are fully met, but they are less in demand than in clinical research, where these functions are required because of the research person's vulnerability, but where the preconditions are not met. Thus, the key question becomes whether informed consent is as relevant for biobank research as it seems, or if other measures should be applied in order to fulfill the aims of science and as well as obtain appropriate moral and legal regulation. To address this question, we will investigate whether it is so that one size fits all, or whether the nature of the subject matter makes a different approach necessary.

Special Research: Special Requirements

It has been argued that biobank research raises special issues which informed consent is not equipped to deal with (Chadwick and Berg 2001: 318). For example, the risk for the research person is so low and that the potential of a valuable outcome (to others than the research persons) is so high that it is legitimate to lower

[2] In the same way as the imperative "administer no poison" in the Hippocratic Oath was not only a matter of ethical concern, but even more so a matter of gaining reputation and trust (Faden and Beauchamp 1986: 62), the institution of informed consent can alleviate social mistrust and instill confidence and trust in researchers and research organizations. In a clinical setting, where consent was first an issue, the consent had another important function, that was, to protect against malpractice lawsuits (Faden and Beauchamp 1986: 81–2).

the requirements for participation in research (Hansson et al. 2006; Helgesson et al. 2007). A routine, nonsensitive use of (excised) biological material should not require a separate consent or renewed consent. The character of this research legitimates looser regulations, e.g., in terms of waivers, broad consent, or blanket consent.

Accordingly, it has been argued that informed consent implies overprotection of human subjects, hampers research, and jeopardizes the quality of research (Wilcox et al. 1999; Greely 1999), and thus the ultimate public benefit of this endeavor (Wadman 2000).[3] Is biobank research so special that it requires special moral and legal standards? If so, what are the special characteristics of biobank research that render alternative moral and legal approaches justifiable?

Moot Material

Biobanks have an unsettled institutional status which raises ontological, moral, and legal challenges. There appears to be many reasons for this. First, the ontological, moral, and legal status of biological material remain unsettled (Rothstein 2005). Correspondingly, there are disagreements about what counts as biological material: sputum, urine, hair, skin cells, and dandruff. Although debates on the scope of research and its limits take place in many kinds of medical research, this sort of profound disputes on the very status of the subject matter of research is not very common.

Second, biological material can enter a biobank from many sources where it has different moral and legal status, such as from diagnostic biobanks, from biopsies, autopsies and from collections of historical material (skeletons, archeological material etc) (Blatt 2000). This process of transposition and conversion of the subject matter in question appears to be peculiar to biobank research.

Third, the status of the *results* of research on biological material is not clear. For example it is not clear what is the status of the information retrieved from analysis of biological material, whether it should be conceived of as a part of the biological material (as is the case in the Norwegian Biobank act) or what its relation to the biological material is when it is separate from the biological material as such. Correspondingly, the status of products or refinement of biological material remains unsettled as well (Rynning 2003).

Exceptional Information

Additionally, it has been argued that genetic information is quite distinct from other kinds of health information, in terms of predictable and discriminative power,

---

[3] It has been argued that biobank research is of a so special character that asking for renewed consent would be intrusive and an invasion on privacy (Eriksson 2003: 183). Correspondingly, it has been argued that specific consent restricts a person's autonomy, while broad (or even blanket consent) enhances autonomy (Hansson et al. 2006).

as well as in scope: the information gained from analysis of biological material is relevant for family members or population groups as much as for individuals (Mitchell and Happe 2001; Greely 2001; Clayton 2005)

Furthermore, it may be argued that relevant information in biobank research is so uncommon to most people, that to inform them properly will scare them off. For example, research on genes that entail high risk for rare diseases will distract participants from important information and unnecessarily discourage participation, although the chance of them having the gene is extremely low (Ring and Kettis-Lindblad 2003: 197–8).

Hence, biobank research may have unknown consequences, both in terms of risk for the research person involved and with respect to its epistemic outcome. Another feature of biobank research is the coupling between information from analysis of biobank material and other kinds of information, such as health information and genealogical information. This is necessary in order to gain clinically applicable knowledge (Lindberg 2003). This coupling of knowledge may also lead to challenges with regard to confidentiality and privacy. First there are practical challenges: how to handle the information so that a research person is not compromised. Second, there are procedural challenges: the combination of biological material (and some clinical information) makes it possible to trace the research person, even though the sample has been anonymized (Rynning 2003; McGuire and Gibbs 2006).

Special Risk

Contrary to clinical research, the risk for the research person in biobank research is considered to be low (Hansson and Levin 2003: 17) The reason is that the acquisition of the biological material does not in itself involve substantial risk; nor does it result in substantial additional risk with regard to procedures performed anyway. Furthermore, it has been argued that the only risk in biobank research is violating personal integrity and privacy, and this can happen only if the research also entails processing of personal data (Essen 2003: 143, 146). On the other hand, taking into account the health information that can be gained from analysis of the biological material combined with other health information, biobank research is normally considered to represent a non-negligible risk to the research person (Roche and Annas 2001). More specifically it has been argued that there is a risk of informational harm (Reilly 1998; Fuller 1999).

Special Outcome

It has also been argued that the outcome or benefit from biobank research is unknown. Furthermore, this research is nontherapeutic in the sense that its primary objective is increased knowledge. And only through further research can this knowledge be turned into therapeutic means. Hence the results are most likely beneficial to others than the research persons themselves. Consequently, it can also be argued that the benefits for the individual research persons are low (Faden and Beauchamp 1986: 12), even though the combined, overall research outcome may be considerable.

Practical Considerations: Kind of Research

It has been argued vigorously that biobank research is conceptually different from other kinds of research involving research persons, and that different principles apply in its research ethics (Eriksson 2003: 183). Epidemiological and register research are distinct from clinical research with experimental subjects, in that the interest is not the individual person, but the population as such. Hence, the individual person does not need the kind of control and protection provided for by informed consent. Epidemiological and register research have a collective goal, and should be regulated by collective means for common ends.

Moreover, different research groups may have access to the biological material in the future. Research is ever more an international affair, where collaboration, exchange of tests and test results, as well as exchange of biological material is an important premise. This extensive (internationally) collaborative aspect of biobank research makes it difficult to withdraw from research. Because material, analyzed data, and information may be stored in many different places, the deletion and/or destruction of it becomes practically challenging.

To sum up so far: biobank research is characterized by the following features:

- Unsettled definitions and status of biological material and its research products (information and products)
- Unsettled relationship to other sources of information
- Unknown consequences with respect to risks and benefits entailed, including the fact that the biological material is a source of unforeseen information about the research person

One could argue, therefore, that biobank research is so special, that it needs to be governed accordingly. Consequently, some sort of "biobank exceptionalism" seems to be warranted. It is important to notice, however, that the exceptionalist argument in principle works bidirectionally: one can argue that biobank research is so special that either exceptionally strict or especially lenient measures have to be applied in order to regulate it. Within genetic exceptionalim it is mainly argued that genetic information is so special that special measures must be taken in order to protect individuals. However, within biobank exceptionalism the major argument is that the risk in biobank research is so exceptionally low that the strict measures entailed in the informed consent doctrine cannot be justified.

New Names: Same Problems

Nevertheless, one could argue that the issues discussed above are not special to biobank research, or that they are not relevant with respect to bypassing requirements of informed consent. This urges us to investigate (a) how special these characteristics actually are for biobank research, (b) whether they reduce the relevance of informed consent, and (c) what kind of consent that is appropriate in biobank research (if at all).

Biological Material

Moral challenges due to the unknown or unsettled moral status of biological material have been extensively discussed in the literature in relation to one particular category of such material, i.e. fertilized eggs, embryonic stem cells, and fetuses on demand for abortion. With respect to abortion (on demand), informed consent is at the core of the action: the abortion is legitimized by the woman being informed and declaring that she understands the consequences of the act, that she is capable of weighing them against other concerns, that she is free from undue influence or coercion, and that she has given her consent.

The same applies to donors of fertilized eggs and embryos for stem cell research. In jurisdictions where embryonic stem cell research is allowed, informed consent is the moral precondition for allowing the egg/embryo transfer. The unknown or unsettled moral status of the subject matter in question is not necessarily a hindrance for applying the principle of informed consent to morally justify the act. So it can be argued that if informed consent can be applied in the case of abortion and embryo donation, then it is also suitable for regulating the donation, manipulation and eventual destruction of other kinds of biological material in the same way. In both cases the consent is given by the person in question (woman or donor), and the status of the biological material is unknown, that is, we know neither what an embryo is nor what a sample of cells is in terms of ownership, etc. Furthermore, issues of personal identity are at stake in both cases. In the first case the personal identity of the embryo is at stake and in the latter information about a person's ancestry and prospective life can be drawn from the sample. However, this also points to important differences in the two cases. In the case of abortion and embryo donation the moral issue is whether the biological material itself is a (potential future) person with certain rights to life and protection, whereas the moral issue in the case of biobank material is the moral (and legal) rights we have over the biological material. Hence, in the first case the moral challenges are related to issues of personhood and moral agency, whereas they are related to ontological and social issues in the latter. This could be taken to support the quest for different regulations. The moral issues related to biobank material are less challenging, and hence require less strict regulation.

Like a double-edged sword, the argument could also be used for defending having the same regulation for both fields. If informed consent can be accepted in the case of abortion and embryo donation, where the moral (and legal status) of the material is unsettled, and where the issue is much more pressing than the moral issues related to biobank samples, then informed consent could be suitable in the latter field as well.

Moreover, it can be argued that the case of biobank material is similar to cases of organ donation. Consequently, this kind of research should be governed by other research ethics principles (Eriksson 2003: 185). When you donate bone marrow or a kidney, you have restricted rights with respect to this material. The same should apply to biological material given to a research biobank. One of the reasons why this argument is not convincing is that epidemiological and register research is normally subject to informed consent, as is certainly also organ donation. Another reason is that whereas donated tissues for transplantation will be incorporated into another person's body, and cannot be removed without considerable harm to that person, biobank material is simply stored and can either be removed or removed from use without any major harmful effects.[4]

Unknown Consequences

Although it can be argued that the outcome from biobank research is of a general kind, and not likely to be relevant for the research person herself, the knowledge gained from the research may in the long run become of vital importance to the research person and/or his or her family, e.g., better therapy (pharmacogenetics), more convenient classification of diseases, and new explanations such as behavioral disorders explained by biological affinities rather than by moral defectiveness (e.g. alcoholism). However, results may also be knowledge of asymptomatic disease or predictive information, which advances dilemmas of whether the research person or her family should be informed or not, and what to do if the research person refuses to reveal knowledge of vital importance to family members. This means that knowledge resulting from biobank research may have a liberating as well as a challenging or even a suppressive potential. Improved classification of dementia can be of vital importance if it leads to more adequate treatment, but it can also be detrimental if it results in social stigmatization and discrimination due to improved predictability (Murray 1997).

What about the relevance of the principle of informed consent when the unknown consequences of biobank research are taken into account (with respect to both general knowledge and knowledge of particular importance to the research person and his or her family)? The unknown consequences represent a real challenge, as informed consent presupposes that the research person has understanding of the scope and the consequences of the research to be undertaken. One could argue that as long as the risk is low, we could accept that the benefits are unknown. Unknown

[4] It is true that cumulative removal of many samples may eventually create problems for the use of the biobank but this is a different kind and order of "harm" than the one in the organ donation case.

benefits are common in research. That is why we perform research in the first place. As informed consent is applied in other forms of health-care research involving research persons, one could argue that it is applicable to biobank research as well.

The major challenge here is of course that, although we can live with the unknown benefit, it is much harder to accept unknown risks (Article 7, §§ 17 and 22 of the Declaration of Helsinki). Hence, although the unknown consequences in terms of benefits do not hamper the use of informed consent with respect to biobank research, the unknown risks involved seem to jeopardize the possibility of there being an *informed* consent at all.

Information and Risk

It could be argued that the informational harm potential in biobank research is of a special kind, as the information is about identity, as well as hidden and future diseases. Furthermore, the analysis of biobank material may reveal information about the research person or the research person's family that we were not looking for, and which are rare in other kinds of clinical research. But this is not unique to biobank research; other forms of health activity also give rise to incidental findings. For example, investigation of the abdomen in relation to participating in clinical research on *Helicobacter pylori* may reveal in a research person cancer with metastases, and newborn screening may reveal additional information about diseases without proper treatment options. Therefore, medical information of many kinds discovered during research may represent a substantial risk of harm to a person, and whether it stems from a biobank does not seem to make the difference. Consequently, there seems to be little ground for granting genetic information with a status of exception (Murray 1997; Suter 2001; Green and Botkin 2003; Gostin and Hodge 1999; Juengst 1998, 2000).

Even the coupling of biological material (and its research results) to health information may not be that special. In clinical studies data from ECG, EEG, CT, or MRI are coupled to clinical information to address a specific clinical research question. The same data may be used to address new and quite different research questions by analyzing them in a different way (and/or couple the results with other forms of health information). Such coupling is therefore not in itself special for biobank research.

What may appear as a stronger argument in favor of a status of exception with regard to information stored in a biobank is that biological material is traceable to the originator even without coupling of other health information, as is the case in forensic medicine. Hence, biological material could be used to trace a person's identity and to connect related health information to this person. Thus, it may be that this traceability of biological material is one of the few aspects which morally speaking distinguishes biobank research from other kinds of medical research with respect to its coupling with other forms of health information. This assumption, however, does not lead us to any clearer conclusion with regard to the role and relevance of informed consent in this context, because if one knows that biological material can be used for tracing, one could argue that one could give consent to this.

Unexpected Consequences

Biological material may be used for analysis fundamentally different from what was described in the original research protocol. This appears to be of utmost importance to the research person, his or her family, or patients in general, because it may generate both benefits and risks that were not originally envisaged. Furthermore, the possibility of assessing the consequences (beneficence and risk) of biobank research may thus be significantly reduced compared to the situation in other kinds of clinical research. This, one could argue, has implications for the extent to which requirements of informed consent can be fulfilled in biobank research.

On the other hand it is argued that the requirement of informed consent should be reduced. A key characteristic of biobank research is that its unexpected results can trigger new research questions that can turn out to be very important and fruitful in terms of initiating new research. This leads to the practical problem of renewing informed consent for every new study one wants to initiate. However, this is not special for biobank research. If data from a clinical research project is used in another research project that is an offspring of the former but has a new scope, there is a long tradition in clinical research for obtaining renewed consent from the research persons. One could still argue that the practical problems with respect to this in biobank research are significantly larger compared to clinical research. The question then is whether these practical difficulties are of a kind and magnitude to legitimate principal renouncements of informed consent. What makes a difference, it appears, is the assessment of risk. If there is no risk, one could well argue that informed consent is overkill as there is nothing to protect the research person from, and that requiring informed consent in such cases is hampering research and beneficial outcomes to people (Chadwick and Berg 2001; Wilcox et al. 1999; Greely 1999; Hansson et al. 2003).

Different Kind of Research

It may be argued that biobank research is dominated by a different kind of researchers without experience of research involving human subjects. Consequently, they are not as familiar with and used to standard requirements in research ethics. Although this may be a valid argument for a practical differentiation between biobank research and other forms of health research, it would hardly be an argument that could invalidate the principle of informed consent in biobank research. Another argument that could possibly be used to question the validity of informed consent in this context is that consent forms have been used to protect commercial interests more than research persons (Andrews 2005). This is, however, hardly unique for biobank research; also in other forms of health research with strong commercial interests informed consent forms may be abused in the same way. What then about the argument that the biobank field is special because not only have patient groups organized themselves but they also have moved into the biobanking enterprise themselves by initiating biobanks involved in research on the kind of diseases they are themselves afflicted by? (e.g. the CFC Biobank, http://www.cfcsyndrome.org/biobank.shtml). It is difficult to see how these aspects in themselves could be used to question the validity of informed consent in biobank research.

Hence, there appears to be no valid argument in favor of biobank exceptionalism. However, although there are no aspects of biobank research that are substantially different from other forms of medical and health-related research, one could still argue that in sum biobank research is quite different, because it features so many special challenges that, although they exist in other kinds of research as well, they either are not that visible or do not present themselves in that combination or magnitude. Were this the case, one would still have to indicate what makes these combinations of features and their related magnitudes so special, that it would bring into question the relevance and validity of the principle of informed consent in biobank research.

Alternatives to Informed Consent in Biobank Research

Where does this leave us with respect to regulating biobank research? If there is little or no principal difference between biobank research and other forms of research, one could argue that informed consent could be used in biobank research in the same way as in for example clinical research. The practical and theoretical challenges will not be worse with respect to biobank research than in other forms of (complex) health-care research. Hence, if informed consent can be applied in other complex forms of health-care research, it could and should be applied in biobank research as well.

However, demanding express informed consent does not address the practical challenges of acquiring consent in the biobank context, e.g., in acquiring renewed consent: dead donors, large drop-out rates, expenses, and the potential of upsetting the research persons by recontacting them (Hansson and Levin 2003: 13; Helgesson 2003: 162–3; Eriksson 2003: 180; Greely 1999: 740; Hansson et al. 2006). Moreover, several empirical studies indicate that research persons do not think informed consent is necessary for genetic research (Wendler and Emanuel 2002) and that those who have given their informed consent, do not appear to remember crucial information about the study nor the fact that they had consented (Moutel et al. 2001; Wendler et al. 2002). Finally, people appear generally to be prone to delegate decisions about the use of their samples to the research ethics committees (Ring and Kettis-Lindblad 2003: 204; Kettis-Lindblad et al. 2006, 2007).

Undetermined Consent

Many of the fundamental and practical challenges raised by express consent have been sought resolved by developing a whole range of different conceptions of undetermined consent, such as implied (or implicit) consent, presumed consent, and hypothetical consent, or "broad consent", "future consent" or "blanket consent" (open or open-ended consent) (Hansson et al. 2006), or letting people waive consent. The prevailing premises have been that the risks are small, the potential benefits large, the research results will benefit the patient group (or people with the same genetic makeup) in general, and coupling of research results and particular research persons is not necessary.

All these conceptions of undetermined consent generate some theoretical challenges, e.g., whether one can be autonomous not knowing facts that are prerequisite to making autonomous decisions.[5] Against this one could argue that detailed knowledge about the type of research in question is not morally relevant for giving consent to research. What is morally relevant is to know that the storage and use of biological material involves a certain risk of harm and that it may violate the research person's integrity (Helgesson 2003: 161; Wendler 2002: 49–50). To be able to assess this potential harm and violation of integrity, specific knowledge about the research project seems, however, necessary.[6] Besides, using the term *consent* in situations that do not comply with the standard requirements for consent, such as understanding the scope and consequences of research, does undermine the concept of consent.[7] It could therefore be argued that the eagerness to stick to the term "consent" even though the premises for consent are not met, is much less because of the interest of protecting research persons, than covering moral challenges by means of a device.[8] This changes the objective of the protective aspirations of consent from the research person to the researcher.

Finally, neither informed consent nor "broad consent" or "waived consent" addresses the challenges discussed above, in particular not the challenge of risk assessment (Maschke 2006). If understanding the scope and consequences of research is a sine qua non for consent, then consent to biobank research of a general and unspecified kind cannot be obtained, neither of a narrow brand nor of a broad one. But if there can be no informed consent because the research person cannot

---

[5] It is argued that to abstain from autonomy is an acceptable option to the autonomous agent (Dworkin 1988: 118). This would make biobank research acceptable within the framework of waived consent. However, this is controversial (Harris and Keywood 2001).

[6] Furthermore, it breaches with a long tradition of research ethics that a person involved in medical research should be informed about the aims and methods of the research (Declaration of Helsinki, article 22).

[7] It is interesting to note that even those who argue that biobank research is not research on human beings, and thus do not fall under the Declaration of Helsinki, recommend the use of consent in biobank research (Helgesson 2003: 160).

[8] See also chapter "Users and uses of the biopolitics of consent: a study of DNA banks", and chapter "Trust, Distrust and Co-Production: The Relationship between Research Biobanks and Donors" or a moral spell ("consent").

know important aspects of future research with respect to the biological material, what alternatives are we then left with?[9]

As the challenges with suboptimal criteria for informed consent have become clear, several alternatives have evolved. Many of them relate to various theories of autonomy, especially theories based on alternatives to individual autonomy: they appeal to evaluative agency (Jaworska 1999), semantic agency (Jennings 2001), or relational agency (Tauber 1999). However, these alternative approaches have evolved from challenges in therapy and not in research. Furthermore, they address reduced competence to consent or strong (social) influence, which are not core challenges in biobank research. In biobank research the key challenge appears to be reduced understanding, and not lack of intentionality or voluntariness. Another alternative would be to argue that biobank research should be based on trust rather than individual self-determination (Williams and Schroeder 2004). However, consent is a crucial prerequisite for trust-based approaches as well (O'Neill 2002; Stirrat and Gill 2005).

Yet another alternative is group consent (Juengst 1998, 2000). As many of the risks and benefits of biobank research concern groups more than individuals, one could argue that group consent is required. However, it is not clear what group consent is, and its moral status with respect to individuals' rights to opt-out or withdraw is unclear (Juengst 2000). Group consent violating individuals' interests in the name of group majority can be conceived of as hard paternalism. Hence, it seems we have to look elsewhere for viable solutions.

Not Autonomy, but Privacy and Confidentiality

One alternative is to acknowledge the reduced autonomy that participation in nonspecific biobank research necessarily will imply, and instead of trying to circumvent this fact, address other normative issues that actually are more pressing, and which contribute to the reduction of autonomy in the first place, such as issues of privacy and of confidentiality (Murray 1997; Greely 2001; Rynning 2003; Gostin and Hodge 1999; Roche and Annas 2001).

The problem for a research person may not be that she or he does not know what biological material will be used for in the future, but that it could violate his or her confidentiality and privacy. We should therefore put our effort in securing these aspects of biobank research, rather than tampering with autonomy and informed consent issues. The point is that we can accept not being able to make an autonomous decision with respect to contributing to biobank research as long as we know that measures with respect to privacy and confidentiality will be taken.

One problem with this approach is to legitimize people's entrance to biobank research in the first place. If we are nonautonomous with respect to unspecific

[9] If an unknown or uncontrolled risk makes informed consent inapplicable in a research project, then one could argue that informed consent is never applicable to clinical research, as there will always be (unknown) risks and benefits involved.

biobank research, we could only enter such research by way of paternalism. There are defendable versions of paternalism, but most of them are based on the principle of beneficence: an intervention in a person's life can be justified (without consent) because its benefits outweigh its burdens for this person. The problem with paternalistic enrolment is that we do know neither the risks nor the potential benefits, so a paternalism based on a trade off between benefits and burdens seems not to be justifiable. Furthermore, this alternative approach is challenged by the potential of traceability of biological material (McGuire and Gibbs 2006). If the material itself is in principle a source of identity, then measures to secure privacy and confidentiality appear to be a remaining challenge.

Consent to Submit to the Researcher

As other parts of this book show, people enter biobank research even if they are skeptical or negative, something which illustrates substantial elements of the power of *virtual trust*.[10] A radical alternative to address such issues would be to apply a Hobbesian perspective to consent in a health research setting. To escape constant attacks of diseases and struggles with illness (in the state of nature), man will develop more in accordance to his higher reason by the way of science, i.e., by contributing to biobank research. In order to do so, he has to submit to the sovereign researcher through a social contract. By freely submitting oneself to the researcher who acted as an agent of the Hobbesian sovereign, a person would preserve a flourishing life in peace. Accordingly, the research person would not only contribute to the general benefit of efficient research, but would become part of a more basic unity of men.[11]

Research ethics committees would then be like "the assembly of men,"[12] and each individual would thus have the right to opt out (to a condition of war and ultimate freedom). The researcher like the sovereign for whom he or she acts would

[10] For this, see chapters "Users and Uses of the Biopolitics of Consent: A Study of DNA Banks" and "Trust, Distrust and Co-production: The Relationship Between Research Biobanks and Donors".

[11] "... and therein to submit their wills every one to his will, and their judgments to his judgment. This is more than consent or concord; it is a real unity of them all in one and the same person, made by covenant of every man with every man, in such manner as if every man should say to every man, I authorize and give up my right of governing myself to this man, or to this assembly of men, on this condition, that you give up your right to him and authorize all his actions in like manner" (Hobbes 1651/1958: 142).

[12] "A commonwealth is said to be instituted, when a multitude of men do agree, and covenant, every one, with everyone, that to whatsoever man, or assembly of men, shall be given the major part, the right to present the person of them all, (that is to say, to be their representative) every one, as well as he that voted against it, shall authorize all the actions and judgments, of that man, or assembly of men, in the same manner, as if they were his own, to the end, to live peaceably amongst themselves, and be protected against other men" (Hobbes 1651/1958: 234).

be obliged to pursue the interest of the people under the external threat of disease.[13] Every individual is autonomous in conferring power to the sovereign by consent (Hobbes 1651/1958: 267). Correspondingly, there is no reduction in autonomy, as the interests of the individual are taken care of by the sovereign, as by the researcher or the research ethics committee.

Since the researchers are selected by the sovereign, and research persons transfer the right to act for them to the researcher, the researcher cannot possibly breach the covenant. "Every [research] subject is author of the acts of the sovereign: hence the [researcher] cannot injure any of his subjects, and cannot be accused of injustice" (Hobbes 1651/1958: 146). Because the purpose of research is health, and the researcher has the right to do whatever he thinks necessary for the preservation of health and security and prevention of disease. Hence, by autonomously committing oneself to research one's good life will be preserved. One consents to the researcher, and not to particular research.

This approach represents a controversial interpretation of informed consent. Nevertheless, political scientists and philosophers trace basic ideas on trust back to Hobbes, Locke, and Hume (Dunn 1988; Hollis 1998; Möllering et al. 2004). This also goes for informed consent. "Nothing done to a man by his own consent can be injury: Whatsoever is done to a man, conformable to his own will signified to the doer, is no injury to him." (Hobbes 1651/1958: 124).[14]

According to Hobbes, consent is seen as a way to transfer power (Hobbes 1651/-1958: 163). In fact consent in one person is the greatest power (Hobbes 1651/1958: 78). Consent transfers power to the sovereign as consent in biobank research transfers power to the researcher and the research ethics committee. Moreover, consent requires knowledge (Hobbes 1651/1958: 214) and voluntariness (Hobbes 1651/-1958: 163). Dominion over others can be obtained by consent or by argument (Hobbes 1651/1958: 163–4).

Improper as it may seem, the Hobbesean perspective on consent addresses some of the important power issues, it displays the development of the concept of consent, and it illustrates that consent is not an absolute static formula that can solve all challenges in research ethics.[15]

[13] We acknowledge that the Hobbesian sovereign is only obliged to pursue the interests of the people in relation to their protection from internal and external violence. Otherwise he is sovereign and can do whatever he likes for whatever reasons, while the sovereign's duty and corresponding rights in this context is limited to the external threat of disease.

[14] This is simply a restatement of the Common Law maxim "Volenti non fit injuria" ("no injury is done to a willing person") which has roots in Roman Law.

[15] Other obligation-based approaches to informed consent have recently been promoted (Rhodes 2005), and trust-based (or duty based) approaches, such as Onora O'Neill (2002), could be conceived of as submissive to moral sovereignty (e.g. universal principles and moral law). Hence, a depersonalized version of a Hobbsian concept of autonomy and informed consent may be of relevance.

Conditional Authorization

Another alternative would be to take into account the challenges both to autonomy, privacy, and confidentiality, and argue that we live in a complex world where we have to make decisions that can threaten or reduce our autonomy, privacy, and confidentiality. Signing contracts, purchasing insurance, marrying, and having children are complex issues of this kind. In practical life we trust people and hope for the best. Although we cannot regulate research in accordance with trust, hope, and love, we can do what we do when we have to formalize such matters, that is, in contracts. Hence one way of putting this into work would be by giving a permission or authorization to biobank research. Acknowledging the challenges with informed consent, Greely suggested the use of permissions to regulate biobank research (Greely 1999).[16] Corresponding to Greely's permission, Árnason proposes an explicit written authorization for entering data from health databases into genetic research (Árnason et al. 2004; 2007).[17] According to Árnason, the written informed authorization implies that the individual has to be informed about, that he or she has understood which information will be stored, how privacy will be secured, how the data will be connected to other information, the manner in which the information will be used, how research on the data will be regulated, and that the individual has a right to withdraw.

Although Árnason's (and Greely's) approach is context-specific (Health Services Databases), the conception of authorization could freely be applied in a broader context, such as including biological material in biobank research. This would be more in line with The Council of Europe's draft of "Recommendation on Research on biological Materials of Human Origin" suggesting that consent *or* authorization have to be given before biological material enter biobanks for research (Council of Europe 2005). An authorization regulating the entrance of biological material into biobank research has to be specified with respect to the conditions of entrance and research, and we therefore suggest the term "conditional authorization." Although a conditional authorization involves many of the same requirements as informed consent, such as understanding, competence, and voluntariness, the requirements for a conditional authorization are less strict than for informed consent, in particular with respect to what the research person has to understand in order to participate in research. As Árnason points out, the "authorization is in the spirit of informed consent, but it is far more general and open and should, therefore, not be confused with it." (Árnason 2004: 45).[18]

---

[16] It is also worth noting that Tristam Engelhardt discusses the "principle of permission" in his work (Engelhardt 1996). His principle of permission appears to be closer to the principle of autonomy, than the contractual authorization discussed here.

[17] It is worth noting that Greely, who suggests the use of permissions, analogous to Árnason's authorization, suggests using individual affirmative informed consent for entering health data into databases for genetic research such as in the Icelandic health services database (Greely 2001). See also Árnason 2004: 44).

[18] For this argument, see also chapter "Scientific Citizenship, Benefit, and Protection in Population-Based Research".

However, due to general and open approach, one could argue that permissions and authorizations are as unspecific and unsuitable to protect individuals and groups and to protect their interests, as is waived consent or uninformed blanket consent. However, in conditional authorization specification requirements are increasing with the generalness and openness of its content. The reason for this is that the extent and limits of authorization have to be specified. This task becomes more challenging as its extension increases. What has not to be included in an authorization (its limitation, and one could say, what differs from an informed consent) has to be specified.

In addition to ordinary requirements with respect to research involving human subjects, conditional authorization should include explicit clauses on the following:

1. Moral and legal status of biological material
   - Who has property rights of the material in the biobank?
   - Who has intellectual rights stemming from work with the biological material?
   - Who has access to the biological material, and under what conditions, including external analysis and export?
   - Remuneration in case of commercial potential
2. Consequences with respect to risks and benefits
   - Measures to secure confidentiality with respect to the biological material and information that stems from it
   - Specific procedures if the research results are of vital importance for the person or his/her relatives
3. Unsettled relationship to other sources of information
   - What other health information is allowed to apply, under what conditions
   - Who is responsible for handling data, and possible breaches of confidentiality?
   - How privacy will be secured
4. Conditions for initiating further research on basis of existing biobank material
   - IRB/REC assessment of further research
   - Approval of patient organization/patient representatives
5. What happens if any of the parts breach the conditions?

Conditional authorization can be based on various contract theories. As in all contractual situations one can trust or distrust, read all of the text of a contract or only read part of it, depending on the situation, e.g., how confident you are with the situation and how much you trust the persons involved. The point is that the situation where one gives a conditional authorization does not give the impression of being consent with an ethical status beyond its actual content.

Moreover, conditional authorization is much more open, e.g., to remuneration and royalties in the case of substantial economic income resulting from research on the biological material, than varieties of consent.[19] It is not obvious that persons contributing to biobank research resulting in substantial incomes for private investors should be without reward, and conditional authorization is one manner of taking this into account.

Conditional Consent?

As conditional authorization is a new alternative to informed consent, it becomes necessary to explain some of the differences to informed consent. In particular as conditional authorization can be a highly detailed contract requiring substantial amount of information, are the requirements regarding understanding by the research person/biobank donor, their voluntariness, and competence the same as with informed consent?

Conditional authorization is specific with respect to spelling out known risks and uncertainties, as well as stating strategies to handle the unknown; e.g., how persons will be contacted in case of information about their health. It is important to note that the problem is not uncertainty as such, because all research involves elements of uncertainty, but how uncertainty is communicated; i.e., we appear to live with well-informed uncertainty every day in all other aspects of life.

The conditions of authorization have to be explicitly assessed by the IRB/REC. Furthermore, conditional authorization would be less dependent on a theoretical framework, such as a theory of autonomy, and can address challenges with autonomy, privacy, and confidentiality in a practical manner. As such, it is a pragmatic approach to attain self-control, security, and protection, and to increase trust.

In this manner conditional authorization avoids hiding behind a general concept of "consent," (Hofmann 2008) but addresses normative challenges explicitly. Hence, conditional authorization can hinder a superficial use of consent, and preserves the content of the concept of consent from erosion. It is not the fact that one has gained consent that makes research acceptable, but the content of such consent. The term *consent* can be applied to cover up important moral challenges, and as long as there is a consent, research is acceptable. In this manner consent becomes a piece of moral technology used in order to promote biobank research and avoid interference and claims from research persons. Hence, instead of protecting the research persons, consent ends up becoming a technical device to protect researchers' interests. The point is not to promote a naïve conservation of an ideal static concept of consent, but rather to avoid that the concept of consent becomes so broad and vague that it looses all applicability.

Specifying what is meant by consent therefore is important and, if it turns out from the specification that it does not qualify as consent, we have to call it something else. However, not obtaining consent does not mean that research is not acceptable. Informed consent is neither a necessary nor a sufficient condition for research on humans, and should not become a "buzzword" covering important moral issues or

[19] However, if the individual consumer-model of autonomy is applied, then one could also argue that biological material could be sold to research.

be used as a spell to make them vanish. Instead we should look for other conceptions better suited to regulate biobank research, and leave the moral concept of consent with some content. Conditional authorization may represent one way of achieving this.

Conclusions

Easy solutions to complex questions are seldom good. Specific solutions to specific contexts are more likely to prove successful. Depending on the specific kind of biobank research, the arguments and approaches discussed above will bear different weight, and should be applied accordingly. One size does not fit all. Biobank material has a variety of bodily sources and statuses (urine vs. heart and brain),[20] stems from persons in a variety of situations (patients, healthy persons; collected by their family doctor, in hospitals, by special institutions, such as blood banks, or by researchers in specific research organizations or enterprises), with a broad variety of moral statuses (fetuses, minors, noncompetent persons, deceased), and are used for different purposes (gaining new general knowledge, obtaining information relevant for the research subject, quality assurance and improvement). Finding one moral formula that addresses all these aspects may be overly optimistic, and even morally dangerous.

Hence, it appears to be difficult to find a general approach to respect for a person's autonomy, which complies with all challenges in biobank research. The point is that none of the alternatives discussed above is flawless. Their suitability varies with context: we have to find an approach depending on the kind of biobank in question, the related risk, and the possible outcome.

Although there are many special challenges in biobank research, none of them appear to justify moral or legal exceptionalism – neither in terms of more stringent nor laxer regulation than with respect to other kinds of medical research involving human subjects.[21] The combination of features and challenges may urge special solutions in practice, although not in principle. The four (traditional) alternatives discussed above appear to be approaches that try to cope with basic and common challenges for certain types of biobank research (including informed consent), and may provide guidance in struggling with the question of whether to consent or not to consent to biobank research.

To consent or not to consent is not the key moral question in biobank research. Other issues such as promoting beneficial research, respecting confidentiality and privacy, protecting against harm, and promoting the interest of the research person are equally important. The crucial challenge is therefore not to settle the issue of ownership of biological material (in general), but to protect the individual providing

[20] It is interesting to note that the moral challenge with respect to the status of the biological material appears to be independent of the source (blood, tissue, autopsies, historical collections), and thus appears to be more than a superficial challenge.

[21] The revealed challenges make it abundantly clear that a wholesale transfer of a set of ethical rules designed for other kinds of biomedical research is also hugely problematic.

the material and secure that his or her interests are taken care of (including the interest of contributing to research). Informed consent is only one way of doing this.

Altogether, it is difficult to justify biobank exceptionalism in the sense that biobank research is so special that lower moral standards could be applied. Rather biobank research highlights some general moral challenges in research, and urges us to find solutions more suitable than those provided by the informed consent doctrine. Hence, to the extent that biobank research can point to alternatives where informed consent stops being a moral guide, it can contribute to refining research ethics in general, by pointing to general weaknesses of applying informed consent formula in areas where there cannot be any consent unless the concept of consent is stripped of any relevant content. If everything becomes consent, nothing is, and then it may well be that we have to turn to other conceptions, such as for example conditional authorization.

Acknowledgment We are most grateful to Jennifer Harris for valuable comments to an early draft of this paper.

References

Andrews LB (2005) Harnessing the benefits of biobanks. Journal of Law, Medicine, and Ethics 33:22–30

Árnason G et al. (Eds.) (2004) Blood and Data: Ethical, Legal and Social Aspects of Human Genetic Databases. University of Iceland Press, Reykjavík Árnason V (2004) Coding and Consent. Bioethics 18:39–61

Blatt JRB (2000) Banking biological collections: Data warehousing, data mining, and data dilemmas in genomics and global health policy. Community Genetics 3:204–211

Chadwick R, Berg K (2001) Solidarity and equity: New ethical frameworks for genetic database. Nature Review Genetics 2:318–321

Clayton EW (2005) Informed consent and biobanks. Journal of Law, Medicine, and Ethics 33: 15–21

Council of Europe (2005) Draft recommendation on research on biological materials of human origin. Steering Committee on Bioethics. Council of Europe, Strasbourg

Dunn J (1988) Trust and agency. In: Gambetta D (Ed.). Trust: Making and Breaking Co-Operative Relations. Basil Blackwell, Oxford, pp 73–93

Dworkin G (1988) The theory and practice of autonomy. New York: Cambridge University Press

Engelhardt HT Jr (1996) The Foundations of Bioethics. Oxford University Press, New York

Eriksson S (2003) Mapping the debate on informed consent. In: Hansson MG and Levin M (Eds.). Biobanks as Resources for Health. Research Program Ethics in Biomedicine, Uppsala, pp 165–190

Essen U (2003) Focusing on personal integrity violation – legal guidelines for ethical practice. Biobanks as Resources for Health. Research Program Ethics in Biomedicine, Uppsala: 129–148

Faden RR, Beauchamp TL (1986) A History and Theory of Informed Consent. Oxford University Press, New York

Friedson E (1970) The Profession of Medicine. Dodd Mead, New York, pp 368–372

Fuller BP et al. (1999) Privacy in genetics research. Science 258(5432):1359–1361

Gostin LO, Hodge JG (1999) Genetic privacy and the law: An end to genetics exceptionalism. Jurimetrics 40:21–58

Green M, Botkin J (2003) Genetic exceptionalism' in medicine: Clarifying the differences between genetic and nongenetic tests. Annals of Internal Medicine 138:571–575
Greely HT (1999) Breaking the stalemate: A prospective regulatory framework for unforeseen research uses of human tissue samples and health information. Wake Forest Law Review 34:737–766
Greely HT (2001) Informed consent and other ethical issues in human population genetics. Annual Review of Genetics 35:785–800
Hansson MG et al. (2006) Should donors be allowed to give broad consent to future biobank research? The Lancet Oncology 7:266–269
Hansson MG, Levin M (Eds.) (2003) Biobanks as Resources for Health. Research Program Ethics in Biomedicine, Uppsala
Harris J, Keywood K (2001) Ignorance, information and autonomy. Theoretical Medicine 22:415–436
Helgesson G et al. (2007) Ethical framework for previously collected biobank samples. Nature Biotechnology 25: 973–976
Hobbes T. (1651/1958) Leviathan. The Liberal Arts Press, New York
Hollis M (1998) Trust within reason. Cambridge University Press, Cambridge
Hofmann B (2008) Broadening consent and diluting ethics. Journal of Medical Ethics 35:125–129
Jaworska A (1999) Respecting the margins of agency: Alzheimer's patients and the capacity to value. Philosophy and Public Affairs 28(2):105–138
Jennings B (2001) Freedom fading: On dementia, best interests, and public safety. Georgia Law Review 35:593–619
Juengst ET (1998) Groups as gatekeepers to genomic research: Conceptually confusing morally hazardous and practically useless. Kennedy Institute of Ethics Journal 8:183–200
Juengst ET (2000) What 'community review' can and cannot do. Journal of Law, Medicine and Ethics 28:52–54
Kettis-Lindblad A et al. (2006) Genetic research and donation of tissue samples to biobanks. What do potential sample donors in the Swedish general public think? European Journal of Public Health 16:433–440
Kettis-Lindblad A et al. (2007) Perceptions of potential donors in the Swedish public towards information and consent procedures in relation to use of human tissue samples in biobanks: A population-based study. Scandinavian Journal of Public Health 35:148–156
Lindberg BS (2003) Clinical data – a necessary requirement for realising the potential of biobanks. In: Hansson MG and Levin M (Eds.). Biobanks as Resources for Health. Research Program Ethics in Biomedicine, Uppsala, pp 21–31
Maschke KJ (2005) Navigating an ethical patchwork – human gene banks. Nature Biotechnology 23:539–545
Maschke KJ (2006) Alternative consent approaches for biobank research. The Lancet Oncology 7:193–194
McGuire AL, Gibbs RA (2006) Genetics. No longer de-identified. Science 21(312):370–371
Mitchell GR, Happe K (2001) Defining the subject of consent in DNA research. Journal of Medical Humanities 22:41–53
Moutel G et al. (2001) Bio-libraries and DNA storage: assessment of patients' perception of information. Medical Law 20:193–204
Möllering G et al. (2004) Introduction: Understanding organizational trust? – Foundations, constellations, and issues of operationalisation. Journal of Managerial Psychology 19(6):556–570
Murray T (1997) Genetic secrets and future diaries: Is genetic information different from other medical information? In: Rothstein (Ed.). Genetic Secrets: Protecting Privacy and Confidentiality in the Genetic Era. Yale University Press, New Haven, pp 60–73
O'Neill O (2002) Autonomy and trust in bioethics. Cambridge University Press, Cambridge
Reilly PR (1998) Rethinking risks to human subjects in genetics research. American Journal of Human Genetics 63:682–685
Rhodes R (2005) Rethinking research ethics. American Journal of Bioethics 5:7–28

Ring L, Kettis-Lindblad Å (2003) Public and patient perception of biobanks and informed consent. In: Hansson MG and Levin M (Eds.). Biobanks as Resources for Health. Research Program Ethics in Biomedicine, Uppsala, pp 197–206

Roche PA, Annas GJ (2001) Protecting genetic privacy. Nature Review Genetics 2:392–396

Rosoff AJ (1981) Informed Consent: A Guide for Health Providers. Aspen Systems Corp., Rockville

Rothstein MA (2005) Expanding the ethical analysis of biobanks. Journal of Law, Medicine and Ethics 33:89–99

Rynning E (2003) Public law aspects on the use of biobank samples – privacy versus the interests of research. In: Hansson MG and Levin M (Eds.). Biobanks as Resources for Health. Research Program Ethics in Biomedicine, Uppsala, pp 91–128

Schneider CE (1998) The practice of autonomy. New York: Oxford University Press

Stirrat GM, Gill R (2005) Autonomy in medical ethics after O'Neill. Journal of Medical Ethics 31:127–130

Suter SM (2001) The allure and peril of genetics exceptionalism: Do we need special genetics legislation? Washington University Law Quarterly 79:669–748

Tauber AI (1999) The breakdown of autonomy. In: Tauber AI (Ed.). The Confessions of a Medical Man. MIT Press, Cambridge MA, pp 49–70

Wadman M (2000) Geneticists oppose consent ruling. Nature 404(6774):114–115

Wendler D et al. (2002) Does the current consent process minimize the risk of genetic research? American Journal of Medical Genetics 113:258–262

Wendler D, Emanuel E (2002) The debate over research on stored biological samples: What do sources think? Archives of Internal Medicine 162:1457–1462

Wilcox AJ et al. (1999) Genetic determinism and the overprotection of human subjects. Nature Genetics 21:362

Williams G, Schroeder D (2004) Human genetic banking: Altruism, benefit, and consent. New Genetics and Society 23:89–100

What No One Knows Cannot Hurt You: The Limits of Informed Consent in the Emerging World of Biobanking

Arthur L. Caplan

Abstract In thinking about the ethics of biobanking many argue that the core protection of individual dignity and privacy is informed consent. But this doctrine will not and cannot be made to handle the ethical load it is being asked to bear in the realm of biobanking. It is time to shift ethical emphasis to general, broad consent to linked-anonymization through trusted third parties as the best way to ensure the ethical practice of biobanking.

The Boom in Guidelines for Research Involving Biobanks

Biobanks have not only created extraordinary enthusiasm among biomedical scientists (Kaiser 2002; Andrews 2005; Knoppers and Chadwick 2005); the boom toward utilizing biobanks, both new and old, in the age of genomics has spawned a boomlet in ethical guidelines. The World Medical Association's "guidelines on health databases" (WMA 2002) regulate research involving health data, including biobanks. Other guidelines address issues related to biobanks under the categories of "human tissue UK and biological samples" (UK Medical Research Council 2001), "human biological materials" (US National Bioethics Advisory Commission 1999; Council of Europe 2006), "biological specimen" (OHRP 2004) or "genetic material" (Australian National Health and Medical Research Council 2005). Associations of geneticists are producing guidelines for biobanking at a rapid clip (ESHG 2003; ACMG 1995; UNESCO 2003). UNESCO adopted an "International Declaration on Human Genetic Data" in October 2003.[41] A parade of nations has followed with their moral frameworks, including France (Comite Consultatif 2003), Germany (Nationaler Ethikrat 2004), Canada (Commission de l'éthique 2003), Switzerland

A.L. Caplan
Center for Bioethics, University of Pennsylvania, Philadelphia, PA, USA
e-mail: caplan@mail.med.upenn.edu

J.H. Solbakk, et al. (eds.), *The Ethics of Research Biobanking*,
DOI: 10.1007/978-0-387-93872-1_2,

(Schweizer Akademie 2005; Petersen 2005; PRIM&R 2007; Simon et al. 2007), and the USA (NCI 2006), to name but a few.

If one looks closely at the various guidelines that have been issued two problems quickly emerge. There is no standardized language in use for referring to either problems or solutions pertaining to biobanking (Knoppers 2005; Knoppers and Saginur 2005; Elger and Caplan 2006). And there is a powerful difference of opinion about the utility of informed consent as the best tool to use to protect the interests, rights, and dignity of those persons and groups, living or dead, whose biological materials are utilized in biobanks (Greely 2007).

Relying on Informed Consent

A core value guiding all of the regulations issued by groups and organizations in the USA, Europe, and Asia governing biobanking is respecting the right of the individual to voluntarily participate in the inclusion of their genetic and biological materials or data about them in a biobank. This means that much attention is devoted in the regulations to informed consent.

Why should participation in a biobank be a matter of voluntary choice? After all, for nearly every biological and tissue sample there is little additional physical risk involved in procuring tissues or genes from persons beyond either simple modes of donation (i.e., cheek swabs, blood samples), through natural means (urine, umbilical cord post-birth), or what is going to be done for medical reasons anyway (removal of tumors, removal of spleen, limb, or thyroid for treatment purposes, etc.).

A key reason for the emphasis on voluntary choice is to protect the liberty and privacy of persons. Most cultures believe that nothing ought be removed from the body of a person without their express permission regardless of the benefit that might accrue to others. Even if I do not legally "own" my body, I control, under the fundamental rights of privacy and bodily integrity, who may touch it and remove materials from it.

Another reason for the emphasis on informed consent is somewhat less lofty. Americans in particular worry that they need informed consent to protect the confidentiality of data and materials in biobanks from unauthorized examination by third parties lest they lose health insurance coverage or employment. Underwriting and risk assessment are omnipresent elements of health, life, disability, and retirement benefits for Americans and thus they have a much stronger reason to be concerned about who it is that might have access to information that might prove disadvantageous to them (Clayton 2005; Greely 2007).

Others are worried about the possibility of penalties being leveled against entire groups who supply materials or data which are subsequently linked to propensities to disease or disability risks. Although this has not yet occurred with genetic information in the United States in any serious manner, it has occurred in the past with other preexisting conditions in such areas as employment, health insurance, and life insurance.

While Americans pin their hopes for the protection of persons' control over what others know about them on informed consent, and while those in other nations do so as well, existing guidelines do not agree about the weight that ought be accorded or even the meaning of informed consent (Deschênes 2001; Hoeyer et al. 2005; Greely 2007).

Consent in Retrospective and Prospective Biobanks

One problem facing informed consent as it is used in American regulations and guidelines is that consent must handle two very different situations. Many biobanks are "prospective." That is, they are banks in which tissue, data, or both are being actively collected for the first time for future use. There are many companies, foundations, and local, regional, state, and national government agencies in the process of building all manner of prospective biobanks.

There are also "retrospective" biobanks. These are banks which use already existing samples and data that have been collected, sometimes many decades before. Many such biobanks exist as individual researchers, surgeons, public health departments, and other groups have collected and stored tissue samples for their own research, teaching, or epidemiological purposes.

The National Cancer Institute of the NIH in the USA has funded many projects in which tissues have been retrospectively collected. However, even in this one relatively small area of biobanking NCI officials have no idea how many such biobanks built with any of their funds actually exist. Nor is it immediately obvious who owns and controls such banks (Washington University v. Catalona et al. 2006).

The use of data and tissues from retrospective biobanks has proven to be a very difficult challenge when it comes to informed consent. Some American regulations, guidelines, and regulatory opinions presume consent based on the reality of possession of tissues or data from tissues and the fact that those from whom they were taken are long since dead or cannot be easily identified. There is a strong presumption that the dead can no longer be harmed by genetic analysis while the living can greatly benefit. In some cases reliance is placed on a consent given decades ago to the collection of tissue, although since this often occurred before the advent of modern molecular biology and any consensus about what needs to be disclosed to those being recruited to engage in research, it is hard to credibly associate the word "informed" with older consent documents. Many doctors and most hospitals treated the material collected as biological waste not as something of value for medical research.

Sometimes guidelines call for authority to provide retrospective consent to be assigned to a committee. They are to act as a surrogate decision-maker on behalf of those whose tissues were collected but can no longer consent either on the basis of what they believe the sources would have said at this point in time or by consenting to what it deems to be in the best interest of the biobank, the public health, or society. However, the legitimacy of the composition of such committees and the basis for their surrogate judgments for persons they in all likelihood never met remains uncertain.

No amount of conceptual gyrating can disguise the fact that informed consent is not a principle that can be readily used to support the use of data from retrospective banks. Often there were no consents obtained when tissues were collected. Or if consent was utilized it was blatantly inadequate. The existence of a bank hardly reveals anything about what a person would agree to today in terms of the future disposition or commercial use of their biological materials or data derived from them. Nor can a committee completely ignorant of the wishes of those whose tissues and materials were collected make any reasonable judgment about what all, many or most persons would consent to in terms of current and future uses in the name of corporate profit, the public good, or any other standard (Schneider 2005).

In one notorious instance in the United States a plan to utilize commercial funding to greatly expand the biobanking power of the Framingham Heart Study collapsed when the descendents of those in the study objected that their parents and grandparents would not have been comfortable with the commercialization of their "gift" which is how they saw their participation in the heart study (Caplan 2006). If this case is any evidence then many now deceased persons might object strenuously to the commercialization of their gift.

Informed consent is a fiction when it comes to retrospective biobanks. It may be a convenient fiction but it is, nonetheless, a fiction. It would seem more reasonable to resolve disputes about retrospective banks by having a committee, broadly constituted to include many public members, to assess what they believe best serves the public interest. Surprisingly, consent is not all that much better as a principle to protect the autonomy and privacy of those involved in prospective biobanking.

A characteristic of most prospective biobanks is that samples and data are collected for long-term use. They are rarely collected only for a single project. Typical examples are the UK Biobank, the Marshfield Personalized Medicine Project in Wisconsin, USA, the Western Australia Genome Health Project (P3G 2007), and the newly formed Framingham Heart Study Project (http://www.nhlbi.nih.gov/new/press/06-02-06.htm). These biobanks aim at studying gene–environment interactions in populations over many years. They try to correlate genetic markers observed in the DNA of the participants, which has been extracted either from their blood or other types of tissue, with data from the participants' present and future medical records. They may also go further and correlate biological information with data from screening questionnaires and routinized physical and laboratory examinations along with data about lifestyle and environmental factors.

In building such banks it is often impossible to anticipate in advance what research studies will evolve. Nor is it possible to predict how studies or biobanks may be merged or integrated with other biobanks both private and government-sponsored. This leaves the prospect for valid consent very much adrift since it is impossible to obtain specific consent to a purpose or commercial activity not envisioned at the time tissue and data collection began.

For consent to truly be meaningful, subjects whose tissues are taken or data accessed should be recontacted for every new research project and sign a new consent form after having been informed about the details of the latest proposed project. If this is the level of consent required then this approach is doomed to fail. It is not only prohibitively expensive, it is impractical since it is highly probable that a large percentage of long-term prospective biobank participants will not be able to be located, will die, or simply will not bother to respond to requests for reconsent. This is especially so for biobanks involving children where initial consent is given by parents but where studies may require decades to carry out.

Informed consent is a doctrine carried into biobanking both from clinical research and therapy. But, it does not fit. Looking backwards, as retrospective biobanks must do, it is too late to invoke it. Looking forward, as prospective banks do, the uncertainty of where biobanking will go and the need to protect the revelation of sensitive data to third parties makes it of limited practical use.

Consent: Other Problems

Many European guidelines take the view that general or broad consent is acceptable. They attempt to avoid American problems with consent by doing away with the specificity of consent with respect to prospective biobanks. German guidelines endorse general consent as do regulations or laws in Sweden, Iceland, Estonia (Kaye et al. 2004).

General consent is considered acceptable in Europe but only if two conditions are fulfilled: the approval of all future projects by a research ethics committee and the right to withdraw samples by subjects at any future time (Trouet 2004). To quote the European Society of Human Genetics (ESHG):

> "... individuals may be asked to consent for a broader use. In that case, there is no need to re-contact individuals although the subjects should be able to communicate should they wish to withdraw". Withdrawal, sometimes also called "opt out" means that those in biobanks "should be given the right to withdraw at any time from the research, including destruction of their sample".

Does this approach to consent really make sense? Is it any more practical than the more detailed form of consent evident in American regulations?

Broad consent with independent committee review and the right to withdraw may make those involved in prospective biobanking feel better but it is not likely to pass muster as any sort of true consent. The risks and benefits, options and alternatives, purposes and funding sources that are all crucial to any truly informed consent in health care cannot be achieved by broad, vague affirmations. It is hard to imagine a court accepting as adequate a consent to any and all purposes that the person soliciting the consent can conceive of or imagine. Even assigning the right to decide to an ethics committee is suspect in that it cannot be known who will sit on this group or what standards they will follow if and when a reconsent is required.

Nor is it really possible to ensure the right of persons to withdraw from a biobank. How will biobanks keep each subject fully informed about emerging projects, studies, and opportunities especially as they themselves merge, evolve, and are bought and sold? And how will biobanks be able to trace the whereabouts of all of those persons whose tissues or data are entered into biobanks especially if as is very likely to happen biobanks move and relocate, as researchers and companies move or merge or the banks begin to engage in studies which correlate findings from other biobanks not in existence at the time of the initial consent?

Time to Abandon Consent in Favor of Anonymization

In Europe weaker standards of consent prevail. American policy still attempts to follow a rigorous standard of consent. Neither will work.

The obstacles to consent that must endure over long periods of time are enormous. Add to this the reality that comprehension and understanding reflected in informed consent in the easiest of circumstances – consent to therapy – is still poor and the continued effort to make informed consent hold the ethical weight of the practice of biobanking makes little sense.

In 2004, the US Office for Human Research Protection (OHRP) proposed a different solution of the problem: an enlargement of the definition of nonidentifiable. This proposal provides a way to protect those in biobanks without creating the illusion that consent can do so.

The US Code of Federal Regulations (45 CFR 46) allows research involving nonidentifiable samples without the obligation to obtain informed consent. This is because there is no way harm or wrong can be done to individuals. It is of course still necessary to obtain general consent in order to gain access to tissues or DNA samples simply to respect each person's right to privacy but if the information garnered from the acquisition of biological materials will remain unidentifiable to those doing the biobanking research then no further consent ought be required since there is no prospect of harming someone by the release, intentional or inadvertent, of sensitive medical information about them.

In their guidance from August 10, 2004, the US OHRP enlarged the definition of non identifiable in the following way: "OHRP considers private information or specimens not to be individually identifiable when they cannot be linked to specific individuals by the investigator(s) either directly or indirectly through coding systems." This is the case if "the investigators and the holder of the key enter into an agreement prohibiting the release of the key to the investigators under any circumstances...."

The advantage given by this definition of nonidentifiable is obvious. If high standards of anonymization can be created and strictly enforced in the biobanking community then the need to invoke informed consent for either retrospective or prospective biobanking can be eliminated.

If researchers are required to hand all data to trusted third parties that, following international standards, can encrypt, anonymize, and link them, then biobanking can be put on a firm, universal, and practical ethical foundation. Researchers must agree that they will not have access to the codes used to anonymize data. Third party organizations can maintain identifiable links to specific persons. But researchers themselves will only receive coded, anonymized information unless the trusted third party entity agrees that there is a reason so powerful as to break the code (i.e., discovery of a drug that can benefit those with a certain genotype).

Implementing anonymization as the core requirement of all biobanking will not be easy. It will take international agreement concerning the standards to be followed both for data encryption, linkage, and security. It will require a transparent audit process of third party data holders to ensure they are adhering to all requirements concerning anonymization, privacy, and confidentiality. Public participation in the governance of third party data holding entities is critical. The holders of identifiers must receive the full protection of the law against compelling disclosure to any private or public entities. If all this is done it should be possible to arrive at a single set of ethical guidelines for biobanking both retrospective and prospective anywhere in the world.

Conclusions

Biobanking can move forward, not by continuing to try to coax flexibility out of the rigid doctrine of informed consent but by turning instead to linked anonymization through the creation of trusted third party data-holders (Elger and Caplan 2006). All data must be stripped of all identifiers and then recoded. The algorithms for linking coded data back to anonymized biological materials, samples, and data must be held by absolutely trustworthy third party sources. If safe havens can be constructed for storing identified information, if researchers in biobanking never know anything directly about the identity of those whose tissues and data they study, then the promise of biobanking is likely to be fulfilled. If ethical regulation proceeds as it is now, with a cacophony of voices trying to fit a doctrine – informed consent – on to a practice – biobanking – that it cannot possibly fit and for which it is a very awkward form of protection of the key risk – loss of personal privacy – then there is a very real danger that the promise of biobanking in advancing knowledge as well as curing and preventing disease will remain only that.

Acknowledgments Thanks to Bernice Elger, MD, for the hard work that established the wide variety of rules and regulations referred to in this paper. I acknowledge the support of the National Science Foundation grant DMR-0425780 in preparing this manuscript.

References

American College of Medical Genetics Storage of Genetics Materials Committee (ACMG) (1995) Statement on storage and use of genetic materials. http://genetics.faseb.org/genetics/acmg/pol-17.htm

Andrews LB (2005) Harnessing the benefits of biobanks. J Law Med Ethics 33(1):22–30

Caplan AL (2006) Smart Mice, Not-So-Smart Humans. Rowman & Littlefield, Lanham, MD

Clayton EW (2005) Informed consent and biobanks. J Law Med Ethics 33(1):15–21

Comité Consultatif national d'Ethique pour les sciences de la vie et de la santé. Avis No. 77 (2003) Problèmes éthiques posés par les collections de matériel biologique et les données d'information associées: "biobanques", "biothèques":http://www.ccneethique.fr/francais/pdf/avis077.pdf

Commission de l'éthique de la science et de la technologie (2003) Avis. Les enjeux éthiques des banques d'information génétique: pour un encadrement démocratique et responsable. Quebec. http://www.ethique.gouv.qc.ca/fr/ftp/AvisBanquesGen.pdf

Council of Europe (2006) Recommedation (REC 2006, 4). http://www.coe.int/t/e/legal_affairs/legal_cooperation/bioethics/texts_and_documents/Rec_2006_4.pdf

Deschênes M et al. (2001) Human genetic research, DNA banking and consent: a question of "form"? Clin Genet 59:221–239

Elger BS, Caplan AL (2006) Consent and anonymization in research. EMBO Rep 7:661–666

European Society of Human Genetics (2003) Data storage and DNA banking for biomedical research: technical, social and ethical issues. Recommendations of the European Society of Human Genetics. Eur J Hum Genet 11 (Suppl 2):S8–S10

Greely HT (2007) The uneasy ethical and legal underpinnings of large-sclae genomic biobanks. Annu Rev Genomics Hum Genet 8:343–364

Hoeyer K et al. (2005) The ethics of research using biobanks: reason to question the importance attributed to informed consent. Arch Intern Med 165(1):97–100

Kaiser J (2002) Population databases boom. From Iceland to the U.S. Science 298:1158–1161

Kaye J et al. (2004) Population genetic databases: a comparative analysis of the law in Iceland, Sweden, Estonia and the UK. TRAMES 8:15–33

Knoppers BM. (2005) Biobanking: international norms. J Law Med Ethics 33:7–14

Knoppers BM, Saginur M (2005) The Babel of genetic data terminology. Nat Biotechnol 23:925–7

Medical Research Council, UK (2001) Ethics series. Human tissue and biological samples for use in research. Operational and ethical guidelines. http://www.mrc.ac.uk/pdf-tissue_guide_fin.pdf

National Bioethics Advisory Commission (NBAC), United States (1999) Research involving human biological materials: Ethical Issues and Policy Guidance, Report and Recommendations, Vol. 1. Rockville. Available at http://www.georgetown.edu/research/nrcbl/nbac/hbm.pdf

National Cancer Institute, NIH (2006) http://biospecimens.cancer.gov/biorepositories/guidelines_full_formatted.asp#background

Nationaler Ethikrat, Germany (2004) Biobanken für die Forschung. Stellungnahme. Berlin. http://www.ethikrat.org/themen/pdf/Stellungnahme_Biobanken_04-03-17.pdf

National Health and Medical Research Council – NHMRC, Australia (1999). Guidelines for gene tic registers and associated genetic material. http://www7.health.gov.au/nhmrc/publications/_files/e14.pdf

Office for Human Research Protections – OHRP, Department of Health and Human Services – HHS (2004) Guidance on research involving coded private information or biological specimens. http://www.hhs.gov/ohrp/humansubjects/guidance/cdebiol.pdf

P3G http://www.p3gconsortium.org/waghp.cfm

Petersen A (2005) Securing our genetic health: engendering trust in UK Biobank. Sociol Health Illness 27(2): 271–292

PRIM&R (2007) Report of the PRIM&R Human Tissue/Specimen Banking Working Group

Schneider CE (2005) Some realism about informed consent. J Lab Clin Med 145(6):289–291

Schweizer Akademie der Medizinischen Wissenschaften (2005) Biobanken: Gewinnung, Aufbewahrung und Nutzung von menschlichem biologischem Material für Ausbildung und Forschung. Medizinisch-ethische Richtlinien und Empfehlungen. http://www.samw.ch

Simon J et al. (2007). A legal framework for biobanking: the German experience. Eur J Hum Genet 15:528–532

Trouet C (2004) New European guidelines. J Med Ethics 30(1):99–103

UNESCO (2003) International declaration on human genetic data

Washington University v. Catalona. See 2007 WL 1758268 (8th Cir. June 20, 2007)

World Medical Association – WMA (2002). The World Medical Association declaration on ethical considerations regarding health databases. http://www.wma.net/e/policy/d1.htm

Users and Uses of the Biopolitics of Consent: A Study of DNA Banks

Pascal Ducournau and Anne Cambon-Thomsen

Abstract In this chapter we intend to examine from a sociological perspective the view of a number of participants in a biobank project on the informed consent procedure they were asked to go through. Having carried out observations, conducted interviews and collected questionnaires as part of an empirical survey, we have concluded that a number of participants feel somewhat suspicious where the procedure is concerned. At least they express caution on its ability to actually serve their autonomy and freedom of choice. As they attempt to detect its potentially perverse effects in terms of power asymmetry and the consequences of diverse responsibilities being devolved to them, their perception of it is far from idealized even if they do not contest it radically. This circumambulatory tour of the users' point of view, which will prove useful to improve communication with the general public, can also be of help in understanding how the contemporary evolutions of biopolitics are perceived.

Introduction

In light of the contemporary evolution of health management systems run by states or medical institutions, it has been widely observed that a new approach in managing public health has emerged: there is greater emphasis on education, consensus and seeking the consent of individuals and populations. This trend is what Foucault called "biopolitics", and one of its main characteristics is to encourage individuals to exercise self-control over behaviour that might affect their health, and over how they make use of the body and its constituent parts, rather than to resort to direct constraints or impose sanitary obligations and health rules: proactive prevention policies, community health initiatives, health education, coaching and patient counselling, gathering informed consent of individuals in various situations (medical

P. Ducournau (✉)
CUFR JF Champollion - Albi/Unit 558, INSERM - National Institute of Health and Medical Research, Toulouse, France
e-mail: ducourna@cict.fr

J.H. Solbakk, et al. (eds.), *The Ethics of Research Biobanking*,
DOI: 10.1007/978-0-387-93872-1_3,

acts, medical research, donations of biological substances: gametes, bone marrow, DNA, etc.).

The concept of biopolitics was first coined by Foucault in order to describe the advent, in the course of the eighteenth century, of a mighty shift in the exercise of power, from conquest and ruling over people's lives to a political technology aimed at fostering life and boosting its yield (Foucault 1976). The exercise of power, which had been exclusively grounded in the right to take lives (bringing death upon some and allowing others to live), as reflected by the lords and monarchs who disposed of the lives of their subjects for their defence (in time of war or an attempt against their person, etc.), slowly evolved into a form of power that no longer "let live" but "gave life" through multiple initiatives aimed at managing, increasing, multiplying, exercising control over and applying regulations to life at an individual as well as a populational level. Such concern for life led to the development of a dual control system:

- On the one hand a growing number of disciplines to be exercised on the human body in order to expand its abilities, draw out its strength, increase its docility
- On the other hand the introduction of the concept of "populational biopolitics", centred on the body as a representative of the species, and designed to pace and monitor the biological processes that affect the population at large, such as births and deaths, states of health, and life spans.

The implementation of this biopolicy has produced new elements of knowledge and new practices that are at the core of the modern state, such as demography, hygiene, urbanism, and public health. The state must now "take charge of the 'bodies', no longer in order to claim rights over their lives or protect them against the enemy, nor simply to exercise punishments or extort taxes, but to help them, and maybe even to compel them to preserve/take care of their health. The obligation to stay healthy is both everyone's duty and the general goal" (Foucault 1994: 16).

Several observers in the "post-Foucault" line of thinking have been struck by the connection between the historic shift described in his works and the many initiatives that have been taken recently to obtain the support, consent, or consensus of the population regarding the management of their health and their bodies. According to Dozon (2001), these measures show that a new public health paradigm is emerging, one no longer based on constraints but on a contract. Alternatively, they could mark the advent of a new body-monitoring system that goes together with the fading away of disciplines, "corporeal liberalism" and an "individualized or delegated biopolicy" (Memmi 2001, 2003). Without indulging in a fairy-tale vision of the new paradigm of health and body management, the existing literature has also stressed that it entails the potential occurrence of complex power relations. Dozon notes that the constraint paradigm is always likely to resurge. Memmi in turn remarks that the ideal form of government, based on the self-control which patients exercise over their biological destiny, often conceals, between the latter and their doctors, "objective power imbalances such as the unequal availability of information, technical knowledge and language skills, which leave the patient with no other option than to give in to the requests of medical professionals" (Memmi 2003: 303–304). On a different note, Berlivet (2004) draws attention to the quite paradoxical manner in which modern-day prevention schemes are presented to individuals: it appears that, while no longer "blaming the victims" and while seeking freedom from addiction and social pressure which typically induce risky behaviour, prevention and health education schemes lead the targeted individuals to adopt preset identities that they are expected to identify with (the "nonsmoker", the "moderate drinker", etc.). The mode of subjectification which emerges is bipolar: subjugation and empowerment. Following these studies on the paradoxical nature of the new enforcement tools for health policies, it has been suggested (Martucelli 2004) that all contemporary initiatives concerning prevention (against cancer, cardiovascular diseases, HIV, etc.) coincided with the development of a modern form of domination based on making individuals liable for their health. Domination traditionally relied on the concept of dependence. However, contemporary forms of domination encourage subjects to become active. In doing so, they prompt them to take on a number of responsibilities that have been devolved to them. Through this devolution of responsibilities, the actors are not expected to abide by a set of norms. Instead they are instructed to face facts that are presented as being the logical and unavoidable consequences of their actions, past or present. To sum up, "it is less a matter of dictating what needs to be done than to get individuals to realize that they are the 'authors' of their own lives" (Martucelli 2003: 491) and, more important, the "authors" of the consequences of their actions.

This brief look at the existing critical literature provides an overview of the subject, and gives reasons to believe that we are currently witnessing a subtle refinement of power relations in the field of biopolitics. Power has acquired the ability to renew itself through the subjectivity of the actors, to further interfere with their desires, choices and actions, even to be heard through the assertion of the ego. But are the actors truly being deceived by the new configuration of power relations? Could the case be that the implementation of this sort of power relations, based on an original acceptance of constraints and the individualization of the latter, is taking place without the individuals concerned (the publics targeted through prevention campaigns, users of health institutions, patients) knowing and in a manner so inconspicuous that they are bound to adhere to it without formulating any criticism? This question deserves careful consideration as two diverging versions of the modern forms of domination can be derived from its answer: one would set forth its imperative, unavoidable dimension, and the other would expose its relative nature, one potentially superseded by the interplay of the actors, who might use their skills to keep it at bay, not letting it influence their behaviour. The second alternative could bring back the notion that the actors always have the possibility of calling into play the "quant-à-soi" (keeping to oneself) in the new context of developing a biopolicy which, it seems, no longer has anything in common with the era of disciplines described by Foucault.

Field Approach and Method

The opinions of the actors who most openly welcome the modern-day health and medical mechanisms set up by health institutions provide a privileged empirical starting-point to answer the question raised. They enable us to approach the topic from a perspective which a priori rules out the "quant-à-soi" or ability to keep oneself at a distance, bypass, become aware of or keep at bay the forms of domination that might accompany the implementation of these mechanisms. We compiled empirical data collected through a survey based on ethnographic observation, interviews and questionnaires involving individuals who agreed to participate in a DNA bank within the framework of an epidemiological survey, using a procedure designed to gather their free and informed consent as required by the existing judicial and ethical norms.

This epidemiological survey was based on a comparison between one "case group" and one group of "control subjects", and was intended to assess the prevalence of cardiovascular pathologies among the "general population", to carry out a follow-up and to identify the risk and protection factors – both genetic and environmental – of the said pathologies. The originality of this survey consisted in the creation of a DNA bank, which concerned the general population as well as a sample population recruited in a hospital. In total about 1,800 individuals – exclusively men aged between 45 and 75, i.e., an age group being considered most "at risk" for coronary and vascular diseases – were recruited in southern France by a laboratory belonging to the Institut National de la Santé et de la Recherche Médicale (Inserm) to participate in the study.

The creation and expansion of DNA banks during the last decade has triggered a prolific output of normative and speculative texts[1] about their stakes and the manner in which they should be set up, governed and used. It is worth noting that these texts often put emphasis on the definition of and respect for the rule of informed consent as a fundamental, sine qua non condition for the realization of DNA bank projects. Thus, a kind of transnational consensus in favour of the application of this rule has emerged, while biobank projects not complying with this rule have been met with harsh criticism both in the specialized press and in the mass media. DNA banks constitute therefore a selected platform for expressing the modern forms of biopolitics. So, to repeat the expression coined by Memmi, we are looking at the implementation of an "individualized or delegated biopolicy" linked to DNA bank projects. Individuals are prompted, in a way that is undoubtedly new and unprecedented, to exercise through the rule of informed consent control over what can and cannot be done with the constituent parts of their bodies.

Thus, the proactive involvement of citizens is vested in this rule, which might become the keeper of sound judgment against the dangers of human genome research (Caze de Mongolfier 2002). "Giving participants the opportunity and freedom to participate in a research protocol puts researchers under the obligation to respect their choices" (Caze de Mongolfier 2002: 67). Or, as specified in one of

[1] For this, see review in Cambon-Thomsen et al. (2007).

the recommendations of the Human Genome Organisation dating back to 1996 (HUGO-ELSI 1996):

> The HUGO-ELSI Committee recommends [...] that any choices made by participants with regard to storage or other uses of materials or information taken or derived therefrom be respected. Choices to be informed or not with regard to results or incidental findings should also be respected. Such choices bind other researchers and laboratories. In this way, personal, cultural, and community values can be respected.

This recommendation originates from a time period – the 1990s – which was marked by controversies about informed consent, with certain actors claiming that "a hijacking of the concept of informed consent" was to be feared leading to the use of DNA samples and the associated medical and social background data for purposes that were not specified at the beginning of the research, or to a commercial use of genetic sample banks (Bungener et al. 2002). The fear of such "hijackings" has given rise to an "era of suspicion" around DNA banks, which the press has pointed out over the past few years. The fear associated with the use of genetic data to analyze behaviour in a strictly biological way (violence, homosexuality, alcoholism, schizophrenia, etc.) has contributed to stressing the importance of describing the medical and scientific aims of genetic research projects when seeking the participants' consent. It is hoped that keeping the public informed will make it possible to achieve the respect of "personal, cultural and community values".

In current practice, the application of the concept of informed consent encapsulated in the protocol for the establishment of DNA banks is meant to put the individual participating in a research project in a position of "informed decision-maker". Consequently, the informed consent procedure must provide clear guidelines for the biomedical actors involved in the protocol as well as for the individuals recruited, through the following sequence of actions:

- Information must be provided to the subject by the research team in the form of an explanatory note and through oral communication.
- The team must strive to make the written and oral explanations as clear as they possibly can so as to ensure correct understanding on the part of the individual concerned.

The physician is formally requested to evaluate the subject's understanding of the project, and to encourage them to ask any questions that might be relevant. The subject is then asked to sign a consent form indicating that his or her decision to participate in the project is informed and free. This procedure was carefully followed with regard to the DNA bank that we observed. We were given the opportunity to observe how the participants' consent was gathered over the 3 years during which the project was carried out. At the same time we assembled a collection of quantitative data using a questionnaire filled out by the physician who interviewed the participants. In the questionnaire the physician documented items regarding the interview, indicated if the participants had asked questions and whether or not they had signed the informed consent form after having read it. The physician was also asked to give to each participant another questionnaire asking for the reasons why they had chosen to take part in the study. We also carried out about 60 interviews with individuals participating in the study in the days following their inclusion, some of whom were known to us as we had observed their interactions with the physician. These interviews were meant to examine more closely the reasons for their participation, to analyze the account of the events which led the subject to agree to participate in the project, the meeting with the physician and the procedure of informed consent. The observations were performed with an ethnographic methodology. The analysis was both qualitative and quantitative. Interviews were recorded, transcribed and submitted to qualitative textual analysis of content; answers to questionnaires were statistically analysed in univariate analysis. The results described below allowed deriving a typology of interactions related to informed consent.

Types of Physician–Subject Interactions Observed

We were able to identify five main types of interaction between the recruiting physician and the subjects whose consent for participation was targeted. We only had the opportunity to carry out these observations in a systematic manner on the group of control subjects. These individuals were sent a letter asking them to participate in the creation of the DNA bank and were offered a check-up on their level of risk for cardiovascular diseases (they were not given the test results pertaining to their genetic characteristics). The potential participants were invited to go to a health centre for a blood test, undergo a number of medical examinations (electrocardiogram, doppler, body fat measurement), and fill out a questionnaire about their lifestyle and their medical history.

The group of subjects recruited in a hospital setting was more difficult to approach. Although we found it relatively easy to carry out interviews with them after their inclusion in the protocol, we could only observe in situ the admission into the research protocol of a very limited number of cases (hardly a dozen), not enough for the data collected to reach an adequate level of information saturation.

During the inclusion procedure, the informed consent form that the participants were expected to sign was placed by the physician on the desk, on the side where the participant was supposed to sit. The form was thus clearly displayed and put apart from any other documents lying on the desk (labels, questionnaires, cards filled out by the physician, etc.). The form repeated the main topics from the explanatory note previously provided to the patient, to which he had already given his preliminary consent. It included two or three additional items for which the participant had to indicate if he agreed to

- Participation in a study on coronary diseases
- Participation in the creation of a DNA bank by giving a blood sample
- Willingness to be interviewed about his experience as a participant

Our own survey was thus included in the very procedure of informed consent, following the intention expressed by the physician in charge of the DNA bank project to declare the sociological survey to an ethics committee and to include it in the ethical framework of the project itself.

The formulas used by the physician to present the form were usually quite repetitive: after going through a short oral presentation of the study, which sums up the explanatory note previously sent by surface mail, the physician asked the participant to "read the document, sign it and feel free to ask questions if anything remains unclear". Most of the time, at this point, the physician took the form from the subject and signed his name in the intended space before he gave it back to them.

The first type of interaction occurred in relation to the consent form itself. That is, the form as such did not seem to play a prominent part in the exchanges between the physician and the participant, in the sense that none of them made particular use of it. Obviously, it was eventually signed by the participant, but this did not require any question–response interplay, nor did it trigger any remark from one or the other. The participant signed the document without uttering a word. The most direct way of neglecting the "consent form" consisted in adding the signature automatically without reading it first.

The second type of interaction was characterized by a greater significance being accorded to the signing procedure. The presentation of the consent form prompted a certain number of participants to consider it and read it, sometimes with great attention. The form then became a useful aid for discussing the conditions of participation in the project. More than a simple aid for discussion, this procedure could therefore be perceived as a third party which provided the participants in the DNA bank project with a coherent framework for ethical decision-making and the reduction of uncertainty.

The third type of interaction that emerged did not result in a reduction of uncertainty through the exchanges that the form brought about. On the contrary, it resulted in a significant increase of uncertainty. In those cases the procedure was not called upon to expose the goals and modes of operation of the study or the uses of the DNA bank. The procedure of informed consent was instead perceived as a judicial device aimed at concealing something ("Why do I have to sign? Are there any risks?" one of the participants asked). The dialogue then revolved around the potential risks that the participants thought they might be exposed to if they gave their consent to the study: "Are you going to put me under the knife to get my DNA? [...]. Then why should I have to sign? I don't even know where you want to take me to...." Questions were also raised about the use that could be made of the genetic samples. "I hope you're not going to make GMOs or stuff like that, are you?"

As an extension of the third pattern of interaction, the fourth type exposes what could be interpreted as a "game of noncommitment" on the part of the participant, who ticked the items but did not sign the form, or signed it but failed to fill it out thoroughly. In such situations, the participants did not exactly refuse to sign or fill out the consent form but, faced with a situation which they seemed to perceive as uncertain and risky, they brought into play what we might call "avoidance tactics". "The surest way of averting danger is to avoid encounters where it might arise" (Goffman 1974: 17).

Finally, the last pattern we observed arose when the participants argued that their word was as good a guarantee for their commitment to participate in the study as their written consent. This pattern could be interpreted as the sign of a clash between two conflicting modes of trust, one based on a written document and the other on the "given word". Some of the remarks these participants made lead us to believe that they perceived the procedure as a sign of mistrust of them and of the genuineness of their commitment to the project.

Judging from the five types of interaction described above, it appears that the procedure of consent is not self-evident or commonly accepted. It may generate distrust or different forms of "avoidance", induce tensions in the interactions between the participants and the physician and even be openly criticized by the former. These observations show what could appear to be a certain form of distrust on the part of the participants towards a possible transfer of responsibilities which they see as the consequence of having added their signature to formalize their consent. The responsibility involved could cover several aspects: obviously, a judicial and administrative aspect is involved, even though the wording on the consent form clearly states that giving one's consent in no way exempts the researchers and medical doctors involved in the study from their responsibilities. But a moral aspect is also involved, since the request for a signature seems to be perceived as the sign that the medical personnel question the genuineness of the participant's commitment. These observations seem to point out that the frames of perception of the actors have reached a raised level of awareness intended to detect a potential power asymmetry that could be the result of a system that grants them greater autonomy and protection. From a quantitative point of view, the patterns of interaction which indicated that these situations were likely to arise (the third, fourth and fifth type) were not in the majority compared to the more common situations which, on the contrary, seemed to show a lack of awareness (the cases where the signature was added in silence, without reading the explanatory note first or apparently without arising mistrust).

Consent as Seen by Its "Subjects"

The interviews that we carried out with the participants in the days following their inclusion in the research protocol shed light on the patterns of interaction that we have briefly described above, as they gave us the opportunity to record their account of their participation in the informed consent procedure, and consequently their points of view. These interviews show that, even though in the majority of cases the procedure was not truly condemned, the participants had not necessarily felt compelled to adhere to it wholeheartedly, even in the case of those who had signed the consent form without any apparent hesitation. Surprisingly, the way consent is perceived by the "subjects" exposes a set of relatively systematized perceptions – and even points of view – that contradict a number of statements pertaining to medical ethics in general, and to the constitution of biobanks in particular. Notably these perceptions highlight the "*ambiguous*", paradoxical nature of the informed consent procedure, since in the eyes of some, it grants individuals a much greater autonomy, whereas for others it induces heteronomy in the relations between individuals and the medical profession. Thus, far from developing a fairy-tale vision of the biopolitics of consent, the points of view that we recorded point to the fact that biopolitics might also carry elements of constraint and power that are relatively unforeseen. They appear all the more insidious as they are embedded in procedures with the objective to expand the subject's autonomy.

Responsibilities

The relatively high incidence of interaction patterns characterized by the absence of questions about the study on the part of the participant, or by the fact that a considerable number of them added their signature without reading the consent form, can partly be explained by the trust factor. The interpretation of a bond based on trust generally falls under one of two types of analysis: one highlights the fact that trust is an act of self-giving and submission to the power of another individual, and the other emphasizes that the trust manifested by a partner always implies the opening of a debt that must be paid off through some form of reciprocity.[2] Obviously the second alternative does not exclude the possibility of there being an unbalanced relationship between the two partners. However, when giving their trust, individuals put the recipients in a position where they have to comply with a moral obligation. It is precisely this obligation that certain participants call upon in their accounts, as we observed in the course of the interviews. The participants are encouraged to take an interest in the information provided to them, to read the explanatory note and to eventually make what may be described as an enlightened decision; yet, on the one hand, they feel incapable of acting in an "informed" manner, and on the other hand they express the trust they have invested in the medical actors. This indicates that there is, in practice as well as in theory, a profound gap between participants and a procedure that is intended to empower them to act as informed and autonomous subjects.

Although according to the accounts of certain participants, there is no way of knowing whether the DNA samples will be used in accordance with the existing

[2] As Karpik notes quoting Benveniste (1969), the relation of trust is an indication of an existing relation of exchange governed by the rule of reciprocity: "one who receives trust is in fact granted an open line of credit by his partner, and therefore his partner holds a letter of credit so the debt can only be cancelled by an equally important compensation in the form of protection or guarantee" (Karpik 1996: 528). Furthermore, the relation of exchange is also grounded in unequal conditions, as shown by the secondary meaning of *fides* (credit or trust): "putting one's fides in another person brought in return their guarantee and their support. But this very notion clearly shows that the conditions are not equal. Therefore there appears to be a power of authority that is exercised concurrently with a power of protection over the individuals that submit to it in exchange for their submission and as far as it extends. This relation implies the existence of a power of constraint on the one hand, and obedience on the other hand." (Benveniste 1969: 118–119).

legislation or will serve ethically "condemnable" purposes ("anyway if they want to do cloning they aren't going to tell us, so there is no way we can find out, and it's not even worth asking because they are not going to tell us"), the relation of trust called upon by the participants places the biomedical actors in a situation of holding a moral debt that they must pay off since failing to do so will bring about "punishment", and this is where we find one of the first attempts at describing the question of responsibility:

> I don't give a damn about knowing what this research is about. It takes too long to read [the explanatory note]. [...] Anyway I asked if I was going to feel discomfort or something, of course. That's what I want to know about. If I had been asked to take drugs and come back every month, well, I might have disagreed with that [...]; I don't intend to go and find out if what I think about genetics is true or completely wrong. There are doctors, I say, that's what they're here for. Then it all depends on whether you want to trust them or not. [...] If you want to be informed you're wasting your time, it's no use. [...] See, I'm a good participant, I don't ask questions [...]. I gave what I had to give. Now [it's up to] the doctors, [the] researchers, to do the rest. And I trust them to do what's right. If they put it to a bad use, God will punish them, and then the devil will burn their feet. That's just a way of describing it, but it's what it comes down to, really (Int. No. 35, 2nd age bracket, case subject).

This sort of account contradicts a number of studies carried out in the field of medical ethics – quite often by physicians – that conclude that the state of ignorance of the participants in the study is directly linked to their lack of information. According to these studies, this lack of information is a result of either the elements provided being intentionally partial, or their formulation being inappropriate and incomprehensible for the lay public (Moutel 2003). The ignorance of the participants may also be voluntary (Michael 1996) and associated with a concept of responsibility (of a medical, ethical and moral kind) that should be ascribed to the medical actors. Applying these parameters to the question of responsibility also contradicts all the ethical declarations mentioned in our introduction that intend to turn the biopolitics of informed consent into the keeper of the contemporary regulation system for the development and uses of gene technology.

It may be worth noting that the concept of responsibility invoked by some of the participants we interviewed carries within itself, albeit in an implicit and popularized form, the elements that provide the basis for a contemporary theoretical and critical approach of those institutionalized procedures which value the involvement and participation of citizens. So the attempts made to avert the dangers associated with the potential disengagement of the authorities and the medical institutions really have enabled researchers to identify certain risks of misuse posed by the act of delegating responsibility to citizens through their involvement in decisions concerning health or biomedical research in the general sense (Polton 2000), and more particularly within the framework of the advances in gene technology (Moulin 2000).

Even more explicitly, some of the accounts collected from participants set forth their perception of the intrinsically ambiguous nature of the informed consent procedure. The following remarks were formulated by another participant, who expressed his concern about the use that might be made of the DNA sample, and about the uncertainties that lie heavily over the human genome research at large:

> "Informed", that would mean having the maximum, knowing the real objectives of the experiment, and all the ins and outs, what it is really for, what act is going to be performed... it sounds ambiguous to me! Because practically speaking what do we know about their intentions? [...] If there is a hidden agenda, at least my intention is not to manufacture weapons or ways to oppress the people. If they [the actors of the DNA bank project] use it to make a diabolical weapon, I don't know but anyway, come to think of it, it's not my responsibility. [...] Whatever can be done with this knowledge, it doesn't belong to me... So OK to do the research, but then again, the applications may not be too good! Einstein worked on relativity, but hey, what did Oppenheimer do with it... So we do research... but again, when you think about it, it's the responsibility of whoever takes fundamental knowledge to decide if they want a biological weapon or a vaccine ... (Int. No. 14, intermediary profession, 1st age bracket, control subject).

The question of responsibility arises all the more forcefully now that the participants have become aware that certain risks linked to their involvement in a human genome research project may turn out to be real. The ambiguity of the procedure is linked to the fact that it is presented as a tool for delivering information fully and thoroughly, in a situation where the information can only be partial and biased depending on the conclusion the participant has drawn. From then on, the scheme is potentially seen as investing the subject, who supposedly gave his "informed" consent, with moral and ethical responsibilities that the subject says he does not wish to assume and, more important, that he says he cannot shoulder given his own knowledge of the project and the fact that he cannot possibly know the intentions of others. Although the written dimension of the consent makes individuals aware of the nature of the study in which they have agreed to participate, it also simultaneously leads them to question openly the quality of the information provided, which in turn leads them to fear that their ethical responsibility might be coupled with legal responsibility as well:

> What I mean is that when I sign and I read something, the least I can do is pay a little bit of attention because you never know what might happen next... maybe they'll want to take blood again, right? Now I don't know what you're going to do with my genes; what do I know...[...] In medicine there are things we know about and things we don't know about. Some people even want to set up sperm banks. Personally I've got nothing against it but, see, I for one wouldn't do it. I say! I wouldn't like to know that my children are walking around like that. Maybe there was a risk, who knows... I don't know... but DNA?... what's that?... they can't... the DNA is mine, it is my own, no one can use it... what do I know... what are they going to do with the DNA? Why take all these precautions? They made me sign that I agreed with all that, but why, I have no idea... (Int. No. 30, Company worker, 3rd age bracket, control subject).

Counter-Intuitive Asymmetry

We have until now chiefly addressed disapproving points of view expressed by the participants about the consent procedure. The time has come to point out the existence of more consensual points of view that nonetheless show an underlying critical approach to the procedure. Quite surprisingly, certain participants brought to the fore the inability of our survey to provide "decision-making clarification", given the individualized framework in which the consent was produced and that was part of the scheme itself. Thus, one of the participants interviewed confessed that he did not feel "too well informed", but also pointed out that additional clarification might have been obtained through a meeting designed to inform and promote exchanges between the potential participants and the biomedical actors of the project. Far from criticizing the fact that such a meeting had not been set up in advance in order to help participants understand the ins and outs of the DNA bank project, he justified the absence of it. As a matter of fact, he seemed to think that it was sensible not to set up an information meeting, as the existing procedure seemed to generate higher rates of participation. Informing individuals collectively would doubtlessly increase the level of interaction between the biomedical team and the participants and would produce better informed decisions, but it might make a greater number of potential participants "'reluctant":

> Oh, I don't think I was well informed. What happened was, I was told to read the paper, and I did, sort of, and then... but anyway, I already knew a bit about those things, I could tell more or less what it was about. [...] Informed consent?... They would have had to call everybody, put them all together in a room and teach them for one hour. There. Then you can call yourself informed. But if you don't do that, if you just tell them "you know, look at that, hey, that's what you need to do and then that's it", well then the guy says yes, he tells you yes, but that doesn't mean he's very well informed (he laughs). That's right! You understand? If you take a course, then you're well-informed, "Does anybody have a question to ask?" Fine. That's informed. But you know what: I can see why they don't want to do things like that. –Interviewer: Why? –Interviewee: (silence) because... let me tell you, this is my own experience, the more you tell people, the more they go 'yes, but' and the more reluctant they get in the end. See? Because you make up your own mind naturally, and you say I'll do it. But then you're in a room, and everybody is saying 'of course, there is this and that, you know, there's that, there's that!' and you end up with a bunch of people saying: 'Well, after all, come to think of it, maybe there's some truth to it...' and blah blah blah, and when all's said and done, instead of having thirty people that go for it, you end up with twenty-five. Or twenty. So there. So for silly things like that it's not worth getting people to argue over nothing (Int. No. 44, Shopkeeper, 3rd age bracket, case subject).

These lines of reasoning show two dimensions of the issue that we believe are worth emphasizing. On the one hand, they indicate that in the context of individualized consent-building, underpinned by a procedure where the participants are called in one by one, the situation seems to foster cooperation, and the participants refrain from intervening actively in an exchange process. On the other hand, they point out that the existing scheme, being designed to call in one person at a time, seems to prevent individuals from using the resources that a group might offer when discussing participation and exposing the process involved. As we study the emic objectivation of the recruitment process and the collection of consent, their structure seems to generate a certain number of effects on decision-making and the nature of the act of cooperation involved. These effects, which are of a counter-intuitive nature insofar as the consent procedure is supposed to lead the individual to make an informed decision, might produce asymmetric power relations between the participant(s) and the actors of the DNA bank.

Conclusions

The patterns of interaction that we have observed, as well as the points of view of the participants recorded in interviews in many cases, indicate that the frames of perception and interpretation of the actors have reached a raised level of awareness intended to detect the potential power asymmetry that might result from a scheme originally designed to provide information and grant their autonomy. The actors in the field of medical research regulation will find the elements in this study useful in planning the evolution of their research and design its practical ethical framework, particularly in the case of biobank projects. The informed consent procedure is not self-evident in the interactions that take place, since many of its users prefer to call into play different ways of keeping oneself at a distance ("quant-à-soi") to eschew the potential risks it could lead them to assume: "avoidance" of the signing procedure, refusal to append their signature, pressing demands to the physician that he justifies this request, etc. Furthermore, the scheme may be criticized for its formal layout and the abusive interpretation of the physician/patient relationship that it seems to officialize by making it contractual. In the course of the interviews we found an explanation centred on two important issues. The first one is the diverse potential responsibilities that the participants identified and linked to their full involvement in the informed consent procedure. The second issue is the counter-intuitive asymmetry that might be generated by the individualized configuration of consent-building.

Our survey points out that research participants call upon their critical abilities in the application of the biopolitics of consent: they strive to detect any potential perverse effects and do not hold a fairy-tale vision of it even if they do not contest it radically. Their reactions and analyses, grounded or not, indicate caution in their assessment and a critical approach towards any form of constraint that might result from the contemporary implementation of biopolitics, and notably concerning the procedures for devolution of responsibilities highlighted in the literature over the past few years. These reactions show that the majority of the actors are inclined to adhere a priori to the current process of biopolitics. In this case, DNA "donors" should not be viewed as disappointed by the institutional policy on the subject, which these days highlights the concept of user autonomy, information and participation. It is obvious that these critical points of view must not be seen solely as an agent retarding the progress of biobank projects, but also as a competency likely to prove useful to the development of such projects, which may provide the basis for a co-production process.[3]

The question of the devolution of responsibilities and the reactions it may stir become all the more important since the context of the research that we have studied overlaps with a sphere of knowledge and action which is, for more than one reason, highly symbolic of the "risk society" described by Beck (1992). The advances that enable us to explore the genomes, keep them in banks and preserve them and the potential applications of this line of research contribute to feed the "anguished

---

[3] For more about this topic, see Chapter 9.

conscience" which seems to be the hallmark of the contemporary world of science and technology.

Therefore, the caution exercised by users when it comes to the new biopolitics is understandable given the anguished conscience of the world. This state of mind may well be the reason why the actors involved in a procedure such as informed consent feel the need to preserve their "quant-à-soi". But this should not cause us to forget that the biopolitics of informed consent may also in turn play a role in building this conscience. In fact, it could well be that once confronted with the choice of what can and cannot be done with the body and its parts, put in a situation they never had to face before, the actors can only become more cautious, wary or even at times diffident when they consider the responsibilities with which they may be invested.

Acknowledgments The authors thank the research team members and the participants for observation and interviews. Useful discussions with Emmanuelle Rial-Sebbag, Pierre Antoine Gourraud and Jacques Lefrançois are gratefully acknowledged. This work was performed with the financial support of the project "Mapping the language of Research Biobanking" (Contract No. 159864/S10) of the Norwegian Research Council and the Norwegian Institute of Public Health, the project "Research biobanking and the ethics of transparent communication" (Contract No. 182269) of the Norwegian Research Council, the FP6 EU Coordination Action PHOEBE (Contract No. LSHG-CT-2006–518148) and an Incitative Concerted Action "Internationalisation of HSS" (No. 04 2 533), GenPos (Human population genetics and public health: multidisciplinary analysis and international comparison of the normative contexts and practices) of the French Ministry for National Education, Higher Education and Research.

References

Beck U (1992) Risk Society: Towards a New Modernity. Sage, New Delhi

Benveniste E (1969) Le vocabulaire des institutions européennes, tome 1: économie, parenté, société. Editions de Minuit, Paris

Berlivet L (2004) Une biopolitique de l'éducation pour la santé. In: Fassin D, Memmi D, *Le gouvernement des corps*, Editions de l'Ecole des Hautes Etudes en Sciences Sociales, Paris pp 37–35

Bungener M et al. (2002) Quelle médecine voulons-nous. La dispute, Paris

Cambon-Thomsen A et al. (2007) Trends in ethical and legal frameworks for the use of human biobanks. Eur Respir J 30:373–382

Caze De Mongolfier S (2002) Collecte, stockage et utilisation des produits du corps humain dans le cadre des recherches en génétique: états des lieux, historique, éthique et juridique. Ph.D. Thesis, Université René Descartes, Faculté de médecine de Necker-Paris

Dozon JP (2001) Quatre modèles de prévention. In: Dozon JP, Fassin D (Eds.) Critique de la santé publique. Une approche anthropologique. Balland, Paris, pp 23–46

Foucault M (1976) La volonté de savoir. Gallimard, Paris

Foucault M (1994) La politique de santé au XVIIIème siècle. In: Dits et écrits. Gallimard, Paris, pp 13–27

Goffman E (1974) Les rites d'interaction. Editions de Minuit, Paris

HUGO-ELSI Committee (1995) Statement on the principled conduct of genetics research. Eubios J Asian Int Bioethics 6:59–60

Karpik L (1996) Dispositifs de confiance et engagements crédibles. Sociologie du travail 4:527–549

Martuccelli D (2004) Figures de la domination. Revue Française de Sociologie, 45:469–497
Memmi D (2001) Surveiller les corps aujourd'hui: un dispositif en mutation? In: Sfez L (Ed.) L'utopie de la santé parfaite. Presses Universitaires de France, Paris, pp 173–184
Memmi D (2003) Faire vivre et laisser mourir: le gouvernement contemporain de la naissance et de la mort. La Découverte, Paris
Michael M (1996) Ignoring Science: discourses of ignorance in the public understanding of science. In: Wynne B, Irwin A (Eds.) Misunderstanding Science? The Public Reconstruction of Science and Technology. Cambridge University Press, Cambridge, pp 107–125
Moulin AM (2000) Les sociétés au siècle de la santé. Médecine/sciences 11:1186–1191
Moutel G (2003) Le consentement dans les pratiques de soins et de recherche en médecine. Entre idéalismes et réalités cliniques. L'Harmattan, Paris
Polton D (2000) Quel système de santé à l'horizon 2020? (Rapport préparatoire au schéma de services collectifs sanitaires). La Documentation française, Paris

Information Rights on the Edge of Ignorance

Anne Maria Skrikerud

Abstract In this chapter I discuss whether there is such a thing as a right to information in biobank research. The concept of "information" is discussed and different theories about what it means to have the right to information are presented. The way in which the right to information may influence moral problems connected to epidemiological biobank research is also discussed.

Introduction

In the bioethical literature "information" is frequently imbued with a variety of meanings, partly because "information" is often discussed in the special context of a *right to information*. But what *right to information* means is not always clear in medical research.

In Norway, medical research is regulated by several acts and regulations, in particular the Personal Data Act and the Patients' Rights Act, which give patients the right of access to their medical records. In addition, international ethical guidelines, such as the Declaration of Helsinki, also regulate medical research. What awoke my interest here is the recurrent use of the concept of the "right" of donors or patients to either demand to know about all data that is catalogued or analysed in an identifiable or coded manner, to demand a copy of their medical records, or to be informed in an understandable way about medical procedures related to studies that are about to be conducted, before giving their voluntary consent (Norway, 1999, 2000, 2001, 2003; World Medical Association 2004).

Behind these ideas of informed consent and the right to be informed about all personal data that is filed somewhere, no matter what kind of data and where it is stored, it seems that a common idea of *a right to information* is lurking. My aim

A.M. Skrikerud
Section for Medical Ethics, Faculty of Medicine, University of Oslo, Oslo, Norway
e-mail: a.m.skrikerud@medisin.uio.no

J.H. Solbakk, et al. (eds.), *The Ethics of Research Biobanking*,
DOI: 10.1007/978-0-387-93872-1_4, © Springer Science+Business Media, LLC 2009

in this chapter is to scrutinize the concept "the right to information" in biomedical research and in particular in biobank research.

The Concepts of "Information" and "Data"

In my view, in order to understand the concept "right to information" it is important to understand the meaning of *information* and the difference between *information* and *data*. In fact, "information" and "data" are two quite similar concepts in medical research.

According to Bartha Knoppers, three groups of *data* are of interest in biomedical research. These are personal data, medical data, and genetic data. *Personal data* is a legal term, defined in the Norwegian Personal Data Act as "any information and assessment that may be linked to a natural person". *Personal data* can be any data that can be stored in a filing system, be it school grades or which DVD you rented last week. *Medical data* are identifiable or coded data that are health related in any way. *Genetic data* are identifiable or coded data on individuals' genes that are systematically analysed and filed.[1] However, these legal definitions of data do not necessarily correspond completely with the prevailing understanding in medicine.

Without being a legal or technical term in the same sense as "data" has turned out to be, "information", without claiming to be precise, covers all types of data. Information is a *vague* concept which means that its meanings differ from individual to individual and from context to context (Waismann 1946). But in general and for the most part we can say that in bioethics "information" encompasses facts of any kind that are communicated. Medical data that are filed in a computer that is turned off and never turned on again will still count as medical data. But at the moment Mr. Smith demands to be told the results of some medical test and a researcher tells him, medical data become *information*. Information can also be communicated facts in the informed consent context. In this context, the patient or donor has the right and duty to be informed about forthcoming procedures. The terms *information* and *data* are often confused and used interchangeably in everyday language.

The Concept of "Right"

When we say that somebody has the *right* to something, we usually mean that something is due to somebody. In saying this, I do not aim for an all-embracing analysis of the concept of "right", but will focus specifically on rights pertaining to biomedical research. So, if a research subject has the *right to information*, what does the word *right* mean in this context? Traditionally rights have been evaluated as either fundamental or derivative rights. Fundamental rights are rights that are given and

[1] I refer to a lecture held by Bartha Knoppers at the seminar "Ethical Challenges in gene-epidemiological research and health registry research", Oslo, August 20–21, 2004.

derivative rights are rights that follow from more fundamental rights. One can hardly claim that the *right to information* is a fundamental right, but if the right to information is a derivative right, it is not clear which right *the right to information* is a derivative of. It may be, and it has often been argued, that the right to information is a derivative of the right to privacy (McGleenan 1997; Allen 1997), while others have claimed that the right to information is a necessary derivative from the right to make autonomous decisions (Häyry and Takala 2001; Harris and Keywood 2001). The right to information does exist as a legal right since it is ensured by most western laws and the Declaration of Helsinki, but if the right is a necessary corollary to a fundamental moral right such as the right to make decisions on autonomous grounds, the right also has moral value.

I will consider two theories that may help to explain the importance of a *right to information* as given in international guidelines: rights theory and liberal utilitarianism. The theory that first and foremost justifies the use of "rights" as a moral value and a legal concept is, of course, rights theory, which has evolved since the end of the eighteenth century. This is a theory that has evolved under strong influence from legal jurisprudence and should be distinguished from human rights ethics (Almond 1991). Hohfeld and Sumner have argued that there is a connection between rights and obligations (Hohfeld 1917; Sumner 1987: 18–31), but only some rights are bound to a duty-concept. The following four categories are all possible understandings of the concept "rights" according to Sumner:

- *Claims:* A claim is a right that generates a corresponding duty by someone else
- *Powers:* Power is a right that provides someone with the possibility to affect the rights of others
- *Liberties:* Liberty is the right to act or refrain from acting. A liberty permits an action
- *Immunities:* Immunity is a right to be protected from the actions of others (Sumner 1987: 18–31; Almond 1991)

According to rights theory, the *right to information* is a claim for the subjects of medical research. Influenced by John Stuart Mill, liberal utilitarianism evaluates states of affairs according to what gives most utility. However, there is a set of values that cannot be sacrificed for any type of utility. Values that are especially important are the values of life, health and autonomy (Häyry 1994). Liberal utilitarianism in bioethics will often be interpreted with what gives most health to mankind without sacrificing the autonomy, health or life of third parties.

An example of a right to know that is equivalent to the right to information is given by Matti Häyry and Tuija Takala in their essay "Genetic Information, Rights and Autonomy". On the basis of the right to autonomy, they argue that if A has a right to know, this may mean at least three different things: "When A has no duty to remain ignorant" this is called a licence. "When others have a duty not to interfere with A's quest for information" this is called a negative claim-right. "When somebody has a positive duty to assist A in her quest for information" this is called a positive claim-right (Häyry and Takala 2001). Usually the right to information in a biomedical context will mean a positive claim-right to information.

To decide whether information exchange according to the right to information has taken place, it is important to keep in mind that the subject that possesses the right must be able to fulfil it; in addition, there must be someone or some public service that has the *duty* to inform, and finally the information must be handed over (Almond 1991; Sumner 1987; Hohfeld 1917). Now if we use this definition in various situations where information exchange is taking place, we should be able to identify at least one right-holder and preferably one duty-holder. Since information is communication, the right demands someone that is obliged to inform, unless of course the information is of a kind that is publicly available. But if the information is not available and if no one has a duty to give information, the right turns into a right without a way of fulfilling it (Sumner 1987). Looking at some examples from biobank research, I will try to analyse the *right to information* in a biomedical research context.

The Right to Information in Research Biobanking

When a subject decides by request that he or she wants to be a donor in a research-biobank, the donor has to give his or her voluntary informed consent. The reason why we have informed consent is partly related to the idea of the right to be informed, since the informed consent form is about making an informed free choice. But the way it has developed in modern research ethics is that the donor is presented with an obligatory act. On request the donor uses his or her *power* to voluntarily decide whether to participate in the research or not. But before the blood test can be taken or the mouthwash given, the donor has to be informed about matters required by law and the Declaration of Helsinki, matters that might be of no interest whatsoever to the donor, e.g., whether the biobank has any conflicts of interest. Häyry's and Takala's interpretation will not leave room for a way to distinguish between this right and a right which the subject may opt for having fulfilled or not. So according to their liberal utilitarian view, this will be a *positive claim-right*. On the other hand, Sumner would call this a *mandatory right*; a right that has gone from the stage from being legally permitted to that of being legally prescribed. In this case the donor has the liberty to be informed without the liberty not to be informed: a so-called *half-liberty* (Sumner 1987).

The donor also has the right to be informed about any medical or genetic data about him or herself that are filed in a biobank. In addition, the donor has the right to be informed about these rights. These rights have legal status in all countries that have ratified the "Directive 95/46/EC of the European Parliament on the protection of individual with regard to the processing of personal data and on the free movement of such data", which says: "...any person must be able to exercise the right of access to data relating to him which are processed, in order to verify in particular the data and the lawfulness of the processing...". "Processed" here means being filed in a systematic way. In several countries the laws that have been introduced to comply with this Directive have been formulated in a way that can be interpreted as if the donor has a claim to information, and that the researcher or the director of the department has the a duty to give information to the donor. The Norwegian Personal Data Act reads as follows: "Any person who so requests shall be informed of...the categories of data concerning the data subject that are being processed..." (Norway 2000).

A multinational biobank that opts for the storage of anonymous data has an informed consent form in which donors are asked to agree to the following statement: "I will not get the results of my DNA sent to me from this project".[2] Hence, from the start of a biobank project, the researchers argue for the possibility to withhold all genetic data if they want to. According to rights theory, the claim of the donors is turned into a *nullity*, because the researchers are ensured the *immunity* of the constraint of the duty by the ethical committee. Liberal utilitarianists would say that the *positive right* has turned into a *licence* (Sumner 1987; Häyry and Takala 2001). This example raises the question: Why do biobanks inform their donors that they will not receive any information about their DNA? There are several reasons for this. One reason is that in multinational epidemiological research data from an identifiable donor are not just located in one file in one computer but are used by researchers in several countries and in several studies. Thus, for practical or technical reasons the data may simply be inaccessible.

Another reason why biobanks do not want to report results of personal DNA analysis back to donors is illustrated by the Danish twin registry. In this case the researchers have to give back information about medical data but are not keen to do it and are concerned about the implications of giving the information. Their concern is on behalf of the research subjects because the researchers can only give research data that have not been medically verified to clinical standards, meaning that there is a statistical risk that the data may be wrong. Some of the research subjects will probably give the results to their general practitioner, where it will most probably end in the patients' medical notes without any caveat concerning the validity of the information. If these research subjects later apply for life insurance, in Denmark insurance companies can ask for permission to contact the general practitioner and read the patients' medical records. A lot of Danes automatically tick off yes to this question, but then information concerning some inheritable genetic disease, which has not been clinically validated, ends up in the hands of an insurance company.[3] The dilemmas faced by the researchers because of the duty of informing are so difficult that they are experienced as ethically deeply problematic.

---

[2] Translated from the Norwegian by the author.

[3] Reference to this example has been made by Kirsten Ohm-Kyvik during discussions at the seminar "Ethical challenges in gene-epidemiological research and health registry research", Oslo, 2004.

The Right Not to Be Informed

The right to be informed is a derivative of more fundamental human rights, but the right not to know is often claimed to be a derivative of the right to information. In 1947 the first case on the right not to be informed was ruled, *Breard v. City of Alexandria*. The case was a door seller who had delivered shop catalogues in Alexandria in the 1930s. He claimed his right to do so. The US Supreme court, however, upheld the right of the individual to be protected from unwanted information on the basis of the right to privacy. Since then the right to be informed and the right not to be informed have been considered as derivative rights from the right to privacy (McGleenan 1997). The right to privacy, which is a legal right, is said to protect the right to be left alone when another civilian's "freedom of speech threatens to disrupt one's liberty of thought and space" (McGleenan 1997). An interpretation which Sumner provides may also explain the right not to be informed. In his view, the subject may either have the *liberty to refrain from being informed* or the *claim to ignorance*. The latter generates a duty for others not to inform. The liberty to refrain from information is dependent on others not having the claim or possibility to inform. The claim to ignorance is dependent upon that there is no one other than those who have the *duty not to inform*, which have the *possibility to inform* (Sumner 1987).

That the right to ignorance can be argued on the basis of autonomy has been disputed since autonomy has to do with the possibility to choose and ignorance does not support a person's autonomy (Harris and Keywood 2001). However, Häyry and Takala have developed a Millean interpretation of the right to ignorance of genetic data based on a person's right to autonomy. I will use their conclusion in my analysis of the right not to be informed in the biobank setting. They argue that if B has the right not to know, this may mean three different things: "… [either] B has no duty to know….[or] Others have a duty not to inform B against his will….[or] Somebody has a positive duty to assist B in remaining in ignorance" (Häyry and Takala 2001).

When a researcher in a biobank wants to recruit donors, he usually gets the permission from the ethical committees to look in health registers or birth registers and by means of those registers to identify possible donors to match into the planned cohort. Often several members of a family are recruited. So when several members of a family are contacted in order to be recruited as donors for a particular research project, they may come to suspect that they have a special gene in their family. But maybe some family member does not want to know that the heart attack of her brother might also strike her. The right not to know has not been respected. According to Häyry's and Takala's interpretation, it is not certain that something wrong has happened. If the government has a duty to protect this woman against knowledge, their duty has not been discharged very well, since it is government bodies such as the ethical committee that has allowed the researchers to inform, and it is those in charge of the health registers who have allowed the researchers to look in the registers. But none of these public institutions know or can know that this woman does not want to be informed. The researchers are the ones that have force-fed this woman with information and by doing this not upheld their duty not to inform. They have breached the woman's negative right not to be force-fed with information. But if these researchers were duty-holders to this woman and others like her, it would actually not be possible to conduct any research at all. If her right only is a right to herself, without having the force to impose a duty on somebody, no one on the other hand can be charged of having violated it.

According to Sumner, the woman will have the liberty to refrain from being informed. One can perhaps also say she has the claim not to be informed but it is difficult to point out who is the duty-holder, except for her general practitioner. The researcher is given permission by the ethical committee to inform and thus to override the liberty of the woman. The researcher has the right and liberty to send out information, and the liberty to inform appears to rule out the liberty to refrain from being informed.

Conclusions

The European Convention on Human Rights and Biomedicine states in Section 10.2 that "everyone is entitled to know any information collected about his or her health. However, the wishes of individuals not to be so informed shall be observed". Sumner's rights theory seems to indicate that the right to information is of greater value because it is easier to decide who holds the duty *to* inform compared to who holds the duty *not to* inform. Even a claim-holder of the right not to be informed may easily come in contact with somebody who is well informed about the medical data, but who cannot plausibly be said to be under a duty *not to* inform. If this claim not to be informed is to have real meaning, one has to be sure that all people to whom the information is available are duty-holders to that specific person. This poses an ethical challenge on biobanks as to how to handle information about donors.

However, for it to have any effect the right to information is also dependent on at least one duty-holder. For example I might have the right to be informed about where Atlantis is, but I will never find anyone with the duty to tell me. So if the duty-holder is the same as the one who is in charge of the medical data in the biobanks, I have the right to information and I may have this right fulfilled. But if the one in charge of the medical data is immune against the duty to tell me, as in the case with the anonymous biobank, I may claim my right as much as I want. However, whether I get the information that I have asked for will depend on the goodwill of the holder of the information.

Those who have the power to influence the claims of others are often the ethical committees. They may find studies acceptable or unacceptable and they are aware of the conditions of the different studies. The researchers are given the duty and the right to protect the value of medical research in society. In some circumstances this will override the individual rights of the donors.

Acknowledgments The financial support of the project "Mapping the language of Research Biobanking" (Contract No. 159864/S10) of the Norwegian Research Council and the Norwegian Institute of Public Health and the project "Research biobanking and the ethics of transparent communication" (Contract No. 182269) of the Norwegian Research Council are greatly acknowledged. In addition I thank Professor Søren Holm and Professor Jan Helge Solbakk for supervising my project.

References

Allen AL (1997) Genetic Privacy: Emerging Concepts and Values. In: Rothstein M (Ed.) Genetic Secrets: Protecting Privacy and Confidentiality in the Genetic Era. Yale University Press, New Haven, pp 31–59

Almond B (1991) Rights. In: Singer PA (Ed.) Companion to Ethics. Blackwell, Oxford, pp 259–269

Harris J and Keywood K (2001) Ignorance, information and autonomy. Theoretical Medicine and Bioethics 22:415–436

Häyry M (1994) Liberal Utilitarianism and Applied Ethics. Routledge, London

Häyry M and Takala T (2001) Genetic information, rights, and autonomy. Theoretical Medicine and Bioethics 22:403–414

Hohfeld W (1917) Fundamental legal conceptions as applied in judicial reasoning. Yale Law Journal 26:710–771

McGleenan T (1997) Rights to know and not to know: Is there a need for a genetic privacy law? In: Chadwick R, Levitt M, Shickle D (Eds.) The Right to Know and the Right not to Know. Ashgate, Aldershot, pp 43–54

Norway (1999) The Act of 2 July 1999, No. 63 relating to Patients' Rights

Norway (2000) The Act of 14 April 2000, No. 31 relating to the processing of personal data

Norway (2001) The Act of 18 May 2001, No. 24 on Personal Health Data Filing Systems and the Processing of Personal Health Data

Norway (2003) The Act of 21 February 2003, No.12 relating to Biobanks

Sumner LW (1987) The Moral Foundation of Rights. Clarendon, Oxford

Waismann F (1946) The many-level-structure of language. Synthese 5:221–229

World Medical Association (WMA) (2004) The Declaration of Helsinki

The Dubious Uniqueness of Genetic Information[1]

Anne Maria Skrikerud

Abstract In population research today, special regulations concerning genetic information are based on the view that accidental disclosure of personal information will be more harmful if the information is genetic. In biobanks the data used for epidemiological research will contain both genetic and non-genetic information. In this chapter four conditions are discussed that should be met in order for genetic information to be harmful when compared with non-genetic information. A key question is whether the practice of emphasising genetic information in the informed consent sheet puts the ethical rights of the donors at risk. Little awareness in ethical committees of the rights of the donors regarding non-genetic information may have a negative influence on how biobanks handle this information.

Introduction

With the emergence of huge population biobanks for research, a number of new national and international regulatory instruments and treaties have been developed, which focus on genetic factors in common multifactorial diseases. These regulatory instruments and treaties tend to view biobanks as gene banks and therefore, considering the implications of genetic information being stored in these banks, concerns are raised about whether the generation/production and potential disclosure of genetic information may harm the donor by stigmatization. Thus genetic information is treated differently when compared to non-genetic health information, both practically and legally. Does this constitute a potential risk of harm to the

[1] An earlier version of this paper was presented at the Conference of the European Society for Philosophy, Medicine and Health Care in Helsinki, August 2006.

A.M. Skrikerud
Section for Medical Ethics, Faculty of Medicine, University of Oslo, Oslo, Norway
e-mail: a.m.skrikerud@medisin.uio.no

J.H. Solbakk, et al. (eds.), *The Ethics of Research Biobanking*,
DOI: 10.1007/978-0-387-93872-1_5, © Springer Science+Business Media, LLC 2009

privacy of the donor? I will attempt to analyse the difference in risk between the two kinds of information, in particular with a focus on the issue of accidental disclosure to third parties, that is, the situation where information becomes publicly available by mistake. There is a significant literature on whether employers or insurance companies should be allowed to demand genetic and other forms of health information, but that is not the primary question I am interested in here, neither will I focus on the harm donors or members of their family may experience if genetic information is accidentally disclosed to them.

The Potential Risk to the Donor

In 1995 Annas, Glantz and Roche published the article "Drafting the Genetic Privacy Act: Science, Policy and Practical Considerations". The article was published in connection with the release of a draft for a genetic privacy act that the authors had been engaged in. In this article they debated the issue of "...whether the genetic information so obtained is different in kind from other medical information..., and, if so, whether this means that it should receive special legal protection" (Annas et al. 1995: 360). They maintain that genetic information is unique in its kind. They give three reasons why they consider genetic information to be "uniquely private or personal information" (Annas et al. 1995: 360). First, "it can predict an individual's likely medical future for a variety of conditions" (Annas et al. 1995: 360). The "likely medical future" is elaborated in the following way:

> [t]he information in one's genetic code can be thought of as a coded probabilistic future diary because it describes an important part of a person's unique future and, as such, can affect and undermine an individual's view of his/her life's possibilities (Annas et al. 1995: 360).

In saying so they implicitly deny the same potential to other health information. They take into account the fact that genetic testing techniques of 1995 did not cover all genes and what that might imply for genetic information:

> ... even if one concludes that genetic information that can currently be derived from DNA analysis is like other sensitive medical information, the DNA sample itself, with its ability to yield far more information in the future remains unique (Annas et al. 1995: 360).

The second reason why they consider genetic information to be unique is that "it divulges personal information about one's parents, siblings, and children" (Annas et al. 1995: 360). They describe the concern for family members as follows: "Deciphering an individual's genetic code also provides the reader with probabilistic health information about that individual's family, especially close relatives like parents, siblings, and children" (Annas et al. 1995: 360). The third argument for uniqueness is its reference to discrimination: "... genetic information and misinformation has been used by governments ... to discriminate viciously against those perceived as genetically unfit to restrict their reproductive decisions" (Annas et al. 1995: 360).

Phrases such as "coded probabilistic future diary" and "likely medical future" may, of course, seem naïve with the knowledge of 2008, since it seems natural to interpret this in a way that genetic information is more deterministic than today's research has been able to confirm. However, I understand the view of Annas, Glantz and Roche when they say that, with genetic tests, it is currently possible, or will in the near future be possible to give a more precise evaluation of an individual's current and future health status, compared to other health information that is used, and that this may have unwanted consequences for families and population groups. This special characteristic of genes and genetic testing, therefore leads Annas, Glantz and Roche to define genetic information as *unique*. In their view the *uniqueness* has at least two aspects. On the one hand genetic information is *unique* because it is presumed to be very precise in its ability to give information about an individual's present and future health status, a characteristic that other forms of health information do not have. On the other hand genetic information is *unique* when compared with other health information, in the sense that it is not information ascribable to any individual, but only to one particular individual. With their three arguments and their view on the special uniqueness of genetic information they conclude that "the genetic information is uniquely powerful and uniquely personal, and thus merits unique privacy protection", suggesting special privacy regulations concerning genes in order to "protect human rights *before* the technology is in wide use" (Annas et al. 1995: 365).

Lately the uniqueness of genetic information when compared with non-genetic information has been questioned, partly by invalidating the arguments of the proponents of special regulations, and partly by showing that most concerns relating to genetic information have non-genetic counterparts (Murray 1997; Holm 1999; Green and Botkin 2003; Kakuk 2006), thus assuming that the risk of discrimination is real, but not greater than for non-genetic health information. Thomas H. Murray argues directly against the claims of Annas, Glantz and Roche in his chapter "Genetic Exceptionalism and 'Future Diaries': Is Genetic Information Different from Other Medical Information?" He admits that, "[g]enetic information does not have to be unique in order to warrant special protection, but it does have to be distinctive and especially sensitive" (Murray 1997: 64). With this in mind he questions whether the arguments of Annas and his co-authors actually show that genetic information is distinctive or especially sensitive. Murray argues in the following way against the first argument of Annas, Glantz and Roche that genetic information gives information about the future of the owner:

> [g]enetic information is neither unique nor distinctive in its ability to offer probabilistic peeks into our future health. Many other things afford equally interesting predictions ... examples include asymptomatic hepatitis B infections, early HIV infection, and even one's cholesterol level. These have implications for future health that are every bit as cogent and sensitive as genetic predispositions (Murray 1997: 64).

The second argument of Annas, Glantz and Roche is the concern that genetic information is also information about the close family and relatives. Murray does not approve of this argument either:

> [t]hat one member of a family has tuberculosis is certainly relevant to the rest of the household, all of whom are in danger of infection, along with everyone who works with or goes to school with the infected individual. Or suppose the main wage earner in the household showed early signs of heart disease that could bring disability and death. Wouldn't the other family members have a profound important stake in knowing this? (Murray 1997: 65).

The claim that genetic information is unique in the sense that it may be used for discrimination is also not a convincing argument:

> [i]nstitutions and individuals can and have used all sorts of information, both visible and occult, as the basis for discrimination .... But it is difficult to make the argument that it is fair to discriminate on nongenetic factors but unfair to discriminate on genetic ones (Murray 1997: 65).

However, Murray admits that "concern for genetic information and discrimination may help explain some of the interest in genetic privacy because it broadens and sharpens important perceptions" (Murray 1997: 66), it "broadens the pool" of factors that may be used for discrimination and individuals who may suffer from discrimination, and it "sharpens" the moral intuition that one should not "be punished for things beyond our control" (Murray 1997: 66). Murray criticizes Annas, Glantz and Roche for basing their argument on what Murray calls, "the 'two-bucket theory' of disease. According to this model, there are two buckets – one labeled 'genetic,' the other labeled 'nongenetic' – and we should be able to toss every disease and risk factor into one of the two" (Murray 1997: 67). In Murray's view this theory, apart from a few exceptions, does not capture the complexity of causes and risk factors of diseases.

He concludes that even if there is no significant difference between genetic information and non-genetic forms of health information "[g]enetic information is special because we are inclined to treat it as mysterious, as having exceptional potency or significance, not because it differs in some fundamental way from all other sorts of information about us" (Murray 1997: 71). Therefore one should not undertake actions that underline the assumption that genes are special, but treat genetic information in the same way as other health information.

Murray illustrates the weaknesses of the assertion that genetic information needs especially strict privacy regulations. In Murray's view there is no extraordinary risk of harm to privacy compared with other health information. He disagrees with the view of Annas, Glantz and Roche that genetic information is *unique* and questions whether genetic information is particularly distinctive or sensitive when compared with other health information. He even questions whether there is any possible way to distinguish between genetic information and other health information. I understand him as a proponent for a privacy regulation that treats both *genetic information* and *non-genetic information* simply as *health information*.

However, there are few signs that this critique has been taken into consideration in European privacy and research regulations. The view of Annas, Glantz and Roche seems to prevail as the dominating influence on the legislation and regulation of biobanks. Thus more focus is directed at genetic information at the expense of non-genetic health information when assessing the risk to the donor of participating in research. Since 1995 a number of national and international regulations and treaties have emerged in this area. These policy documents emphasize that genetic information is in need of strict regulation acknowledging a high potential for discrimination. In the *Universal Declaration on the Human Genome and Human Rights*, UNESCO demands special regulations for the human genome and for research subjects participating in research on the genome (UNESCO 1997). Generally, the harm caused by genetic information being made public, thus influencing the donors' future job and insurance prospects, is considered to be the most severe risk (Wolf 1995; Annas et al. 1995). For example while health insurance companies may not use genetic information to refuse or enhance payments for health insurance (Mathiessen-Guyader 2005), they are entitled to use any non-genetic health information for the same purposes.

The consequence of the lack of regulations for non-genetic information is that this aspect of relation to biobanks receives less attention from ethical committees and in informed consent forms for the donor. The question I want to pursue here is whether there is a risk of abuse and discrimination of non-genetic health information in research biobanking, and whether the regulations as they are today reflect the needs of the donors. But first I want to take a closer look at what the difference between genetic information and other health information effectively means when it comes to epidemiological research biobanking.

Genetic and Other Health Information as Research Data in Biobanks

The main type of research data in population biobanks is genetic information. In addition, biobanks for epidemiological purposes make use of non-genetic health information. These two types of information are treated differently. First, when genetic information is obtained by DNA analysis the focus of research is unknown factors related to the genes themselves. A research biobank for epidemiological population studies is typically used to address the association between genetic factors, environmental factors and diseases. It is thus assumed that genetic factors may increase or decrease the risk of disease. Second, the methods used to obtain genetic data vs. other type of health data are quite different. Genetic information is obtained from laboratory analysis of a blood sample or biological material from a mouth swab, while non-genetic health data come, to a large extent, from health questionnaires, health surveys, or access to the donor's health record. Third, because non-genetic forms of health data often originate from health questionnaires, this allows donors the possibility of guarding their privacy with respect to the kind of information they wish to divulge. For example, the donor may consider body mass index (BMI) too shameful or may refuse to acknowledge drinking habits. In this way the donor can give incorrect answers or can choose not to respond to certain questions. Fourth, many countries in Europe have legislations that treat genetic information differently from non-genetic forms of health information.[2] In Norway and Sweden the biobank acts define biobanks as containing identified biological material. Genetic information in a biobank is regulated by special regulations, while non-genetic health information is regulated by general health data acts. Any other research data are not legally a part of the biobank (Norway 2003; Sweden 2002). In Denmark the *Act on Biomedical Research Ethics Committee System* (Denmark 2003) defines a research biobank in the same way as in Norway and Sweden, so research committees primarily evaluate biobank research on genetic and biological information, while the law is unclear about whether processing of non-biological information needs to be evaluated by a research ethics committee. The same lack of clarity is to be found in the UK, since the Medical Research Council demands ethical evaluation of research on biological material, but not of the same information gathered from health questionnaires (Medical Research Council 2007). In addition to this there are specific acts or regulations that restrict the use of genetic information once released. An example of this is the prohibition for employers to ask for and use genetic information in Denmark, Finland, France, UK and Norway. A similar prohibition for insurance companies to request and use genetic information exists in Denmark, France and Norway. A voluntary or government-based moratorium for insurance companies to use genetic information is found in Finland, Sweden and the UK. These examples illustrate that genetic information is highly protected against abuse for discriminatory purposes (Mathiessen-Guyader 2005).

The Interests of the Donor

In the Declaration of Helsinki (WMA 2004) it is emphasized that the risk of research must be weighed against the estimated benefits, especially when healthy individuals are involved as research subjects (§ 18). This means that if it is considered that biobanks may threaten the personal privacy and/or integrity of the research subjects, the benefits must be estimated to be more important. In addition, the research should in some way also benefit the research subject. Many population biobanks have a research protocol that mainly looks for reasons why some people and not others get common multifactorial diseases. In a population health perspective this is quite important. But the Declaration of Helsinki also states that "... considerations related to the well-being of the human subject should take precedence over the interest of science and society" (WMA 2004: § 5). In this respect, the Council for International Organizations of Medical Sciences (CIOMS 2002) has a more permissive view than the Declaration of Helsinki. The Council states in its *International Ethical Guidelines for Biomedical Research Involving Human Subjects* that

[2] I have chosen to take into consideration both legislation that is specific to biobank research and legislation that in some way affects the genetic privacy of the donor in society. The reason for this is that some legislation regulates how genetic information may be used in society if the information should accidentally be released.

> [r]isks of interventions that do not hold out the prospect of direct diagnostic, therapeutic or preventive benefit for the individual must be justified in relation to the expected benefits to society (generalizable knowledge). The risks presented by such interventions must be reasonable in relation to the importance of the knowledge to be gained (CIOMS 2002).

The subject may undergo research if it benefits science and society provided the risk is small. CIOMS chooses to understand the Declaration of Helsinki as not to "... preclude well-informed volunteers, capable of fully appreciating risks and benefits of an investigation, from participating in research for altruistic reasons ...." (CIOMS 2002).

Thus, I interpret the CIOMS guidelines as meaning that when the donor does not have any personal benefit from the research and he/she is well-informed, and when there is a foreseeable benefit to society, then biobank research is conducted in accordance with international standards of risk-benefit calculations, provided that the risk is small.

Generally, one could say that the prime interest of research subjects in medical research is not to suffer any substantial harm. In biobank research the subject will not undergo any painful treatment or risky drug trials that may give rise to physical harm. So any harm that may constitute a risk to the donor to a biobank is harm connected with personal and sensitive information and the handling of such information, and that there is a risk of breaching *the donor's right to privacy*.

Health data processing in medical practice is usually strictly regulated in order to protect the privacy of the patient. When biobanks emerged the setting was so different that ordinary health regulations were not considered good enough. Several European countries brought in acts and regulations to cover the different aspects of biobanks, especially issues concerning the privacy of the donor. In several of these acts the health data protection focus is on genetic information or on human biological material, leaving non-genetic information in a sort of legal vacuum.

The Dubious Uniqueness of Genetic Information

The various legislations and regulations on research biobanking and genetic privacy, as I understand them, imply that there is a real risk of genetic information being released to third parties against the donors' wishes and interests. If genetic information about a donor is released to the public, then the right to privacy is breached, something which is wrong. But apart from that, how harmful would this really be? Genetic information, defined as information derived from analysis of genetic material of a donor, has been viewed as having a high potential for causing harm. Since a lot of information about our genes, such as sex and hair colour, is available to the public, the harm that is in question must be the harm that can be caused by publishing genetic information that is *not* available by other means, and considered to be private in nature.

Let us assume that an exchange of data among researchers has gone wrong and the variables connected to one individual were sent to a wrong e-mail address. From the data we can understand that the individual is: "... male, born on February 2, 1979.... He is 176 cm high and lives in Dalvik [Iceland]"[3] (Árnason 2002). This information is enough to uniquely identify the individual, in this example, as Helgi. I consider that *four* conditions need to be met in order for this scenario to be harmful to the identified donor.

First, if information is considered to be known to somebody, it must be *intelligible* to that person in one way or another. If genetic information is translated and interpreted in everyday language, then it may be understandable to a sufficient number of people to cause damage. However, data in biobanks are not kept in everyday language. So the information released may look something like "ApoE*4". Genetic data are kept as symbols, and if these symbols are not understood, there is no information that can be obtained. Most people do not understand this information, since they are not educated in genetics. A few letters and digits accidentally released together with identifiable demographic data could mean nothing to the public. The harm that could be done to the donor, in this case the spread of knowledge that Helgi has an allele associated with a slightly increased risk of Alzheimer's disease (Lahiri et al. 2004), is reduced to those situations where the receiver understands the information. So the potentially harming person must possess *background knowledge in genetics*.

Second, not every allele of every gene in the human genome is *fully investigated* with regard to its implications. In a research biobank the focus will typically be on *uninvestigated alleles*. This means that even an educated geneticist may not necessarily understand the implications of a particular piece of information. Therefore in order for the genetic information to be harmful the *implications* of the allele must also be known or believed to be known to some degree. If the information is not understood as meaning that the identified individual has a particular allele, which may lead to specific consequences for that individual, then the information is not specific information and thus unlikely to be harmful. Presently, the phenotypic expressions for the majority of human genes are completely or partly unknown even to experts.

Third, if genetic information is released to someone who can read the information, and if the implications of the genetic information are fully investigated, then the *content* of the genetic information must also have the *potential to cause harm*. Most genetic information is not harmful, as the information describes traits for characteristics not necessarily considered as sensitive or private information. For example this could be a gene coding for the number of fingers. The knowledge that originates from genetic information must be associated with information that in some way is considered to be private or sensitive, for instance disease susceptibility, and if released to the public is likely to cause (considerable) disadvantage to the donor. On the other hand, genetic information about a protective allele may have a favourable effect to the donor if released to the public. So, while the gene variant ApoE*4 of Helgi is connected to an increased association with Alzheimer's disease, other

[3] Here I borrow an example from Einar Árnason's article "Personal Identifiability in the Icelandic Health Sector Database" (Árnason 2002).

variants of the same gene are associated with a decreased risk (Lahiri et al. 2004; Bento et al. 2004; Davis 2004).

Fourth, the *receiver* of genetic information must be in a position *to cause* a *harmful effect* so that, if sensitive information about a donor is released to a person, this person must have an interest in using the information to affect the donor's future life in a harmful way. This means that the receiver of the wrongly sent e-mail must be in a position to transform the information into a positive action that in some way or other causes unwanted harm to the donor. This will only very rarely be the case.

The Generalizability of Non-genetic Health Information

The donor has more control over non-genetic information than over genetic information. He or she will typically fill in health questionnaires and may choose to leave out certain facts, in particular facts perceived as private. He or she will know what kind of non-genetic information the biobank has obtained, but will only be vaguely informed about exactly which genes are analysed and for which specific purposes. Nevertheless, this informational privacy does not make the information given less sensitive or less harmful if accidentally released.

Since the information is often collected by means of health questionnaires, it is stored in a form that lay people understand so that contrary to what happens with genetic information, if it is released, the general public will be able to understand it. If Mary Anderson is asked about her alcohol consumption, and if she ticks off "5 drinks a day", this information is highly sensitive but also more informative about future risks for cancer in the oesophagus than the genetic code ALDH2*2 would be (Poschl and Seitz 2004).

Also when information is stored in a less cryptic way it is not so important to know the context of the information. Even though the identity of a donor and the information that she suffers from a specific disease is not necessarily the same as that the information is true, one cannot expect the truthfulness of the information to be questioned either. A receiver can easily interpret the information and is likely to believe it to be true. But the information is actually quite likely to be incorrect, since the health questionnaire on which it is based may have been answered several years ago, and *the information may be obsolete*. Genetic information is not incorrect in this way.

Common multifactorial diseases are characterized by many factors that are associated with increased risk of an illness, and one risk factor will usually not be enough. Another characteristic is that gene variants associated with common multifactorial diseases are common in the population, but the increased risk is considered to be low. An example of the latter is the allele GSTM1 null associated with an increased risk of lung cancer. This variant is carried by approximately 50% of the white population. But a history as an asbestos textile worker is considered to be a far more important risk factor, and thus more interesting to know about. So, on the one hand, GSTM1 null status is fairly probable, and is thus something that employers and insurance companies have to expect. On the other hand, asbestos exposure is rare but associated with relatively high risk of lung cancer (Vineis et al. 2001; Pira et al. 2005).

Not all sensitive non-genetic information is actually *harmful* to the donor if it is released to the public. For instance, information that the donor had otitis as a child is sensitive but hardly directly harmful if made public. On the other hand, detailed information about mental illnesses may be harmful to the donor if people around him learn about it, and more harmful than a single susceptibility gene.

Conclusions

In my view the risk associated with harmful information being released to the public is higher the more people are able to interpret and understand the information. Considering the way genetic information is stored in biobanks, release of non-genetic information is potentially more damaging. Existing biobank regulations aim at protecting donors' rights, but their focus of attention is on genetic or biological material. CIOMS demands that donors must be fully capable of appreciating any risks if the research is not of direct interest to the donor (CIOMS 2002). The problem is that neither the law nor research ethics have captured the meaning of genetic information and non-genetic information in the regulations relating to epidemiological biobank research. To set a standard that demands full information for the donor is therefore quite ambitious, especially when the regulations and the everyday situation of the researchers are contrasting realities. So, the processing of non-genetic health information constitutes a risk to the donors by being to some extent overlooked, in academic as well as in public debates and in the national and international regulations relating to research biobanking.

Acknowledgments The financial support of the project "Mapping the language of Research Biobanking" (Contract No. 159864/S10) of the Norwegian Research Council and the Norwegian Institute of Public Health and the project "Research biobanking and the ethics of transparent communication (Contract No. 182269)" of the Norwegian Research Council are greatly acknowledged. In addition I thank Professor Søren Holm and Professor Jan Helge Solbakk for supervising my project.

References

Annas G et al. (1995) Drafting the genetic privacy act: Science, policy and practical considerations. Journal of Law, Medicine and Ethics 23:360–366

Árnason E (2002) Personal identifiability in the Icelandic health sector database. Electronic Law Journals – JILT 2: http://www2.warwick.ac.uk/fac/soc/law/elj/jilt/2002_2/arnason

Bento JL et al. (2004) Association of protein tyrosine phosphatase 1B gene polymorphisms with type 2 diabetes. Diabetes 53:3007–3012

CIOMS (2002) International Ethical Guidelines for Biomedical Research Involving Human Subjects
Davis J (2004) Type 2 Diabetes Gene Discovered. WebMD Health News: http://diabetes.webmd.com/news/20041028/type-2-diabetes-gene-discovered
Denmark (2003) Act on the Biomedical Research Ethics Committee System
Green MJ, Botkin JR (2003) Genetic exceptionalism in medicine: clarifying the differences between genetic and nongenetic tests. Annals of Internal Medicine 138:571–575
Holm S (1999) There is nothing special about genetic information. In: Thompson AK, Chadwick R (Eds.) Genetic Information – Acquisition, Access and Control. Kluwer/Plenum, New York, pp 97–103
Kakuk P (2006) Genetic information in the age of genotype. Medicine, Health Care and Philosophy 9:325–337
Lahiri DK et al. (2004) Apolipoprotein gene and its interaction with the environmentally driven risk factors: molecular, genetic and epidemiological studies of Alzheimer's disease. Neurobiology of Aging 25:651–660
Mathiessen-Guyader L (2005) Survey on national legislation and activities in the field of genetic testing in EU Member States. European Commission
Medical Research Council – MRC (2007) Human tissue and biological samples for use in research–operational and ethical guidelines. Ethics Series
Murray TH (1997) Genetic exeptionalism and future diaries: Is genetic information different from other medical information. In: Rothstein M (Ed.) Genetic Secrets: Protecting Privacy and Confidentiality in the Genetic Era. Yale University Press, New Haven, pp 60–73
Norway (2003) Norwegian Act Relating to Biobanks
Pira E et al. (2005) Cancer mortality in a cohort of asbestos textile workers. British Journal of Cancer 92:580–586
Poschl G, Seitz HK (2004) Alcohol and cancer. Alcohol and Alcoholism 39:155–165
Sweden (2002) Biobanks in Medical Care Act
UNESCO (1997) Universal Declaration on the Human Genome and Human Rights
Vineis P et al. (2001) Misconceptions about the use of genetic tests in populations. Lancet 357:709–712
Wolf SM (1995) Beyond genetic discrimination: toward the broader harm of geneticism. Journal of Law, Medicine and Ethics 23:345–353
World Medical Association – WMA (2004) Declaration of Helsinki

Duties and Rights of Biobank Participants: Principled Autonomy, Consent, Voluntariness and Privacy

Lars Øystein Ursin

Abstract In this chapter the notion of principled autonomy is presented, and the perspective enabled by this notion is applied in the field of biobanking. Some consequences of the perspective of principled autonomy on aspects of biobank recruitment are discussed in relation to concepts of voluntariness, consent, and privacy. These discussions aim to focus on the fruitfulness of the notion of principled autonomy in bringing out the interconnectedness of the duties and rights of biobank participants – both in general, and in a context of taking part in a research-based universal health care system in particular.

The Assumption of Rights

The discussion of how biobank participants are to be respected deals with a certain assumption about an individual, namely that a person should – in some sense or another – have control over herself because she *owns* herself. Such an assumption harks back to the thinking of the principle of respect for the individual in terms of rights rather than of laws, as described by Charles Taylor: "The notion of a right, also called a 'subjective right', as this developed in the Western legal tradition, is that of a legal privilege which is seen as a quasi-possession of the agent to whom it is attributed. (...) Law is what I must obey. (...) By contrast, a subjective right is something which the possessor can and ought to act on to put it into effect" (Taylor 1989: 11).

The perspective of rights – or even "natural rights" – fits in nicely with a picture in which every individual is the possessor of her body and information about herself. On the one hand, such a picture downplays the interpersonal aspect of moral obligations: If I am asked to provide blood samples for a biobank research project,

L.Ø. Ursin
Philosophy Department and Bioethics Research Group, Norwegian University of Technology and Science, Trondheim, Norway
e-mail: lars.ursin@hf.ntnu.no

J.H. Solbakk, et al. (eds.), *The Ethics of Research Biobanking*,
DOI: 10.1007/978-0-387-93872-1_6,

I can make a decision all by myself, since I happen to *own* the blood in my veins. On the other hand, this picture emphasises the interpersonal aspect by making it the duty of the individual to govern her involvements with others and vice versa: Since I own my blood, rather than it just being a part of my existence, I am supposed to control the uses to which it might be put. This way of answering the question of *why* biobank participants are to be respected leads to a discussion of *how* they are to be respected in terms of the aptness of different notions of ownership and control.

Another way of answering this question is in terms of a picture of a community of rational agents, in which the manipulation and coercion of any person would deny her rationality, and as such is incompatible with such a community. In this picture, individual control is linked to moral impartiality rather than to personal property. On the one hand, this picture emphasises the fact that every person should be respected as self-governing: If I am asked to provide blood samples for a biobank research project, I might make a good decision on my own, since I am an individual able to make reasonable choices. On the other hand, this picture downplays the personal – or private – aspect of governing oneself by linking respect for an individual to the exercise of rationality in the sense of impartial decision-making. It is the aptness of such a picture for biobank research which is the subject matter of this chapter.

Principled Autonomy

The perspective of Onora O'Neill on the assumption of rights is to emphasise that duties precede rights.[1] She argues that you cannot claim anyone's rights, without stating who has the duty to fulfill these rights. For instance, in order to claim the right to (better) health care, there might be a real sense that one ought to take part in sound health research, unless there are good reasons not to. Current epidemiological research is a way to better health care tomorrow, from which anyone can benefit. O'Neill argues that the importance placed on autonomy in the bioethical debate, and the use of informed consent in medical practice, might "encourage ethically questionable forms of individualism and self-expression, and may heighten rather than reduce public mistrust in medicine, science and biotechnology" (O'Neill 2002: 73). O'Neill thinks that a better approach to securing sound ethical standards and the rights of the individual is to focus on obligations because "(...) a right that nobody is required to respect is simply not a right" (O'Neill 2002: 78). Rights and obligations are two sides of the same coin. Rights without corresponding obligations are illusions at worst, ideals at best. To focus on obligations also brings out the relational nature of individual rights. It sheds light on how our autonomy is embedded in social settings and institutions, and on how these can enable and disable the exercise of our autonomy.

O'Neill bases her account of autonomy on the Kantian definition of the concept. Kant defined the notion of autonomy as *ethical*, in addition to and distinct

[1] For this, see O'Neill 2002.

from its *political* origins.[2] The Kantian notion of autonomy is based on obligations, O'Neill points out, and for her it negates the notion of individual autonomy: "For Kant autonomy is *not relational, not graduated, not a form of self-expression*; it is a matter of acting on certain sorts of principles, and specifically on principles of obligation" (O'Neill 2002: 83–84). O'Neill calls this Kantian notion *principled* autonomy, which links autonomy to an adherence to principles instead of an attainment of independence.

According to O'Neill, principled autonomy connects to a kind of self-legislation – to oblige oneself to be led by ethical reasoning. O'Neill quotes Allen Wood, suggesting that this will lead us to a dilemma: If we are somehow obligated by ourselves, does such an obligation amount to much? Is it logically possible to obligate oneself to anything? This seems to be just an illusion, on a par with the invention of a game where I am the only one who ever knows the rules. Or is it a description of the ideal of authenticity – our moral obligation to be true to ourselves? If we, on the other hand, say that this self-legislation is an obligation towards principles of reason that are somehow independent, does this not oblige us to accept the prevailing rationality, rather than one's own will?

This dilemma will be avoided, however, if autonomy is neither a private obligation nor a commitment to common thinking but rather the fundamental principle of reason itself. We are reasoning if we make it possible for others to follow us – in thought and in action. In that case, autonomy is the principle by which it is possible to give reasons at all. In O'Neill's view, the fundamental requirements of an account of reason are "the necessary conditions that anyone who seeks to reason with others must adopt. As Kant sees it, principled autonomy is no more – but also no less – than a formulation of these basic requirements of all reasoning. (...) we *must* act on principles others *can* follow. So there is no gap between reason and principled autonomy, and specifically no gap between practical reason and principled autonomy in willing" (O'Neill 2002: 92).

Principled autonomy, then, requires us to act on principles that can be understood and acted on by anybody – in principle. Individual autonomy is a necessary, but not a sufficient, condition for principled autonomy. The notion of freedom involved here is freedom to act, independent of irrational influences. For Kant causal independence – freedom from controlling impulses and plain coercion – is a more prominent condition for autonomy than social independence and self-expression. Principled autonomy requires mutual, and not just individual, understanding of the principles by which we guide our actions.

Universal moral principles and principles of reasoning are the essence of O'Neill's conception of autonomy. Realised principled autonomy implies a common understanding between researcher and participants, and thereby promotes involvement and non-maleficence in medical research. Demanding independence rather than reasons might have the paradoxical consequence of weakening rather than strengthening the ability of the individual to autonomously pursue her own interests.

[2] For this, see Schneewind 1998, esp. Chaps. 22 and 23.

Justifying Consent by Principled Autonomy

Giving consent to become a patient or research participant based on principled autonomy is thus about preventing coercion and abuse, rather than about promoting personal autonomy. Autonomy should be seen to be a matter of adherence to moral principles, which is grounded in the autonomous recognition of these by the people concerned, and their mutual *trust* in each other to adhere to these principles. Autonomy as self-expression puts the emphasis on independent decision-making with reference to the (rights of the) individual, while principled autonomy puts the emphasis on finding and acting from commonly accessible and assessable reasons. To justify informed consent requirements as the promotion of autonomy-based trust consequently seems to fit the principlistic conception better than the individualistic one. Informed consent justified by principled autonomy thus makes for legitimate biobank research recruitment in meeting both the demand of participants by promoting trust, and the demand of the Helsinki Declaration by securing informed and voluntary participation.

The adequacy of justifying informed consent under a Kantian conception of autonomy is also argued for in *Autonomy and informed consent: A mistaken association?* by Sigurdur Kristinsson. Kristinsson takes the *Belmont Report* as his point of departure. The shift towards protecting the research participant against undue paternalism and not just harm from the voluntary consent of the Nuremberg Code of 1947 to the informed consent of the Helsinki Declaration of 1964 is even more apparent in the Belmont Report of 1979. The Belmont report states in Sect. B1 that respect for persons is a basic ethical principle of vital significance to research ethics. The report explains the notion of respect for persons in this way:

> Respect for persons incorporates at least two ethical convictions: first, that individuals should be treated as autonomous agents, and second, that persons with diminished autonomy are entitled to protection. The principle of respect for persons thus divides into two separate moral requirements: the requirement to acknowledge autonomy and the requirement to protect those with diminished autonomy. An autonomous person is an individual capable of deliberation about personal goals and of acting under the direction of such deliberation (The Belmont Report 1979: Sect. B1).

In section C1 the requirements set by the principle of respect for persons in medical research is given: "Respect for persons requires that subjects, to the degree that they are capable, be given the opportunity to choose what shall or shall not happen to them. This opportunity is provided when adequate standards for informed consent are satisfied" (The Belmont Report 1979: Sect. C1). The report states, in other words, that informed consent is required in order to respect persons by respecting them as autonomous beings, understood as individuals "capable of deliberation about personal goals and of acting under the direction of such deliberation" (The Belmont Report 1979: Sect. B1).

Kristinsson now questions the moral justificatory power of the Belmontian concept of autonomy in general, and in relation to informed consent in particular. Why should respect for people's "deliberation about personal goals" be of basic moral significance, rather than just a fashionable idea, wonders Kristinsson. An attempt to link it with Kant's Formula of Humanity[3] will fail to capture Kant's intention, if humanity is thought to be something more than the mere ability of rational agency. And, if autonomy is rightly understood as the Kantian *duty to oneself to be rational*, autonomy as the justification of elaborate informed consent procedures designed to secure the *personal* – but not necessarily rational– deliberation *of others* disappears. Kristinsson joins O'Neill in holding that justifying informed consent by way of Kantian autonomy means that "the ultimate point of informed consent policy is not to increase the incidence of personal deliberation but rather to decrease the incidence of manipulation, deception and coercion" (Kristinsson 2007: 257).[4]

To argue that the justification of informed consent should be viewed in terms of avoiding harm rather than as promoting the personal autonomy of the individual means that its main function is to waive specific rights of the individual. This means that the norms grounding these rights rather than the exercise of individual autonomy are the real basis for the normative significance of informed consent requirements, as argued by Manson and O'Neill: "Consent (...) can be used to waive important norms, rules and standards, and so has considerable ethical importance. But since its use always presupposes whichever norms are to be waived, it cannot be basic to ethics, or bioethics" (Manson and O'Neill 2007: 149).

This view emphasises the relationship between the negative obligations of researchers not to manipulate participants or violate their bodily integrity and the rights which correspond to these duties.

Autonomy, Perfection, and Neutrality

Another aspect at play here is whether the justification of informed consent requirements as a promotion of individual autonomy is compatible with the basic principle of liberalism, namely that of securing the equality of all citizens by letting the right precede the good. A *neutralist* understanding of this principle would be that the state always should act in ways that are neutral between rival conceptions of the good, rather than to promote any(one's) particular and controversial conception of the good. The question is whether the emphasis on individual autonomy indeed can be given such a neutral justification, or whether it is the promotion of the substantial conception of the good that is controversial.

A Millian kind of justification for advancing the autonomy of biobank participants seems to violate such a principle of neutrality. Rather than to respect a participant's right to handle his or her involvement with medical research as he or she wishes, it seems to impose an ideal of personal autonomy that involves an obligation to approve the relevant research. According to Mill, it is crucial that

[3] The formula of humanity as an end in itself is the version of the categorical imperative that dictates that you should act in such a way that you always treat humanity, whether in your own person or in the person of any other, never simply as a means, but always at the same time as an end.

[4] Cf. Kristinsson and Árnason 2007.

people are left alone to be able to exercise their liberty, because it is essential to self-realization (Mill 1832/1977: 277), to promoting self-esteem and the ability to exercise mature choice (Mill 1832/1977: 277), and even essential to developing a more prosperous State (Mill 1832/1977: 310).

Kristinsson argues that the promotion of individual autonomy not only fails to fulfill the ambition of identifying how participants in medical research should be respected as individuals, but that the promotion of individual autonomy in a liberal society is *in opposition to* respecting participants as individuals. The reason for this is that personal autonomy is a substantive moral ideal that is not compatible with the liberal principle of neutrality, according to which state regulations such as informed consent requirements "should be acceptable to all citizens, regardless of their comprehensive conceptions of the good" (Kristinsson 2007: 258). Therefore, Kristinsson concludes, in order to respect individuals and the liberal principle of neutrality, a Kantian rather than a Belmontian conception of autonomy is called for.

To respect individuals and to treat them as equals does not necessarily mean treating them without favouring any particular notion of the good, however. It can also be argued that it should take the form of treating them according to the notion of the good that is thought to be superior. Liberal states often carry out attitude campaigns and economic incentives, numerous non-coercive but also non-neutral state policies in the form of public education. This aspect of the liberal state can be brought in accordance with the principle of neutrality if we distinguish a narrow neutrality principle from a comprehensive one.[5] In opposition to the comprehensive principle, which holds that state neutrality should extend both to the basic framework and the specific policies of the state, the narrow principle holds that neutrality is restricted to the constitutional structure of the state. According to a narrow conception, the state can legitimately promote an ideal of individual autonomy in non-coercive ways, even if such an ideal is controversial.

Research participants and policymakers all agree that the relationship between the state and its citizens in a liberal society should not be based on blind trust and/or unrestricted rights to intervention and access to information about citizens, as this would open the door to totalitarianism and the loss of citizens' freedom from paternalism and domination. The notion of principled autonomy does not promote blind trust, but it might nevertheless be susceptible to being regarded as a conceptual variant of positive liberty; "as soon as the autonomous self of the individual begins to be equated with the rational self as such (shared by all rational agents), a slide into paternalism begins" (Kristjánsson 1996: 142). A *liberal perfectionist* like Mill would argue that the promotion of a citizen's *personal* autonomy is essential to a liberal state. For the liberal perfectionist, respecting citizens as individuals comprises enabling the individual to deliberate on personal goals. It is not just to respect citizens through the shared obligations of principled autonomy and to restrict state policies by the principle of neutrality.

In the Kantian conception of autonomy promoted by Kristinsson and O'Neill an act is justified by the ability to back it with coherent rational and moral principles.

[5] For this, see Wall and Klosko 2003: 6.

In a Millian conception of autonomy, the moral obligation *comprises* the promotion of people's ability to develop and express their own character. The principlist emphasis on moral justification thus misses an important aspect, and makes it restrict itself to analysing the role of rationality in ascriptions of autonomy.[6] And the formal character of principled autonomy hands us a concept of autonomy that tends to presuppose rather than bring forth the way personal autonomy has certain substantial empirical conditions.[7]

Trust and Negatively Informed Consent

The perspective of principled autonomy emphasises the genuine trustworthiness of the institutions. Sometimes an individual might want to give up her personal autonomy and to leave decisions concerning herself to others. Ulrik Kihlbom, however, argues that an individual's ignorance of the specifics of a research project does not have to compromise her autonomy. He aims to show that there is a way to leave decisions to others without giving up her autonomy. To do this he asserts the falsity of a common assumption of what the autonomous decision to give an informed consent requires, namely that "it is necessary that she has positive belief in the methods, means and risks concerned" (Kihlbom 2008: 147). The assumption alluded to, which Kihlbom finds illustrated in Beauchamp and Childress' *Principles of Biomedical Ethics* (Beauchamp and Childress 2001), is that "to exercise one's autonomy is a matter of being the direct and intentional cause to what happens to oneself, and, in turn, the presupposition that this can only be the case if you understand and are aware of what is happening to you" (Kihlbom 2008: 146).

For Kihlbom the exercise of my autonomy does not necessarily depend on *positive* knowledge about the research project. It might be more important for me to have some crucial *negative* knowledge as to what the research project does not and will not include. *Negatively informed consent* in a clinical setting, accordingly, requires that the patient have a clear understanding of the aims of the treatment but not of the methods, difficulties, and risks involved. He knows that the treatment is voluntary and that he can be given more information about and withdraw his consent to the treatment at any time. The patient who gives negatively informed consent not to receive more information regarding the treatment then explicitly chooses to trust the physician to promote the best possible treatment for him. This trust should be well-founded. For Kihlbom "this rules out negative IC in situations where the physician and patient know little about each other."[8]

[6] For this, compare Hill 1991.

[7] See Guyer 2003.

[8] In requiring that personal acquaintance between patient and physician, and knowledge about the patient's personal values, be a necessary condition for negatively informed consent, Kihlbom does not argue for Kristinsson's view that principled autonomy is the justification for informed consent requirements that best secures respect for individuals.

In Kihlbom's view, a relaxation of the specificity of informed consent requirements might even turn out to *enhance* a patient's personal autonomy. It is a bit unclear as to how this might work but it seems that Kihlbom is thinking of a patient's ability to reach his personal aims,[9] which indeed might be enhanced by someone else. He does not argue that autonomy in the end is merely of instrumental value and as such might legitimately be overridden by the physician to promote a patient's real interest – namely his well-being. But holding that trusting others to promote your ends might enhance your autonomy implies that the instrumental value of autonomy is important.

The crux of the matter, however, is the tenability of Kihlbom's distinction between giving up autonomy and trusting others to make decisions for you. Kihlbom is correct in pointing out that "many of the means we use we are not familiar with. These states of ignorance do not threaten our autonomy" (Kihlbom 2008: 148). Indeed, instead of saying that I must know all the health consequences of drinking the tea you are offering me, in order to make the autonomous choice to have tea with you, it is better to say that it is enough to have the well-grounded belief that you are not trying to poison me.

But this tells us something about the situations in which we employ the concept of autonomy, rather than about the relationship between autonomy and positive vs. negative knowledge. Autonomous choices must be significant, and related knowledge – positive or negative – must be relevant to making such choices. Thus, neither general nor negatively informed consent is grounded in personal autonomy, *if* it entails that you leave significant decisions to others. If it does not, the consent is specific, since there are no further significant choices for the individual to autonomously decide.[10]

Now, if personal autonomy is promoted as an ideal, an individual should learn to see the personal significance of more choices. This creates a paradox in which participants see no problem in giving general consent, while the government pushes for more specific consent – because the participants should recognise that there are still significant choices to be made.[11]

Authorisation and Voluntariness

The perspective of principled autonomy emphasises the voluntariness of participants. In the case of biobank research, the unknown nature of future research projects and the significance of the findings for participants have, for instance, led

---

[9] See Kihlbom 2008: 148: "If I, as the patient, choose to let you, as the physician, determine my treatment, and I have well founded beliefs that you will choose the treatment that best promotes my values, and that the risks of the treatment you will choose are in accordance with my attitudes towards different kinds of risks, I will exercise my autonomy, not waive my right to exercise it."

[10] For an elaboration on this point, see Ursin 2009.

[11] For an elaboration of this point, see Ursin et al. 2008.

the HUNT[12] biobank in Norway to make a kind of general consent with continuously updated information of the ongoing research projects available to participants. In this way the dichotomy between specific and general consent is transcended through the introduction of the dimension of *time*: Consent becomes a continuous rather than a one-off decision. This kind of consent could be called *processual*, or – as argued by Sigurdur Kristinsson and Vilhjálmur Árnason – an *authorisation*.[13]

According to Kristinsson and Árnason, an authorisation will, in a system of trustworthy and transparent institutions, safeguard participants from manipulation, and make voluntary participation possible. They argue that to require specific consent would be ineffective, burdensome, and even present a privacy risk, while general consent fails to meet the moral motivations for consent.[14] Kristinsson and Árnason argue that the safeguarding of the voluntariness of biobank participants is a major moral motivation for informed consent requirements. The problem is, however, that "in contexts where relevant outcomes are foreseeable without being commonly known, potential subjects need to be informed in order for their participation to be voluntary. In contexts where possible relevant outcomes are poorly understood by even the researchers themselves, it is hard to see how participation can be voluntary" (Kristinsson and Árnason 2007: 206).

For Kristinsson and Árnason, both intentionality and control are conditions for voluntariness. Concerning intentionality, they state that "voluntary participation in research must be based on the subject's awareness of all aspects of the participation that are relevant to describing the act" (Kristinsson and Árnason 2007: 205). And, since "the only specific ingredient that could possibly be explained [when giving an informed consent to take part] is the right to withdraw from the database at any time" (Kristinsson and Árnason 2007: 212), the use of informed consent does meet the requirement of securing the voluntariness of participants.

The possibility of declining to be included – and if included, to be given the permanent possibility to opt out – is a *necessary* condition for the voluntariness of research participation. But is it not, in contrast to the view of Kristinsson and Árnason, also a *sufficient* condition for the voluntariness of participation?

Rather than defining the concept of voluntariness in terms of intentionality, Serena Olsaretti defines voluntariness negatively in terms of options in this way: "A choice is non-voluntary if and only if it is made because the alternatives which the chooser believes she faces are unacceptable" (Olsaretti 2008: 114). For Olsaretti the existence of *acceptable alternatives* is essential to voluntariness. This distinguishes voluntariness from freedom: I might be free to leave an island, but since I will die if I try to get away by swimming, my decision to nevertheless stay on the island is non-voluntary, since no acceptable alternative is available.

---

[12] HUNT is an acronym for *Helseundersøkelsen i Nord-Trøndelag*, which translates as *The Nord-Trøndelag Health Study*. For a further description of HUNT, see Chap. 15 of this book.

[13] Cf. Kristinsson and Árnason 2007 and also Árnason 2004.

[14] It seems, however, more accurate to say that, because of its nature (or more precisely because of the unknown future nature of the protocols), biobanking is unsuitable for the use of any form of *one-off kind* of consent.

Olsaretti's concept of voluntariness makes moral responsibility depend on voluntariness rather than freedom. I am not responsible for handing over money to a robber pointing at me with a gun, even if I am free to do so. But does the linking of moral responsibility to acceptable alternatives make acts done out of duty non-voluntary? If I am in a position to prevent a robbery in such a way that this is the only morally acceptable thing to do, do I do this non-voluntarily?

If that is the case, and moral responsibility depends on voluntariness, I am not responsible for acting in a morally laudable way. I just *had to* do it. Moreover, according to Ben Colburn, since I am not responsible for my way of acting, it makes my act ineligible for moral praise. This is contra-intuitive. This account, however, fits with the intuition that I am eligible to claim some kind of compensation for any damage to myself from the victim pointed out by the robber, again since I had to do it – I did not do it voluntarily.[15]

To account for our intuitions in terms of her view, Olsaretti distinguishes between moral and substantive responsibility. I am morally responsible if I act deliberately and in a morally reasonable way, while I am substantively responsible if I act from moral obligations. Thus, if I prevent a robbery I am acting voluntarily in the sense that I recognise other acceptable alternatives. I choose, however, to act morally responsible in a deliberate way, and thus I am praiseworthy. On the other hand, I see that I have an obligation to act in a certain way – no other choice is morally possible – so I act as substantively responsible and in this sense non-voluntarily. I might be acting voluntarily from the perspective of moral responsibility, while the same act is non-voluntary from the perspective of substantive responsibility.

Olsaretti's notion of voluntariness brings out vital elements of consent based on principled autonomy. On the one hand, I should be voluntary in the sense of not being coerced, which means that not taking part is a real and acceptable alternative. On the other, I might perceive of my participation as a substantive moral obligation, which makes it involuntary in this sense. Olsaretti's notion of voluntariness as "acceptable alternatives" thus seems to offer a more intuitive way of describing the important element of non-coercion of consent based on principled autonomy than the notion of voluntariness as "awareness of all aspects of the act" used by Kristinsson and Árnason.

Patients' Duties and Privacy Rights

The main potential for harm to biobank participants is not in terms of inappropriate physical invasions but in terms of inappropriate use of personal information. Such inappropriate use of information might be a matter of breaches of confidentiality

[15] For this, see Colburn 2008. Colburn's solution is to say that I act voluntarily even if the alternatives are *morally* acceptable. So, if I refuse to rob a bank purely on moral grounds, this does make me responsible for the continuation of my poverty. In this way my choice is eligible for praise, and I am responsible for the consequences. Colburn's view, however, seems to draw an unwarranted distinction between alternatives which are morally and prudentially unacceptable. He also seems to lead us back to an account of moral responsibility in terms of freedom rather than voluntariness.

that might lead to stigmatization, discrimination, and existential or familial complications. In the biobank context, it might also be a matter of proprietary privacy concerns regarding the kinds of research to which the information is put by the biobank researchers. It might be ethical concerns of the participants such as avoiding research contrary to human dignity or political concerns such as promoting research for the benefit of special groups. It might also be economic concerns, if profits might be gained by the biobank in selling the information, or from products or services developed from research on the biobank information provided by the participants.

According to David Wendler "(...) involvement in research includes three distinct elements: (1) exposure to risks; (2) performance of research mandated behaviours; (3) contribution to answering a research question. (...) To consider a specific example, the standard drug trial involves individuals facing risks (risk element) as a result of taking an experimental drug (performance element) in a way that helps investigators determine whether the drug might be clinically useful (contribution element)" (Wendler 2002: 33–34).

In biobank research, Wendler's case in point, risk elements "involve unwanted information flow" (Wendler 2002: 35). Performance elements involve medical tests, giving biological samples, and completing questionnaires. Finally, contribution elements involve the participation in pursuing specific research aims. In biobank research with de-identified information, both the risk and the performance elements are negligible, according to Wendler. He then goes on to ask "what reason could there be to solicit sources' informed consent for research that poses no risk for them, and does not affect them personally? What is left for sources to consent to?" (Wendler 2002: 38). The answer is that the contribution element is the sole part left to consent to here. This calls for balancing individual and collective interests: First, our interest in enabling the participants' autonomous decisions on which projects to contribute to; and second, the burdens of obtaining such consent – which may hamper our interest in make beneficial research done.

Biobank legislation in different countries qualifies the individuals' right to autonomy over biobank information by making specific uses of biobank information permissible in ways stated by law or with approval from research ethics committees.[16] Most often, however, the right to autonomy over biobank information must be waived by the individuals themselves in terms of giving their informed consent to placing the information at disposal for research purposes. When participation is

[16] In a liberal society the individual's rights to autonomy and privacy also extend to biobank information, albeit in a qualified way. The current biobank legislation in Norway, for instance, states that in order for research biobanks to be established in a legitimate way, there are three acceptable ways of relating to the participants, legally speaking: First, biobank information may be used if the individuals taking part waive their right to keep the relevant information private, and exercise self-determination by giving their informed consent. Second, biobank information may be used if the scientific goal and benefit clearly exceeds any inconvenience caused to the individual. Third, biobank information may be used if this right is specifically founded in the Biobank Act for the biobank concerned. This means that the interests of the individual, of society, and of the researchers involved must be balanced. It is up to the law to state principles by which such a balance is achievable. In either case, privacy is a good which is protected by the relevant body.

mandatory, the element of immediate autonomy is no longer an issue, but the element of privacy remains. When consent is required, participants are asked to entrust their interest in or right to privacy to the biobank.[17]

What unites these ways of governing biobank research is the premise that the individual's contribution to the biobank is a private concern – even if its usage in biobank research is to generate de-identified data about various groups for the benefit of public health in general, with any feedback in terms of personal health information being given to the individual. The discussion about the legislation of biobank research in Norway illustrates this. The participation of every citizen receiving treatment at a Norwegian hospital in the Norwegian Patient Register (NPR) is mandatory. In 2006, the Norwegian Parliament decided that entries in the formerly anonymous NPR should be linked with every patient's personal ID number. This was in February 2007 sanctioned by law. By having identifiable (pseudonymised) patient entries, the NPR can now be used for the purposes of research, since it enables cross-references and linkages to other registers. During the round of hearings that preceded the new Act, the Norwegian Data Inspectorate (NDI) argued against the proposal to make the NPR identifiable by person. The NDI argued:

> Such a change will in any case represent erosion of professional secrecy, and of the principle that everyone should be able to control the use of any information about them. The fact that information conveyed in personal communication with the doctor is to be registered centrally, regardless of the wishes of the patients themselves, will create anxiety and insecurity for many patients. In evaluating the need for a change, one should take into consideration the possibilities for and consequences of the fact that some patients will fail to contact the health service, or will give incorrect information, out of fear that information might be passed on elsewhere. The registration might be counter-productive in realising its purpose, and this possibility must be kept in mind in assessing the need for an NPR identifiable by person. There is no doubt that information about persons can be misused and that errors will occur, and the question now must be how soon and how often this will happen. The greater the amount and the greater the collections of data the greater the possibilities of misuse and the consequences thereof (Datatilsynet – The Norwegian Data Inspectorate 2005).[18]

The arguments of the NDI partly echoes the reasons for judgement given by the European Court of Human Rights in the case *Z v. Finland*:

---

[17] Why is the right to privacy more readily claimed in order to protect health information than to protect economic information? Is it because health information is of a more delicate nature than information about personal finances? Or is it more, as Alexander Rosenberg suggests (Rosenberg 2000), that peoples' economic situation is something they have generally earned, both literally and metaphorically, and consequently it is to a large extent something they have to accept or cope with. The make-up of peoples' bodies (health, race, sex, looks, etc.), on the other hand, is something which is more inherited than earned, and leads easily to kinds of personal or structural discrimination which we try to discourage. To have a right to privacy in these matters is then to put up a temporary and local Rawlsian "veil of ignorance", while waiting for these kinds of discrimination to disappear because of new attitudes or through the making of new policies, both of which should do away with certain forms of contemporary discrimination. In this perspective privacy is a good which is instrumental and reciprocal in the sense that its justification lies in a leveling of the playing field if we all grant it to each other.

[18] Translation by author.

> The protection of personal data, not least medical data, is of fundamental importance to a person's enjoyment of his or her right to respect for private and family life as guaranteed by Article 8 of the Convention (Art. 8).[19] Respecting the confidentiality of health data is a vital principle in the legal systems of all the Contracting Parties to the Convention. It is crucial not only to respect the sense of privacy of a patient but also to preserve his or her confidence in the medical profession and in the health services in general. Without such protection, those in need of medical assistance may be deterred from revealing such information of a personal and intimate nature as may be necessary in order to receive appropriate treatment and, even, from seeking such assistance, thereby endangering their own health and, in the case of transmissible diseases, that of the community (European Court of Human Rights 1997).

The arguments of the NDI and the European Court of Human Rights are partly based on matters of principle and partly on matters of empirical consequence. The principled objections are against the weakening of client confidentiality, the loss of control over personal information, and the lack of consent. The empirical arguments that the NDI points to are that potential research participants might decide against participation (involving informed consent) on the grounds that this would be linked to the proposed NPR. Information might go astray and compromise the patient's right to privacy. If enlistment in the proposed NPR is mandatory for all patients, they might choose not to submit relevant information, or to give incorrect information. The violation of the patient's interests in discretion may lead them to have less trust in the health-care system. In addition, the proposed NPR will ultimately not only come to conflict with the patient's interests in privacy, but also impair the quality of the register, and consequently the quality of the research, administration, and therapy based on the register.

David Korn recognizes two primary causes for this anxiety about information and privacy: "One, which I call 'pragmatic', is the concern about such things as loss of health insurance, discrimination in employment, and social stigmatization. The second root is 'ideological' and springs from a strong, deeply held belief in an individual's right to privacy" (Korn 2000: 964). Korn's "pragmatic" root and the "empirical" arguments of the NDI both highlight a right to privacy which is based on a right not to be harmed. Any citizen should have the right not to experience social harm or unjust treatment as a consequence of participating in medical research. Korn's "ideological" root and the arguments of principle of the NDI both highlight a right to privacy which translates into a right to property. Any citizen should have the right to decide what is going to happen inside their own private sphere as well as what can be done with (material from) their bodies and information about themselves.

---

[19] In paragraph 1 of Article 8 of the European Convention on Human Rights the Council of Europe it is stated that "everyone has the right to respect for his private and family life, his home and his correspondence." (The European Convention on Human Rights, available at http://conventions.coe.int). Paragraph 2 of the Convention, however, qualifies the right to privacy of paragraph 1 rather strongly: "There shall be no interference by a public authority with the exercise of this right except such as is in accordance with the law and is necessary in a democratic society in the interests of national security, public safety or the economic well-being of the country, for the prevention of disorder or crime, for the protection of health or morals, or for the protection of the rights and freedoms of others" (The European Convention on Human Rights, available at http://conventions.coe.int).

The questions and concerns of the NDI are highly relevant for the regulation of biobank research. Provided that all research projects are subjected to thorough ethical scrutiny by the relevant ethics committee, the risk the participants most meaningfully can be said to run is that risk of their personal information being accidentally leaked and misused – they do not run the risk of the material being abused in otherwise unethical research projects. It is therefore ethically imperative to minimize the risk for information being leaked or used inappropriately, as this is one way to address both the principled and the empirical arguments mentioned by the NDI in their statement above.

Onora O'Neill's concept of autonomy separates it from privacy. Principled autonomy is not about securing a private sphere of free choice but about partaking in non-coercive intersubjective practices and giving reasons in discussions. This means that justifying informed consent requirements out of respect for autonomy implies that participants should neither be coerced nor be unable to give cogent reasons consenting or declining participation. O'Neill claims that justifying informed consent requirements in terms of individual autonomy aims to secure both the well-being and the reflective choice of participants, but that in failing to connect autonomy with moral reason it secures neither well-being nor reflection. Current informed consent practices based on individual autonomy just promote any choice, not specifically the ones that demand and defend the moral interests of the individual (O'Neill 2002: 38).

To promote the specific privacy interests of participants in biobank research, it is according to the perspective of principled autonomy important to situate and value these interests in their proper contexts. As pointed out by the NDI, this context might be ambiguous for a patient who relates both to his physician and to registry research by mandatory participation in the NPR. The context here might, however, also be viewed as a relation between a participant in a universal health-care system which offers medical treatment based on research. The right to receive medical care could then be argued to correspond to a duty to take part in the maintenance of the system.

In such a nexus of mutual obligations, the relevant health information is private in the sense of confidential rather than in the sense of ownership. Rather than implying a duty to secure an interest of the individual to control this information, it implies a duty to secure that the information should be handled with respect, that it should not be passed on, and that it should be ensured that its usage does not adversely affect or otherwise compromise participants in the system. The personal origin of any information is not a sufficient condition for requiring consent to any use made of it. As argued by Manson and O'Neill: "Where research is non-invasive, as in the case of secondary research using anonymised data that have already been legitimately obtained and stored, nothing is done to the 'research subjects' to whom these data pertain and it may be hard to establish a case for requiring informed consent" (Manson and O'Neill 2007: 82). The relevant research here concerns group level phenomena rather than the health status of the individual. It might thus be viewed to be of no concern to individuals' rights of privacy at all. This would imply that to require the informed consent of the participants in these kinds of research must be justified by other concerns than a right to privacy.

Conclusions

Rather than just to provide the opportunity to promote their personal autonomy, informed consent has been in this chapter regarded also as the means to respect biobank participants on the basis of principled autonomy. The negative purpose of informed consent, to make participants able to avoid harm (or indeed to avoid research participation), has been subsequently emphasised. The main aim of informed consent has in this perspective been argued to be a legitimate way for biobank participants to be voluntary participants and to waive rights to privacy. A crucial question raised by this perspective, however, is when and whether biobank participants have any privacy rights to be waived. The perspective of privacy endorsed here was that the nature of the information depends on the relation it is a part of and how it is put to use. This in turn determines the rights and duties concerning the handling of the information. Regardless of whether the justification of consent is viewed as promoting the control over private information or in terms of avoiding harm, the question becomes whether the information in the relevant context is to be regarded as private. And in the case of certain kinds of biobank research, it is possible to argue that the relevant relation and intention is such that biobank information is not of a private nature, and thus that the need for requiring consent for these kinds of biobank research falls away.

References

Árnason V (2004) Coding and consent: Moral challenges of the database project in Iceland. Bioethics 18:27–49

Beauchamp T, Childress J (2001) Principles of Biomedical Ethics. Oxford University Press, Oxford

Colburn B (2008) The concept of voluntariness. The Journal of Political Philosophy 16:112–121

Datatilsynet – The Norwegian Data Inspectorate (2005) Høringsuttalelse om etablering av Norsk pasientregister som et personidentifiserbart register

European Court of Human Rights (1997) Z v. Finland, paragraph 95: http://www.echr.coe.int/echr/

Guyer P (2003) Kant on the theory and practice of autonomy. In: Paul EF et al. (Eds.) Autonomy. Cambridge University Press, Cambridge

Hill T (1991) Autonomy and Self-Respect. Cambridge University Press, Cambridge

Kihlbom U (2008) Autonomy and negatively informed consent. Journal of Medical Ethics 34:146–149

Korn D (2000) Medical information privacy and the conduct of biomedical research. Academic Medicine 75:963–968

Kristinsson S (2007) Autonomy and informed consent: a mistaken association? Medicine, Health Care and Philosophy 10:253–264

Kristinsson S, Árnason V (2007) Informed consent and genetic database research. In: Häyry M et al. (Eds.) The Ethics and Governance of Human Genetic Databases. Cambridge University Press, Cambridge

Kristjánsson K (1996) Social Freedom. Cambridge University Press, Cambridge

Manson N, O'Neill O (2007) Rethinking Informed Consent in Bioethics. Cambridge University Press, Cambridge

Mill JS (1832/1977) Collected Works of John Stuart Mill, Vol. XVIII. University of Toronto Press, Toronto

Olsaretti S (2008) The concept of voluntariness – a reply. The Journal of Philosophy 16:112–121
O'Neill O (2002) Autonomy and Trust in Bioethics. Cambridge University Press, Cambridge
Rosenberg A (2000) Privacy as a matter of taste and right. In: Paul F, Miller Jr. FD, Jeffrey P (Eds.) The Right to Privacy. Cambridge University Press, Cambridge
Schneewind J (1998) The Invention of Autonomy. Cambridge University Press, Cambridge
Taylor C (1989) Sources of the Self. Cambridge University Press, Cambridge
The Belmont Report (1979) Ethical Principles and Guidelines for the Protection of Human Subjects of Research. The National Commission for the Protection of Human Subjects of Biomedical and Behavioral Research
Ursin LØ (2009) Personal autonomy and informed consent. Medicine, Health Care and Philosophy 12(1):17–24
Ursin LØ et al. (2008) The informed consenters: governing biobanks in Scandinavia. In: Gottweis H, Petersen A (Eds.) Biobanks. Governance in a Comparative Perspective. Routledge, London
Wall S, Klosko G (Eds.) (2003) Perfectionism and Neutrality. Rowman & Littlefield, Lanham
Wendler D (2002) What research with stored samples teaches us about research with human subjects. Bioethics 16:33–54

Biobanking and Disclosure of Research Results: Addressing the Tension Between Professional Boundaries and Moral Intuition

Lynn G. Dressler

Abstract The role of biobanks is changing to accommodate the expanding needs of the research enterprise. In addition to collecting human specimens, many biobanks also collect research results derived from those specimens. With the advent of technologies to screen the whole genome and the inter-relatedness of multiple genes with multiple diseases, research results will increasingly reveal information with health implications for the contributor of the specimen. What is the responsibility of the biobank to communicate these research findings? What are the benchmarks to guide decision-making on a daily basis? Although there is an emerging ethical imperative from international guidelines to communicate research results to the individual, how should these be implemented in practice? The answers to these questions are highly contextual and currently lack standards of reference. This creates tensions between the traditional boundaries of a biobank, as a resource to store specimens, and the moral intuition of the biobank personnel, as gatekeepers to potentially beneficial health information. This chapter explores these tensions and issues of disclosure of research results in the context of biobanking and provides practice recommendations and next steps for policy development.

Introduction

Samples collected and stored in biobanks generate data which may be clinically useful for research subjects and their physicians. Some biobanks, in addition to collecting and controlling use of specimens, also have access to research results, functioning as a databank as well as a biobank. Until recently, most biobanks

L.G. Dressler
Eshelman School of Pharmacy, University of North Carolina at Chapel Hill, Chapel Hill, NC, USA
e-mail: Lynn_Dressler@unc.edu

J.H. Solbakk, et al. (eds.), *The Ethics of Research Biobanking*,
DOI: 10.1007/978-0-387-93872-1_7, © Springer Science+Business Media, LLC 2009

that collect research results, as well as researchers generating research data, have operated on the premise that research results should not be disclosed to research subjects, as it may do more harm than good (Dressler 2005). As genetic and genomic research continues to reveal an increasing array of new associations with health, how does a biobanker or researcher determine whether and when it is appropriate to disclose research results to the individual who contributed the specimen? What factors should guide the assessment of risks and benefits of disclosure of research results? Although many international guidelines promote the return of research results (Knoppers et al. 2006), in the daily practice of biobanking these questions remain a challenge.

Disclosure of Research Results in the Context of Biobanking

Biobanks can be large, commercial, national or academic repositories, or they can be small investigator collections of specimens, where the investigator acts as the "biobank" as well as the researcher conducting the study. In addition to storing and collecting specimens for research use, often the biobank will collect and store clinical and demographic information. Running a biobank includes many quality control and quality assurance practices, including validating that the sample stored in the bank accurately reflects the expected pathology and diagnosis (Dressler et al. 1999; Schilsky et al. 2002; Dressler 2005; Jewell 2006; NCI 2007). For example, when collecting frozen or fixed paraffin blocks of tumor or normal tissues, sections are cut and reviewed by a pathologist before being distributed to a researcher for use. In addition, for some studies, the biobank may need to cut deeper into blocks of tissue to provide sufficient material to the researchers. What happens when new information, which is clinically relevant, is revealed during this process? Consider the following example: a block of tissue labeled as ductal carcinoma in situ of the breast is found, on deeper sectioning, to reveal invasive cancer. This finding is not a research result per se, but it may change the prognosis and course of therapy for the contributor of the tissue, especially if the diagnosis of ductal carcinoma in situ was recent. What should happen when the biobank discovers that a lymph node marked as "negative, not involving cancer" by traditional pathology review, reveals several cancer cells when a special cytokeratin stain is used, a test not yet considered the standard of care? What should biobank personnel do with this information? In the United States (U.S.), there currently exists no standard protocol or policy to guide decision-making in these situations. Many biobanks now also store or have access to the research results derived from the banked specimens and therefore act as "databanks" as well. When the biobank is also a databank, they are placed in a position of determining when it is appropriate to communicate a spectrum of complex, often uncertain, research results to the research subject or biobank contributor.

Tensions Between Professional Boundaries and Social Responsibility

Unlike clinicians who endeavour to follow the Hippocratic oath (Hippocrates) and have clear professional codes of conduct, researchers have no code of conduct to guide them. Researchers are not trained to make decisions in the best interest of the research subject as clinicians do for their patients; they are trained to maintain scientific integrity and prevent scientific misconduct. They are trained to think in broad terms, not in terms of an individual subject. Yet the implications of today's genetic and genomic medical research have significant impact for individuals, families and society, thrusting researchers and biobankers into unfamiliar settings, where issues such as disclosure of research results, reach far beyond the usual, narrower boundaries of scientific misconduct (Pelias 2005). For many researchers and biobankers, the consideration of how their work may affect an individual research subject, the subject's family or a community creates new and often unsettling tensions (Dressler 1998; Pullman and Hodgkinson 2006). In addition to conducting good science, the researcher or biobanker must on the one hand now consider moral issues for which most lack appropriate training, and which may further delay or deter their studies. On the other hand, as more researchers experience situations where their research reveals clinically relevant information about an individual research subject and their family, it is likely that their moral intuition (e.g. to return these individual results) and their profession's boundary (e.g. returning research results is more harmful than good) will be in disharmony. The tensions resulting from these opposing forces will make it even more challenging to develop policies to guide practical decision-making regarding the disclosure of individual research results.

When Do Research Data Constitute Information that Should Be Communicated to the Subject?

This question has moral and scientific implications. In the U.S., federal agencies overseeing human subjects research are clear in their positions that preliminary results do not constitute information and that "even confirmed findings may have some unforeseen limitations" (OPPR 1993). In the context of genetic information, they support the "development of reliable, accurate, safe and valid presymptomatic [clinical] testing". Pathologists, the gatekeepers of tissues in the hospital setting in the U.S., have emphasized that "until genetic research data is confirmed and validated to the extent that it is used in genetic counselling, it should not be considered valid clinical information" [and should not be communicated to research subjects] (Grizzle and Groday 1999).

This chapter will explore the moral and scientific issues of disclosure by first giving an overview of what we mean by "disclosure of research results", especially in the context of biobanking; present the ethical arguments for and against disclosure of individual research results, and return to the question of when does research data become information that should be communicated to the research subject. The chapter will end with recommendations about policy and the next steps to be taken.

What Do We Mean by the Term "Disclosure of Research Results"?

What are we referring to when we speak of *research results* and to whom are we considering disclosing these results? For the purpose of this chapter we will use the term *research results* to refer to the information that is gained *anew* in the course of research. It will not refer to any other information that may already exist (such as family history, medical conditions, treatments received, etc.). The discussion will refer to laboratory data revealed, either through quality control practices of the biobank, data collection or analysis by the primary investigator, i.e. the researcher generating the data, or a secondary investigator, i.e. the researcher who may use the data generated by the primary investigator. When we speak of disclosure of research results, this can include disclosure to the individual research subject, their family, their physician and potentially other third parties or groups (e.g. tribal leaders). The discussion in this chapter will focus mostly on disclosure to the research subject or individuals contributing specimens to the biobank.

Aggregate Data vs. Individual Data

Aggregate data refers to the composite information obtained in a research study. Consider the following hypothetical example: Of the 500 breast cancer patients participating in this study, we observed that 12% have a mutation in their estrogen receptor, and of those with such a mutation, 50% have a polymorphism in their CYP2D6 gene. This may explain why some women respond to tamoxifen and others do not. This type of research result is generally reported in the form of a scientific manuscript, which is published in the scientific literature. It has been the norm to refer research subjects to these publications when requests for information are made by the subject or their family. Recent regulations in the U.S. (NIH 2008) require that these scientific manuscripts be available for public access via the Internet, i.e. PUBMED Central (NIH 2008). The intent of this regulation is to make available aggregate data to the public, whose tax dollars support the research. However, these manuscripts and abstracts are written for scientists, in scientific and technical language, and are not readily understandable by the public. There is a movement afoot, however, where efforts are being made to communicate aggregate results in plain language, summarized in the form of a mailed newsletter, Web site or annual meeting specifically designed for research subjects (e.g. University of North Carolina Specialized Program of Research Excellence in Breast Cancer Research, Carolina

Breast Cancer Study Annual Forum). In the U.S., the disclosure of aggregate results is not routine.

There have been relatively few circumstances that have compelled disclosure of individual research results. In the U.S., the norm has been not to disclose individual research results except in very rare circumstances (NBAC 1999; OPPR 1993). Recent international guidelines, however, have promoted an emerging ethical duty to disclose individual research results, largely driven by the principle of autonomy (Knoppers et al. 2006; MRC 2001; CIOMS 2002; Council of Europe 2005).

Disclosure of Individual Research Results

What kind of information falls under the category of *individual research results*, and how do we define this? For the purposes of this chapter, two categories of individual research results will be described: one is research results *related to the scope of the research study* and the other is results revealed through the course of the research but *incidental or unrelated to the scope of the original study.* As we perform more genome wide association studies and learn more about the inter-relatedness of the biologic function of genes and their impact on disease, this line will become blurred, but for now it serves a helpful purpose for the following discussion.

To describe what is meant by research findings related to the scope of the original research study, it is helpful to use an example. From the hypothetical study described above, researchers were looking at the frequency of variations in the estrogen receptor gene and in polymorphisms in CYP2D6 to explain differential response to tamoxifen in a group of 500 breast cancer patients. Returning individual research results related to the scope of the study would mean that women would be told the results related to their specific specimens, especially women with mutations in the estrogen receptor and those with the CYP2D6 polymorphism.

Results unrelated to the scope of the research refer to information revealed during the course of the research that is not related to the aims or research questions of the original study. Findings incidental or unrelated to the original study can include a variety of scenarios. For example, what if the researchers were using genome wide association techniques, a technology that has the capacity to scan the entire genome of an individual, and although they were conducting a breast cancer study, they found mutations in a series of genes already known to be highly correlated to the development of colon cancer? This information is discovered during the course of research, but is not related to the scope of the original breast cancer study. It is a finding that one may have anticipated using genome wide technologies, but it is a new finding. What does the biobanker or researcher do with this incidental information?

To make matters more complex, let's take the example of pleiotropy, an ever increasing phenomenon where one gene or form of a gene may have many functions and affect many diseases (Wachbroit 1998). Some of these pleiotropic effects are known, but most are being newly discovered each day. The classic example of pleiotropy is the APOE gene, where one isoform of the gene (A4) which is associated with risk of heart disease is also associated with risk of developing Alzheimer's disease (Wachbroit 1998). Although the reliability of predicting Alzheimer's disease from knowing the A4 status is still uncertain, should a researcher who is conducting a study in heart disease and finds that 10% of his patients have the A4 isoform withhold information related to susceptibility to Alzheimer's? What if in the next year it is discovered by another investigator that the APOE A4 isoform is also related to a small change in DNA sequence (i.e. polymorphism) which is associated with a risk of a fatal toxicity if the individual is given a certain class of drugs? Is the biobank responsible to communicate these findings to the contributor of the specimen? What level of certainty should be in place before findings are released? What standards for clinical validity or utility should be used, and who should determine those standards? These questions remain unanswered, yet it is critical that we anticipate these scenarios and analyse the moral, professional and societal implications and factors that should guide the decision to disclose or not. The following discussion gives a brief summary of the ethical arguments for and against disclosure of research results.

Arguments for and Against Disclosure of Results

What are the ethical arguments for disclosure of research results? The two most common ethical principles discussed in support of disclosure of results are respect for persons and beneficence. Respecting an individual research subject's autonomy underlies many of the international guidelines which support release of individual research results (Knoppers et al. 2006; MRC 2001; CIOMS 2002; Council of Europe 2003). Advocacy groups support the opportunity for each individual to decide whether or not they want to have access to the research information that may be learned about them (NAPBC 1997). Some advocate for release of information only if research data are analytically and clinically validated (NAPBC 1997). Others advocate for the release of information regardless of the clinical validity, as long as the individual is informed of the uncertain nature of the information. Wilfond and colleagues recommend a formulary which combines validity of test, value to the individual research subject and closeness of the relationship between the investigator [biobanker] and the research subject as criteria for disclosure (Ravitsky and Wilfond 2006). Most groups advocate that research information should not be withheld when it provides evidence of immediate risk to individual research participants (Dressler 2005). Withholding information would be acceptable only if disclosure would predictably compromise the safety of the participant or third party (NAPBC 1997).

What are the ethical arguments against disclosure of research results? The two most common arguments opposing disclosure of research results include the principle of non-malevolence and the philosophy regarding the intent of research. Most professionals in the research community and research regulators, especially in the U.S., have long held that the intent of research is to provide generalizable knowledge, not necessarily to benefit the individual research participant. Many in the research field consider the return of research results antithetical to the concept of research as it further challenges the existing difficulty to distinguish research from clinical care. This argument may prove more difficult as translational genomic research becomes more integrated into individualized treatment strategies.

The position that release of research results is likely to do more harm than good is predicated on the experimental nature of research: research investigates an unknown, tests a hypothesis. In the U.S., the release of preliminary findings that are not confirmed or validated is not supported (OPPR 1993; NBAC 1999). "Research findings must be confirmed and validated and proven clinically useful before communicating information to anyone, otherwise we risk false assurances, or unnecessary scares" (NBAC 1999). The prevailing position, articulated more than 15 years ago, still guides most researchers today: "The uncertainties of predictive accuracy of future disease development outweighs benefits of disclosure – except when early treatment exists or can improve the prognosis" (OPPR 1993). In the U.S., there also exists a legal component to the non-disclosure position. By law, information that may be used in clinical decision-making must be generated in a laboratory that has received special federal certification called CLIA approval (CLIA 1988).

The arguments described above continue to inform the discussion of disclosure. A larger literature discussing these positions in more detail exists (Bookman et al. 2006; Parker and Lucassen 2003; Manolio 2006; Renegar et al. 2006; Banks 2000; ASHG 1998; Dressler and Juengst 2006; Ossorio 2006; Parker 2006). As more health-related information is revealed in the course of research and contributors to biobanks as well as the public become more aware of the implications of these findings, there will be increasing pressures in the biobanking community to determine the appropriateness of disclosure of research results.

What Motivates the Decision to Disclose or Not?

Numerous factors are involved in the decision to disclose research results. They include but are not limited to the following:

- Professional duty or responsibility
- Level of certainty regarding the association of the result with an outcome
- Magnitude of harm if result is not released
- Magnitude of harm if inaccurate information is released
- Availability of an intervention, medical or lifestyle, to offset the severity or development of the disease or condition
- Standard of care or standard of conduct for the profession
- Legal concern, e.g., negligence, duty to warn
- Relationship with biobank contributor or research subject; e.g., how well does the biobanker know what the contributor would value regarding research results

- Timing of the finding: temporal relationship of the information to the diagnosis or potential onset of the condition
- Availability of a clinical test, outside the research setting (Dressler 2007)

All of these factors contribute to the tension between ill-defined professional boundaries and moral intuition. In the clinical setting there is a clear sense of responsibility – physicians have a fiduciary duty to their patient and are expected to make decisions in the best interest of their patient. Here, one could argue that disclosure of clinically relevant information would be in the patient's best interest, if the researcher was also the research subject's physician. Does this "best interest standard" also extend to the individual's family and other relatives? In the U.S., court cases give mixed answers.

Disclosure of Research Results vs. Duty to Warn

We can draw on the clinical experience and situations of a physician's *duty to warn* to inform our discussion on disclosure of research results (Dugan et al. 2003; Falk et al. 2003; Offit et al. 2004). Because of the physician's fiduciary responsibility to his or her patient, the term *duty* refers to a moral and legal obligation or conduct arising from the physician's professional position. In the legal setting duty to warn cases have involved the determination of whether or not the physician was responsible to warn a family member of their risk of harm, based on knowledge [of a mutation, for example] of the individual the physician is treating. In the U.S. clinical cases have been determined by what is considered "standard of care" at the state level. For example, in Florida (*Pate v. Threlkel* 1995) a mother was diagnosed with medullary thyroid cancer in 1987; three years later, the adult daughter was diagnosed with the same disease. The daughter claimed that the doctor and his colleagues should have known the risk to her, the daughter, and had a duty to warn her [and get screening]. The Florida court held that the physician had a duty to warn the mother of the hereditary nature of the disease, but not the daughter (*Pate v. Threlkel* 1995). The basis for the decision was the prevailing standard of care in Florida and protecting the confidentiality of the doctor–patient relationship. However, only a year later in New Jersey a different decision came down from the courts (*Safer v. Pack* 1996). In this case the court held that there can be a duty to warn relatives at risk as long as there is a foreseeable risk, the individual at risk is easily identifiable and substantial future harm is easily identified or minimized by a timely and effective warning [such as monitoring or screening]. This case involved a father who had multiple polyps and adenocarcinoma of the colon. The father was originally diagnosed in 1950, and died of his disease in 1964 at the age of 45. The colonic polyps and his young age are factors contributing to the inherited nature of his disease. In 1990, 34 years later, the daughter was diagnosed with metastatic colorectal cancer and sued the deceased physician's estate, claiming that the physician knew of the hereditary nature of the disease and failed to warn the mother or the daughter (*Safer v. Pack* 1996).

From the Clinical Setting to the Research Setting

How do these lessons from the clinic apply to the research setting? Is there a "look-back" liability for biobankers or researchers? George Annas stated many years ago: "Where the implications of the research are unclear or where there is no effective therapeutic intervention, there could be no liability. However, if the investigator were also the treating physician, there is a small risk of liability if for example, a colon cancer gene mutation is found and the patient does not undergo routine screening (which could have identified early changes) and subsequently develops colon cancer" (Annas et al. 1995). Researchers, who are not physicians, as well as biobankers, currently have no fiduciary duty to the individual research subject or their family. Researchers and biobankers do not necessarily uphold any type of "best interest standard" in relation to the research subject, often because there is only an indirect relationship with the research subject via their specimen and/or research results. Researchers and biobankers, who are not physicians, have not been concerned with liability issues, at least not until now. In the U.S., beyond needing to be in compliance with federal regulations, researchers' and biobankers' professional responsibilities to the research subject and society have been ill-defined. These responsibilities, however, may be changing as many international guidelines support an ethical duty to disclose research results as a component of biobanking and human specimen research (Knoppers et al. 2006; MRC 2004; Council of Europe 2005). In addition to developing policy for implementing these guidelines, concern is also being raised regarding their legal implications. "It is hoped that fear of potential legal liability will not give rise to protectionist approaches mandating such a duty [to disclose research results] under law" (Knoppers et al. 2006).

An Emerging International Ethic of Duty to Disclose Genetic Research Results

Numerous guidelines have emerged from the international community addressing the issue of disclosure of aggregate or individual research results (Knoppers et al. 2006). Current guidance addresses the right of the individual research subject to decide whether or not they want to receive research information as well as the opportunity to change their minds about receiving such information (Knoppers et al. 2006). In some guidelines, such as CIOMS, there has been strong support that "individual subjects will be informed of any finding that relates to their particular health status" (CIOMS 2002). A similar position has been articulated by the Council of Europe (Council of Europe 2005), as well as the World Health Organization and UNESCO, especially in relation to disclosure of genetic research results (WHO 2003; UNESCO 2003). The 2004 amendment issued by the Council of Europe (Additional Protocol to the Convention on Human Rights and Biomedicine Concerning Biomedical Research) indicated: "If research gives rise to information of relevance to the current or future health of research participants, this information must be offered to them" (Knoppers et al. 2006). The Medical Research Council (MRC) in their Operational and Ethical Guidelines for Human Tissue and Biological Samples Used in Research Report stated: "Where the clinical relevance of research results becomes clear some time after the sample was obtained, or where the results obtained from secondary research may impact on the donors' interest, [a mechanism should be in place] to inform donors that results of potential interest may be available and offer them the opportunity to receive individual feedback or advice if they wish" (MRC 2004).

It should be noted, however, that none of these guidelines support the disclosure of research results, unless they have been analytically and clinically validated, demonstrate clinical significance or utility or in some other way significantly benefit the biobank contributor or research subject. The challenge is that although these conditions are generally accepted, we still lack consensus regarding what the reference should be for determining *clear* clinical validity or utility and who should be at the table to make these determinations (e.g. participants in addition to professionals). There are no consistent criteria to aid a biobanker or researcher in determining when and if a research result may be of benefit to an individual participant. Therefore the practical interpretation and application of these guidelines, on a daily basis, is still a challenge.

When Do Research Data, in General, Constitute Information that Should Be Communicated to an Individual Research Subject?

Most experts agree that interim or inconclusive findings should not be disclosed to the research participant, and only "confirmed, reliable findings constitute information" (Dressler 2003, 2005). They emphasize that early or preliminary data are imprecise and interpretations of such data may change over time. The consequences of disclosing preliminary or uncertain data, including anxiety, stress and possible medical intervention, are serious and can do more harm than good. However, people will act on information whether that information is provided in a clinical setting, using a validated clinical test, or whether that information is provided in the context of research or biobanking. Even in the absence of a valid test, perceptions of risk can cause individuals to change their lifestyle and not always for the better (Lynch et al. 1997). Results of testing, especially genetic testing, have significant impact on individuals and families. For example, in a study performed on families with hereditary breast–ovarian cancer, Lynch and colleagues found significant changes in women who underwent prophylactic surgery, including removal of both breasts and/or ovaries, after disclosure of positive results (Lugogo et al. 2007). Rates of compliance for both breast and ovarian cancer screening recommendations were also significantly increased among mutation carriers following result disclosure as opposed to before result disclosure (Lugogo et al. 2007). Women who were mutation carriers had feelings of guilt about passing the mutation down to their children, were worried about getting cancer, were worried about their children getting cancer and were concerned about health insurance discrimination (Lugogo et al. 2007). Approximately half of the women who underwent prophylactic surgeries before the disclosure of the results, turned out to be BRCA mutation negative, causing another level of significant stress.

As we grapple with decisions to disclose or withhold disclose information revealed in the course of research, it is vital that we keep this in mind: individuals *will act on information* and that the information *will have an impact on their lives*, the magnitude of which will depend on the nature of the information and the value the individual places on that information. It also underscores the need to address the intersection of many external and internal forces acting on the decision to disclose or not – including process, policy, professional boundaries and a researcher's moral compass. What process should be in place to address the growing likelihood that clinically relevant, valid information will be obtained in research studies? Who is responsible for making these decisions? Who is accountable for the outcome and consequence of communicating health-related information to a research subject and their family? What infrastructure and guidance need to be in place to support this process? These questions involve judgments on professional and moral levels. They also underscore the need for the research community to establish a professional code of conduct which addresses social and moral responsibilities.

To date, in the U.S., the over-riding philosophy has been, except in rare circumstances, that there should be no disclosure of individual results in the research setting, unless the information is required for eligibility or randomization in a clinical research trial. Even the disclosure of aggregate results is not a common activity in the U.S. For the U.S., the question remains how we can agree when it is appropriate to violate the non-disclosure rule. To paraphrase bioethicist and philosopher Bernard Gert (Gert 2004): under what conditions would you be willing to let everyone know that it is okay to "break the rule" in this way and not just in your circumstance? What is the reference upon which the decision to disclose should turn? Should biobanks or researchers use the *best interes*t standard? This would be hard to do if the researcher does not know the value which the research subject places on the information to be disclosed. Should the biobank use the *reasonable person standard* – what would a reasonable person want? Empirical research is needed to fully answer this question. Should the *medical intervention standard* be used – only disclose if a medical intervention exists? Should the biobank promote the autonomy standard – the research subject has a right to know and a right not to know the information that is gained about them through the course of research? Or should the approach involve a combination of these standards? Do we need to develop a new standard for decision-making in this context? Most professional, advocacy, advisory and regulatory groups agree that some type of vetting through a research ethics monitoring committee is appropriate (Dressler 2007). The inclusion of community members on such a committee would be an important consideration.

Policy Suggestions and Next Steps

The issue of disclosure of research results will become even more pronounced as the inter-relationship between previously unrelated diseases and genes becomes known. Policy needs to be developed to anticipate the rise of these situations. Practical guidance needs to be available for biobankers, researchers, and ethics review boards which face these challenging decisions and specimen contributors who are affected by these decisions. Although the international community have presented a general guidance which favors disclosure in support of individual autonomy, the practical criteria to guide this decision on a case-by-case basis is still lacking. Guidelines in the U.S., however scarce, continue to reflect the presumption that the disclosure of individual research results represents an exceptional circumstance, although with the advent of whole genome studies, this too may evolve over time.

Requirements for Biobankers and Investigators

Regardless of which guidelines one follows, the resolution of whether or not to disclose individual results still remains a challenging issue. At the very least, biobanks and researchers who have access to research results should be required to

- justify the intent for disclosure of individual research results as part of the biobanking protocol or research proposal.
- explicitly state in the informed consent process the intent for disclosure of research results, either individual or aggregate results, describing options for right to know and not to know and the opportunity for the contributor or research subject to change their mind.
- describe, in the biobank protocol or research proposal, the likelihood for the research to reveal findings related or unrelated to the scope of the research which may have impact on the subject's health.
- describe and justify, in the biobank protocol or research proposal, what circumstances might lead to the decision to disclose individual research results (e.g., validated result and medical intervention available).
- describe, in the biobank protocol or research proposal, a plan regarding how to manage a disclosure.
  - How this information is communicated to the research subject and by whom, i.e., not necessarily by the researcher or biobanker
- review and approve of the disclosure plan by an ethics review committee or an independent advisory board (Dressler 2007).

Elements of the Disclosure Process

If individual research results are going to be disclosed, the following considerations should be part of the disclosure process:

- Opportunity for research subject (or contributor to the bank) to decide whether or not they want to have research results disclosed
- Opportunity for the research subject (or contributor to the bank) to change their mind
- Research results, including appropriate medical advice or referral, should be provided by a medical expert who is also trained in counselling
- Counselling provided both before and after information is communicated
- Details and logistics of counselling, including needs for referrals, costs of counselling, follow-up visits if needed
- Appropriate precautions should be taken to determine the spectrum of information to be disclosed
  - E.g., information needed for diagnosis or treatment of a condition
  - E.g., disclosure of misattributed paternity, would not be warranted, except in rare cases, where disclosure of such information may prevent harm (Dressler 2007)

Next Steps

DNA and other human specimen banking coupled with studies in genetic and genomic research highlight the need to transition to a more socially responsible standard of research conduct in biomedicine. We need a deliberative process to address the roles and responsibilities of biobankers and researchers to inform the development of "codes of conduct". This process must address the tensions between moral intuition and professional boundaries so the resulting codes are broad enough to allow for moral analysis and yet narrow enough to provide some boundary for decision-making. This would require moving toward a collaborative process for decision-making, with a strong involvement by the community and contributors to the biobank, not just the professional or regulatory groups. Case-based scenarios for education and training and moral reasoning will help move this process forward and reveal underlying motivations that drive our decision-making. Deconstructing the arguments for and against disclosure among all the stakeholders is a necessary next step in the development of practical guidance in the U.S. and elsewhere.

Acknowledgments This work was supported by a NIH, National Human Genome Research Institute (ELSI Program), National Research Service Training Award: F32HG004200. This work would not have been possible without the support of Dr. Eric Juengst, mentor and colleague. The author also acknowledges the helpful assistance of Janell Markey, MS, in the preparation of this chapter.

References

Annas GJ et al. (1995) Drafting the genetic privacy act: science, policy, and practical considerations. J Law Med Ethics 23:360–366

ASHG (1998) ASHG statement. Professional disclosure of familial genetic information. The American Society of Human Genetics Social Issues Subcommittee on Familial Disclosure. Am J Hum Genet 62:474–483

Banks TM (2000) Misusing informed consent: a critique of limitations on research subjects' access to genetic research results. Sask Law Rev 63:539–580

Bookman EB et al. (2006) Reporting genetic results in research studies: summary and recommendations of an NHLBI working group. Am J Med Genet A 140:1033–1040

CIOMS International Ethical Guidelines for Biomedical Research involving human subjects (2002): http://www.cioms.ch/frame_guidelines_nov_2002.htm

CLIA (1988) Public Law No. 100–578, C.R.V. Clinical Laboratory Improvement Amendments Stat. 102 (October 31, 1988). Available from: http://www.fda.gov/CDRH/clia/100-578.pdf [Accessed 5-515-2008]

Council of Europe (2005) Additional Protocol to the Convention on Human Rights and Biomedicine, concerning Biomedical Research: http://conventions.coe.int/Treaty/en/Treaties/Html/195.htm

Dressler L (1998) Genetic testing for the BRCA1 gene and the need for protection from discrimination: an evolving legislative and social issue. Int J Breast Dis 10:127–135

Dressler L (2005) Human specimens, cancer research and drug development: How science policy can promote progress and protect research participants. Commissioned paper prepared for the National Cancer Policy Board, Institute of Medicine and Research Council of the National Academies: http://www.iom.edu/Object.File/Master/26/207/IOM_fnl.pdf

Dressler LG (2007) Disclosure of genetic research results: ethical and regulatory perspective. Meeting abstract. Public Responsibility in Medicine and Research (PRIM&R). (October, 2007).

Dressler L, Juengst E (2006) Thresholds and Boundaries in the disclosure of individual genetic research results. Am J Bioethics 6:18–20

Dressler L et al. (1999) Policy guidelines for the utilization of formalin-fixed, paraffin-embedded tissue sections. The UNC SPORE experience. Breast Cancer Res Treat 58:31–39

Dugan RB et al. (2003) Duty to warn at-risk relatives for genetic disease: genetic counselors' clinical experience. Am J Med Genet 119:27–34

Falk MJ et al. (2003) Medical geneticists' duty to warn at-risk relatives for genetic disease. Am J Med Genet 120:374–380

Gert B (2004) Common Morality: Deciding What To Do. Oxford University Press, Oxford

Grizzle W, Groday W (1999) Recommended policies for uses of human tissue in research, education, and quality control. Arch Pathol Lab Med 123:296–300

Hippocrates. The Hippocratic Oath. National Institutes of Health [cited 5–14–2008]: http://www.nlm.nih.gov/hmd/greek/greek_oath.html

Jewell S et al. (2006) Biospecimen banking, standardization and lessons learned from the Cancer and Leukemia Group B Pathology Coordinating Office. Sem Breast Dis 8:93–99

Knoppers BM et al. (2006) The emergence of an ethical duty to disclose genetic research results: international perspectives. Eur J Hum Genet 14:1170–1178

Lugogo NL et al. (2007) Genetic profiling and tailored therapy in asthma: are we there yet? Curr Opin Mol Ther 9:528–537

Lynch HT et al. (1997) A descriptive study of BRCA1 testing and reactions to disclosure of test results. Cancer 79:2219–2228

Manolio TA (2006) Taking our obligations to research participants seriously: disclosing individual results of genetic research. Am J Bioethics 6:32–34

MRC (2001) Human tissue and biological samples for use in research – operational and ethical guidelines: http://www.mrc.ac.uk/consumption/idcplg?IdcService=GET_FILE&dID=9051&dDocNameMRC002420&allowInterrupt=1

MRC (2004) Human tissue and biological samples for use in research – operational and ethical guidelines: http://www.mrc.ac.uk/Utilities/Documentrecord/index.htm?d=MRC002420

NAPBC (1997) Model Consent Forms and Related Information on Tissue Banking and Routine Biopsies, With comments by the PRIM&R and ARENA Tissue Banking Working Group. Letter to Directors/chairpersons and Members, Institutional Review Boards. Public Responsibility in Medicine and Research. National Action Plan on Breast Cancer, National Cancer Institute. Available at: http:/www.primr.org

NBAC (1999) Research involving human biological materials: ethical issues and policy guidance. Vol. 1: Report and Recommendations of the National Bioethics Advisory Commission: http://bioethics.georgetown.edu/nbac/hbm.pdf

NCI (2007) NCI best practices for biospecimen resources: http://biospecimens.cancer.gov/global/pdfs/NCI_Best_Practices_060507.pdf

NIH (2008) National Institutes of Health. Public Access Frequently Asked Questions. Available at: http://publicaccess.nih.gov/FAQ.thm?print=yes

Offit K et al. (2004) The "duty to warn" a patient's family members about hereditary disease risks. JAMA 292:1469–1473

OPPR (1993) Human genetic research: issues to consider: http://ohrp.osophs.dhhs.gov

Ossorio PN (2006) Letting the gene out of the bottle: a comment on returning individual research results to participants. Am J Bioethics 6:24–25

Parker LS (2006) Best laid plans for offering results go awry. Am J Bioethics 6:22–23

Parker M, Lucassen A (2003) Concern for families and individuals in clinical genetics. J Med Ethics 29:70–73

Pate v. Threlkel (1995) 661 So.2d 278 (Fla.), rehearing denied (October 10, 1995)

Pelias MK (2005) Research in human genetics: the tension between doing no harm and personal autonomy. Clin Genet 67:1–5

Pullman D, Hodgkinson K (2006) Genetic knowledge and moral responsibility: ambiguity at the interface of genetic research and clinical practice. Clin Genet 69:199–203

Ravitsky V, Wilfond BS (2006) Disclosing individual genetic results to research participants. Am J Bioethics 6:8–17

Renegar G et al. (2006) Returning genetic research results to individuals: points-to-consider. Bioethics 20:24–36

Safer v. Estate of Pack (1996) 291 N.J. Super. 619, 677 A.2d 1188 (N.J. Super.App.Div.)

Schilsky R et al. (2002) Cooperative Group Tissue Bank as Research Resources: The Cancer and Leukemia Group B Tissue Repositories. Clinical Cancer Research 8:943–948

UNESCO (2003) International Declaration on Human Genetic Data: UNESCO-the United Nationl Educational, Scientific and Cultural Organization (October 16, 2003). Available at: http://portal.unesco.org/en/ev.php-URLID=17720&URLDO=DOTOPIC&URLSECTION-201.html

Wachbroit R (1998) The question not asked: the challenge of pleiotropic genetic tests. Kennedy Inst Ethics J 8:131–144

WHO (2003) Genetic databases: assessing the benefits and impact on human and patient rights [Report]: http://www.law.ed.ac.uk/ahrc/files/69_lauriewhoreportgeneticdatabases03.pdf

Biobanks and Our Common Good

Erik Christensen

Abstract There are many different kinds of biobanks with various scopes and purposes, such as diagnostic biobanks, therapeutic biobanks, and research biobanks. My focus is on research biobanks, which enable us to identify genetic and environmental causes of complex diseases. The hope is that research biobanks of this kind will be able to provide us with new medical knowledge of large public health issues. Research biobanks have the potential to produce knowledge that could hold great value and significance for many. In this mindset, such institutions may be considered a social asset benefiting all. I aim to explore the obligations we have in relation to research biobanks and what we can expect and demand from them. I argue that good reasons for everyone to participate in this type of research can be found in the principles and values that characterise modern societies and that many of us take for granted. To explore the rights and obligations, we have vis-à-vis biobank research. I base my arguments in a communitarian and liberalistic understanding of individuals and communities. These two approaches illustrate in separate ways what is at stake. We shall see that both approaches facilitate arguments claiming that biobank research is part of our understanding of ourselves and of society. This means that biobank research can become part of the kind of society that provides the individual with the opportunity to realise his understanding of a good life. If this is the case, we are not obligated to obtain consent from the individual in connection with biobank research. A central prerequisite is that the research promotes values and benefits we can all support. To safeguard this principle, a wider discussion and debate concerning what to research is needed. From this it follows that it could be unethical not to research certain diseases genetically.

E. Christensen
Department of Philosophy and Bioethics Research Group, Norwegian University of Science and Technology, NO-7491 Trondheim, Norway
e-mail: erik.christensen@hf.ntnu.no

J.H. Solbakk, et al. (eds.), *The Ethics of Research Biobanking*,
DOI: 10.1007/978-0-387-93872-1_8,

Introduction

Biobank research raises a number of questions concerning the relationship between the individual, society, and biobanks. What are the responsibilities of each stakeholder? What are their rights and obligations? What type of resource and asset does biobank research represent? How do we distribute this asset (Austin et al. 2003; Hansson and Levin 2003; Tutton and Corrigan 2004; Cambon-Thomsen 2004; Godard et al. 2004; Ashburn et al. 2000; Winickoff and Winickoff 2003; Sutrop 2004; Häyry et al. 2007; Gottweis and Petersen 2008)? Biobank research combines genetic data with health and lifestyle information (Holmen et al. 2003). If our "collective genetic legacy" can secure good health and prosperity for ourselves and future generations, the issue of whether it ought to be a duty to contribute to the realisation of this common good becomes highly relevant.

In medical research consent is a fundamental principle; nobody can be forced to participate in research if they do not give consent. More generally speaking, we do not accept setting aside individual rights for the greater good of the community. This mindset, in combination with an individualistic ethos, has led to a situation where people do not automatically volunteer for matters concerning the greater good of the community. Instead, it is argued that we have a moral duty to contribute to medical research (Rhodes 2005, 2008; Herrera 2003; Harris 2005; Harris and Woods 2001; Evans 2004; Orentlicher 2005). However, because medical research historically has been a subject for a number of scandals and abuse, this moral argument cannot be made too strongly; i.e., we cannot unduly judge the ones who do not participate nor can we make participation compulsory. It may appear that a moral standpoint is the only basis from which to argue that the individual ought to participate in biobank research. In this chapter it will be argued that good reasons for everyone to participate in this type of research can be found in the principles and values that characterise modern societies and that many of us take for granted. The condition is that biobank research does not constitute a breach with these principles and values. Freedom is such a principle, autonomy or self-realisation is another, and equality is a third. To explore the rights and obligations we have vis-à-vis biobank research, I will take as a point of departure first a communitarian and subsequently a liberalistic understanding of individuals and communities. In separate ways, these two approaches illustrate what is at stake. We shall see that both approaches facilitate arguments claiming that biobank research is part of our understanding of ourselves and of society. This means that biobank research can become part of the kind of society that provides the individual with the opportunity to realise his understanding of a good life.

Communitarianism, Society and Biobanks

Biobank research has brought to the fore the issue of whether the individual has certain obligations vis-à-vis the community. To a certain degree we can claim that all medical research benefits the individual, but the nature of biobank research has made this issue all the more pressing. Because biobank research is believed to constitute a social asset, communitarian values have become central, and several voices argue that we have obligations vis-à-vis the community as a whole, and that these obligations must be balanced with the rights of the individual (e.g. Chadwick and Berg 2001). Communitarian values, however, are based on a certain view of individuals and communities that often goes unexplained. The degree to which we commit to the communitarian values, thus, is dependent on whether or not we share this view. Consequently, it is meaningless to appeal to communitarian values without accounting for the views on which these values are based.

The communitarian view claims that "[o]nce we recognize the dependence of human beings on society, then our obligations to sustain the common good of society are as weighty as our rights to individual liberty" (Kymlicka 2002: 212; see also Sandel 1998; Taylor 1985b). If we are to make an argument for why we ought to contribute to biobank research, we must demonstrate that individual rights cannot be interpreted independently of our obligations to biobank research and our connection to the community in general. It is commonly held that the individual has certain rights. This evokes a sense of a "primacy of rights" and entails that our obligations to society are derived from, or are secondary to, these rights in some sense. Freedom is such a right, which makes it difficult to claim that biobank research is, or constitutes, a social asset which one is obligated to support. However, if it turns out that biobank research conveys central values in our society, values that help define our identity and sense of belonging, maybe the "primacy of rights" should be amended with a set of "politics of the common good".

Whereas liberalism is based on the belief that we are free to choose our goals and the type of life we want to lead, the communitarian understanding claims that the exercise of this freedom only makes sense on the basis that we discriminate between different values and which goals we seek to realise. Our understanding of freedom is based on the belief that some forms of freedom are more significant than others. What enables this understanding is the fact that we view objectives and purposes in light of what is important to us. We thus distinguish between significant and trivial forms of freedom depending on who we are and our understanding of ourselves. That having been said, we cannot determine what is significant by how strongly or how often we desire something. Instead, it seems as though some feelings and objectives are more important than others regardless of their strength or how often they occur. This is explained by the fact that we do not only have first-order desires, but also second-order desires, which are desires about desires. Taylor has dubbed this "strong evaluation": "We experience our desires and purposes as qualitatively discriminated, as higher or lower, noble or base, integrated or fragmented, significant or trivial, good and bad" (Taylor 1985c: 220; see also Taylor 1985a, 1998). In other words, we find that some of our feelings and objectives are intrinsically more significant or valuable than others. We discriminate between different values and feelings. This implies that our freedom is contingent on values and objectives about which we can be wrong.[1] If we act on the basis of these values we will fortify our "unfreedom" rather than our freedom. Freedom (or autonomy or self-fulfilment) cannot be understood independently of the values and objectives that give our actions meaning.

The right of the individual to choose not to participate in biobank research, to not contribute to this social asset, must be seen in light of which values and objectives give our actions meaning. Freedom of choice has no value in itself; the value is in the actions we do choose. The communitarian approach entails that the right of the individual to decide for himself cannot be separated from the objectives held by society. The self comprises objectives that we do not ourselves choose but discover as a result of being embedded in various social practices and contexts. Exercise of freedom thus does not mean that one is free to reject or revise one's objectives and projects, but that one is capable of fulfilling these by realising what the constituent parts of the objectives are. If the community, in part, constitutes our objectives, the common good will be part of the individual good. Our identity and connection with the community means that we are unable to view ourselves as separate from it. If biobank research constitutes a common good or a value from which we cannot separate ourselves, any fulfilment of this common good will mean that we fulfil our own understanding of the good life, because this understanding, in part, is founded on the common good. The best way to promote and serve the interests of the individual would be to limit the options of not participating, hence contributing to a realisation of the good life. The right to self-determination may undermine a central asset in society, and a fortiori undermine the opportunities the individual has to pursue his own projects, because these, in part, are based on society in general.

The communitarian understanding of how best to serve the interests of the individual entails a specific view of which type of society is the best. Consequently, this excludes other views, including the liberal one. We cannot appeal to communitarian values without simultaneously reflecting on the type of society in which we prefer to live. Whether communitarian values ought to constitute an argument in medical research in general, and biobank research in particular, is thus dependent on whether we commit to the communitarian understanding of individuals and communities on which these values are based.

If we look at the liberal understanding, the conception of the common good is based on the different preferences held by the individual. This approach states that the common good is to give the individual an opportunity to realise his understanding of the good life on an equal footing with the opportunities of others to pursue the same goal. Consequently, we can argue that the common good in a liberal society "is adjusted to fit the pattern of preferences and conceptions of the good held by the individuals" (Kymlicka 2002: 220). The communitarian view, however, interprets the common good taking a substantial understanding of the good life as defined by

[1] The alternative is to claim that we can never be wrong in terms of our desires, which is only possible if our feelings are "brute facts". This position, however, is not very plausible. We distinguish between feelings such as pain on the one hand and shame and fear on the other. The former is a case of brute fact; if you do not feel pain, you are not in pain. The latter two, however, might be subject to mistakes; they may be irrational and thus unfounded, cf. Taylor 1985c: 222–227.

the community's way of life as its point of departure. The common good represents the standard by which the individual understanding of the good life is judged. If biobank research represents a common good in society, the individual's understanding of the good life will be evaluated on this basis. And because society's understanding of the common good involves ranking values and goals, the preferences that are in accordance with this common good will be emphasised more than others. The communitarian understanding is perfectionistic "since it involves a public ranking of the value of different ways of life" (Kymlicka 2002: 220). The consequences of this perfectionism could be that the individual is encouraged to participate in biobank research, while efforts are made to restrict his or her opportunities to refrain from contribution.

In the communitarian understanding lies the assumption that the good life may only be realised in a society maintaining the social conditions making it possible to form an opinion as to the kind of life one wants to lead. Biobank research is a common good serving as a foundation for the objectives the individual seeks to realise. By limiting the opportunities the individual has of not participating in biobank research, the social conditions facilitating self-determination are maintained. The communitarian view is that a substantial understanding of the common good is necessary to protect the freedom and rights of the individual. On the other hand, if an individual exercises his or her right not to participate in biobank research, he or she will undermine the conditions for realising his or her understanding of the good life, and you will thus be less free (Taylor 1985c).

Liberalism, Equality and Biobanks

Liberalism advocates a belief in the rights of the individual and the individual's right to make autonomous choices. This would appear to mean that liberalism may not be used as an argument in favour of contributing to biobank research: If liberalism involves the right to freely choose one's values, how can anyone claim that some choices are more right than others? If the value lies in the very freedom to choose and not in what is chosen, it is hard to see how contributing to common assets such as biobank research follows from a liberalistic point of view. However, we shall see that liberalism does not necessarily represent an argument against introducing a form of obligation to participate in biobank research, but rather constitutes an argument in favour of our participating. I argue that if we want to defend the basic values on which liberalism is based, we should also, with good reason, contribute to the kind of common good biobank research represents. The condition is that the biobank research can be said to constitute key values to which the individuals of a liberal society have committed. Which kinds of values are these?

Political theories appeal to various values, such as "contractual agreement", "the common good", "utility", "rights", and "identity". Does this mean that the various theories appeal to ultimately conflicting values? And does this mean that we have to choose one to the detriment of the others? In this case, the choice of theory will determine to which values we appeal. But how do we know which theory to choose in the first place? Which criterion do we use to determine which theory is the right one? There is no simple answer to this question, but Ronald Dworkin presented one possible solution that may show us a way out of the problem. He claims that modern political theories do not have different foundational values but that they rather share the same ultimate value, equality (Dworkin 1977: 179–183, 1983, 1986: 296–301, 2000; Sen 1992, 2004: 22; Cohen 1993). They are all "egalitarian" theories. Dworkin does not, in this context, refer to equal distribution of income or equal distribution of property. They are obviously not egalitarian in this regard. We can, however, interpret the term egalitarian as treating everybody "as equals". This is indicative of a more abstract and fundamental understanding of the concept, entailing that "the interests of each member of the community matter, and matter equally" (Kymlicka 2002: 4). Everyone is entitled to equal concern and respect. The question thus becomes which conditions must be met before this is the case. In this approach, one can argue that each political theory presents a distinct definition of what it means to treat people as equals. We must therefore look at the conditions that, according to liberalism, must be present in order for us to treat each other as equals.

Liberalism is grounded on a belief in the individual's right to freedom. In this lies the right to make autonomous decisions as well as the right to realise the understanding of how one wants to live. The only limitation on this right is another's right to the same freedom. Consequently, an individual has the right to as much freedom as is compatible with the others' rights to the same amount of freedom. Why does freedom have value? One possible answer to this question is the fact that it enables us to obtain the benefits we want. Freedom protects our interests. But as not all forms of freedom are equally important to us, it is common to distinguish between "basic liberties" and "non-basic liberties". Basic liberties include inter alia the "freedom of thought and liberty of conscience; the political liberties and freedom of association, as well as the freedom specified by the liberty and integrity of the person" (Mill 1859/2004: 15–16; Rawls 1999: 53, 2005: 291). By categorising these as basic liberties, Rawls does not only refer to how they are more important to us than other rights; they are necessary for us to be able to exercise our moral ability, and they are inalienable. In this context, however, the first sense is the most relevant: "the basic liberties" are necessary to obtain various understandings of the good life. This means that we must organise our society in such a way that we all have the same opportunities to realise our understanding of the good life, whatever our interpretation of it may be. Society does not determine what the common good is, but the common good is that society provides everyone with the opportunity to realise their understanding of how they want to live. If we distinguish between the means or goods which we need and the types of goals we ought to seek realised with the help of those goods or means, the former is the responsibility of society, whereas the latter is up to the individual. If goods exist that can be said to be necessary in order to realise one's understanding of the good life, goods that "have a use whatever a person's rational plan of life" (Rawls 1999: 54), the idea is that it is fair to give everyone equal access to these goods so that everyone has equal opportunity to realise the life they want to lead.

How do we justify social differences, such as some people making more money than others? One possible answer, and perhaps the most prevalent, is that these differences can be justified, given the existence of equality of opportunity. Under the premise of fair competition for offices and positions resulting in benefits, such as higher income and increased prestige and status, differences tied to these offices and positions may be justified. As long as no one is discriminated against on the basis of gender, race, or social background, economic differences in income are not unfair. What, then, of those worse off? According to this principle they cannot object to these differences, nor are they entitled to any of the benefits. Not everyone agrees with this principle, and as we shall see, there is good reason to expand the principle of equality of opportunity to include those worse off in such a way that they may also benefit from the differences.

What is it that makes equality of opportunity intuitively appear fair and just? This is primarily due to it being based on what people do and not who they are. What we end up as depends on our choices and not on our circumstances. It is not determined by our gender, race, or social background – conditions we are unable to do anything about.[2] Hence, we believe that social differences are justifiable if they are a consequence of the individual's choices and actions. Similarly, we find it unreasonable if differences are caused by social circumstances, such as gender or race. It becomes a question of what you deserve. What, then, of differences in natural abilities and skills? Similar to differences in social circumstances, these are not affected by individual choices, and as such are not up to the individual to determine. No one deserves being born into a particular social class, race, or gender. But the same premise seems to apply to natural differences; no one deserves being born with a disability or an extraordinary talent. Rawls thus concludes that "once we are troubled with the influence of either social contingencies or natural chance on the determination of distributive shares, we are bound, on reflection to be bothered by the influence of the other" (Rawls 1999: 64). The original intuition behind equality of opportunity was that economic differences ought to be determined by personal choices and actions and not by social advantages and disadvantages. If this is true, we must also make allowances for natural differences, as these are as little a result of personal choices and actions as are social circumstances. However, instead of levelling these differences the way we did with social circumstances, Rawls proposes that we permit these differences on the condition that they do not only benefit a few, but also those worse off (Rawls 1999: 54, 87). If some differences benefit all, nobody loses. In other words, we do not have to eliminate all differences, only those that work to someone's disfavour. Rawls calls this approach "the difference principle".

The principle of equality of opportunity and the difference principle are concerned with how we inter alia ought to distribute economic resources and how we can explain differences, such as, for example, differences in income. Rawls does not, however, concern himself much with the issue of health (Rawls 1999: 83–84, 2005: 21),[3] and it has thus been necessary to explain these principles to define

---

[2] What it takes to actually achieve equality of opportunity is controversial, but it does not affect the argument in this context.

[3] On health as a "primary good", see Rawls 1999: 54. See also Rawls 2001: 173–174, 2005: 184–185 and 244–245.

the rights and obligations of individual's vis-à-vis the health benefits related to biobank research.[4] In the context of biobanks, health is interpreted as a result of environmental (social) circumstances and genetic (natural) conditions. One of the key characteristics of biobanks is the connection between genetics and health and lifestyle information, which enables us to explore complex causal connections in the interaction between genetics and environmental conditions. The fact that health is tied to natural differences follows from the roles our genes play. Some have a genetic makeup making them more prone to contract or develop a number of diseases, but it does not necessarily follow that this will happen. Environmental or social circumstances play a role in the risk people have of developing cardiovascular diseases, diabetes and cancer.[5] A person does not pick one's genes or one's social circumstances.[6] If health is a good not only affecting one's well-being, but also the chances of securing other social goods, it follows from the principle of equality of opportunity that social differences related to health must be levelled, whereas natural differences ought to benefit those worst off. Only then do we treat people as equals. In other words, it is unfair that some people benefit from advantages that are not a consequence of their own choices and actions.[7]

What, then, are the consequences for the individual in terms of participation in, and contribution to, biobank research? First, the benefits and values produced by biobank research will contribute to levelling social differences related to health, because many of the diseases which biobanks study are related to socioeconomic conditions. One can compensate for these social differences by other means, but as long as a number of the most common diseases seem to be associated with social factors, new knowledge and insights gained from biobank research will remain one of the most important tools to fighting social differences caused by health issues. In other words, health benefits and assets must be distributed in such a way that we level differences and give everyone equal access. This entails that the individual ought to participate in, and contribute to, biobank research, because this

[4] But Norman Daniels has quite successfully tried to apply Rawls' principles to the issue of health, focusing on health care and health need and allocation of scarce resources; see Daniels 1981, 1985, 2004, 2008, and Pogge 1989: 181–196. Less emphasis has, however, been placed on the application of these principles to problems related to biobank research within bioethics.

[5] A number of studies, surveys and reports have shown that social differences with regard to health are derived from variables such as class, race, gender and geographical belonging; see Black and Morris 1992; Marmot et al. 1978; Marmot 2004: 49–50; Marchand et al. 1998; Mackenbach et al. 2008; Daniels et al. 1999; Daniels 2008, Chap. 3.

[6] Rawls refers to health as a "natural primary good" and not a "social primary good"; see Rawls 1999: 54. In the context of biobank research it is, however, also appropriate to interpret health as a "social primary good"; see Daniels 1981, 2008.

[7] A possible objection could be the argument that health is a personal and not a social responsibility. It will thus not seem fair that differences in health are to be levelled if they are a consequence of personal choices and actions. This objection is not decisive for my argument as biobank research looks at the interaction between genetics and environmental conditions in relation to the risk of developing or contracting diseases based on this complex causal condition. It is also not given that personal choices and actions are enough to explain differences when it comes to health, and the personal responsibility for health argument should thus only play a minor role (Wikler 2004); but see Shapiro 2007.

is a prerequisite for producing the knowledge and insight needed to compensate for diseases caused by social circumstances. Second, biobank research will benefit those most at risk for developing diseases for which one is genetically disposed. As nobody picks their own genes, this means that differences in terms of the risk (disposition) of developing a disease are random. They are expressions of brute luck and not based on what a person may or may not deserve. We can imagine that individuals not particularly at risk for diseases stand to receive a larger share of the social assets, whereas those with a genetic disposition for developing a disease stand to receive a smaller share of the assets. This means that persons with a genetic disposition for developing the diseases biobanks want to research have the most to gain from this type of research, whereas persons who do not have this genetic disposition have little to gain. Differences in the genetic disposition for developing certain diseases should neither be an advantage nor a disadvantage for the individual in terms of access to assets, as these differences are not an expression of choice. However, instead of thinking that nobody should benefit from these differences, we should let the differences benefit all so that those worse off will also benefit. People not genetically disposed to develop certain diseases cannot use this as an argument to evade contribution, because that would mean that they used their advantage in a way that would increase, not reduce differences. This conflicts with the idea of treating everybody as equals, because it would only promote the interests of some groups, not all. However, if one contributes to biobank research, the position of those worst off will also improve in spite of differences in their genetic disposition for developing diseases, assuming that the risks are equally distributed among these groups (or that the worst off gain the most).

Biobank Research and Informed Consent

Biobank research does not accomodate arguments for obtaining consent as easily as other types of medical research. The question then is to what extent it is reasonable to require consent in biobank research. Moreover, if consent is indeed required, what type of consent would be relevant in this context (Hoedemaekers et al. 2007; Hansson et al. 2006; Hansson 1998; Beskow et al. 2001; Kaye 2004; Wendler 2002; Chadwick 2001). In my view, there is good reason to argue that consent is unnecessary if biobank research promotes and preserves values and assets with which we all agree in principle. This does not mean that all types of biobank research fall into this category. We must thus distinguish between the types of biobank research where obtaining consent is necessary and justified and the types of biobank research where this need does not exist. In order for biobank research to be exempt from the requirement of obtaining consent, the research must be entirely uncontroversial. We can assume that this category concerns research into diseases affecting a broad majority of the population. The research must apply to the population at large and not just affect a few select groups. Diseases falling into this category affect a large share of the population, and we must all relate to them in one way or another.

The nature of the values and benefits produced by biobank research in this regard must also potentially be made available to everyone, and the benefits must contribute to levelling social and natural differences.

This approach calls for a descriptive (neutral) rather than normative (perfectionistic) understanding of biobank research. We are talking about an asset necessary for realising the good life, irrespective of what our understanding of what that is.[8] Biobank research may be interpreted as a necessary means to ensuring that the individual attains his goals. Therein lies the claim that biobank research is a common good: it is an asset needed by everyone to realise their understanding of the good life. The usage of the term common here thus refers to it being a means everyone needs, and not that it ought to be used for a joint purpose to which everyone agrees. The problem of referring to biobank research as a common good in the normative sense is that this approach is based on a substantial understanding of the individual, as well as a perfectionistic understanding of society. If, implicit in the appeal to participate, there is an understanding of which is the worthier choice (to contribute to research) and which are the right ideals (communitarian). One could potentially alienate the individuals who do not share this view. Relying on a normative argument will fall short in a pluralistic society if the argument encapsulates values with which only part of the population can identify. Rather than constituting a collective (common) good, it may be perceived as divisive.

Instead of basing biobank research on the requirement of obtaining informed consent, there may be other and better ways of safeguarding the best interest of the individual. One option is to replace the requirement of consent with the opportunity to influence the topic of the research. This would enable the individual to exercise autonomy on a very different level than what is possible through the requirement of consent. In this way, the problems related to normative arguments in favour of biobank research are eliminated by having the individual participate in deciding the topic of research. Part of the problem of normative arguments in favour of biobank research is taking the values appealed to for granted. We can, however, not presume that a consensus as to which values to appeal to exists independently of the values held by individuals. If, on the other hand, we allow individuals to participate in deciding the topic of research, we solve this problem. Hence, we could say that biobank research is made accountable and ethical questions are seen in the light of the topics chosen for research.

One argument in support of requiring informed consent in biobank research is the right to privacy. If the right to privacy is critical, obtaining consent will be a way to safeguard this right. I would, however, like to address the issue of whether this right should be critical. The right to privacy is intuitive. Nobody likes it when it is violated or threatened. But even if a person dislikes actions which infringe on his right to privacy, it does not necessarily follow that these acts are impermissible. Freedom of speech comes to mind. Exercising one's right to free utterance may

---

[8] See Rawls' concept of "primary goods": "Primary goods ... are things which it is supposed a rational man wants whatever else he wants" (Rawls 1999: 79). See also Rawls 1999: 54–55 and 79–81, 2005, lecture V. Rawls also refers to primary goods as "all-purpose means" (Rawls 2001: 57–61).

cause another individual to feel that his right to privacy has been violated. Yet, very few are convicted of defamation. The reason for this, of course, is that we regard the right to free speech as so fundamental that we are willing to go to great lengths to protect it. Consequently, there will be times when we will have to put up with situations we dislike or find uncomfortable. If biobank research is based on information that cannot be traced back to the individual, and offers information at group level, so that the information presented does not apply to the individual, we have to ask ourselves whether the right to privacy ought to be so strong that obtaining consent is necessary. If other, weightier concerns exist, the right to privacy and consent may have to yield, even if the persons involved do not agree. This does not, however, necessarily mean a breach with the fundamental rights granted to the individual. Whether or not this is the case will depend on whether the right to privacy in this context may be fairly interpreted as being part of the fundamental rights granted to all. If the right to privacy constitutes such a fundamental right, the issue becomes whether this right should take precedence in the event it conflicts with other rights. In other words, the right to privacy may conflict with other rights granted to us, and the right to privacy cannot be interpreted as always being opposed to, or conflicting with, social or community interests.

Conclusions

The right of the individual to determine whether or not he wants to contribute to a common good, such as biobank research, comes off as problematic in light of the limited risk this poses to the individual. When the degree of risk for the individual is so low, and the potential gain for society is so great, it is justified/legitimate to ask whether the right to self-determination and the requirement of informed consent should carry as much weight in biobank research compared to other forms of medical research where the risk to the individual is considerably greater (WHO 2003; Helgesson and Eriksson 2008). Genetic research in general and biobank research in particular "has been accompanied by a shift in emphasis towards the ethical principles of reciprocity, mutuality, solidarity, citizenry and universality." (Knoppers and Chadwick 2005: 75). Even though biobank research forces issues related to communitarian values and principles, one should perhaps still question the communitarian understanding of individuals and society. In this context, one should especially emphasise the fact that a communitarian approach may imply a form of paternalism (Berlin 2002; Taylor 1985c: 229), which would make basing the argument for biobank research on communitarian values difficult.

In western democratic societies the position of the liberal egalitarian tradition is strong. It is thus wise to be sceptical of the increased emphasis on communitarian values and principles in the biobank debate. If communitarianism promotes an understanding that the self does not precede but comprises its objectives (given the social context) as well as a perfectionistic (non-neutral) understanding of the common good (as defined by the community's way of life), we must ask ourselves if we are willing to give up the liberal egalitarian understanding. This is founded on the belief that everyone should have equal opportunities to realise their objectives (whatever an individual's understanding of the good life may be), and it presents an antiperfectionistic understanding of society, in the sense that the common good is defined based on the kinds of means and resources needed and not the types of objectives that ought to be sought using these means and resources. It may appear that the concepts of "solidarity", "community" and "benefit" have become the new "buzz-words" of biobank research, but it is important to clarify the theoretical foundation on which these concepts are based. Perhaps the argument that one is obligated to contribute to biobank research could bring about a discussion as to which topics to research (Williams 2005; Williams and Schroeder 2004). In my opinion the interests of the individual are better served not by the requirement of informed consent but by a discussion of what is our common good resulting from participation in biobank research (Weldon 2004). In that case, contributing to biobank research would be a natural extension of the liberal values and democratic principles on which our society in general and medical research in particular are founded.

Acknowledgments My research has been funded by a fellowship from the Research Council of Norway's Programme for Functional Genomics. Part of the paper was written during my stay as a visiting researcher at the Kennedy Institute of Ethics at Georgetown University, which provided me with excellent research facilities. I am grateful to the Kennedy Institute of Ethics for their kind hospitality and to the staff at the National Reference Centre for Bioethics Literature for their invaluable help in finding the relevant literature for my research. I am also grateful to Berge Solberg, and to the editors of this anthology, for their comments and suggestions.

References

Ashburn TT et al. (2000) Human tissue research in the genomic era of medicine. Archives of Internal Medicine 160:3377–3384

Austin MA et al. (2003) Monitoring ethical, legal, and social issues in developing population genetic databases. Genetics in Medicine 5:451–457

Berlin I (2002) Two concepts of liberty. In: Hardy H (Ed.) Liberty – Incorporating Four Essays on Liberty. Oxford University Press, Oxford

Beskow LM et al. (2001) Informed consent for population-based research involving genetics. JAMA 286:2315–2321

Black D, Morris JN (1992) Inequalities in Health: The Black Report and the Health Divide. Penguin, London

Cambon-Thomsen A (2004) The social and ethical issues of post-genomic human biobanks. Nature 5:866–873

Chadwick R (2001) Informed consent and genetic research. In: Doyal L, Tobias JS (Eds.) Informed Consent in Medical Research. BMJ Books, London, pp 203–210

Chadwick R, Berg K (2001) Solidarity and equity: new ethical frameworks for genetic databases. Nature Review Genetics 2:318–321

Cohen GA (1993) Equality of what? On welfare, goods, and capabilities. In: Nussbaum MC and Sen A (Eds.) The Quality of Life. Oxford University Press, Oxford, pp 9–29

Daniels N (1981) Health-care needs and distributive justice. Philosophy and Public Affairs 10:146–179

Daniels N (1985) Just Health Care. Cambridge University Press, Cambridge
Daniels N (2008) Just Health: Meeting Health Needs Fairly. Cambridge University Press, Cambridge
Daniels N (2004) Is there a right to health care and, if so, what does it encompass? In: Kuhse H, Singer P (Eds.) A Companion to Bioethics. Blackwell, Malden, MA, pp 316–325
Daniels N et al. (1999) Why justice is good for our health: the social determinants of health inequalities. Daedalus 128:215–251
Dworkin R (1977) Taking Rights Seriously. Duckworth, London
Dworkin R (1983) In defense of equality. Social Philosophy and Policy 1:24–40
Dworkin R (1986) Law's Empire. Harvard University Press, Cambridge, MA
Dworkin R (2000) Sovereign Virtue: The Theory and Practice of Equality. Harvard University Press, Cambridge, MA
Evans HM (2004) Should patients be allowed to veto their participation in clinical research? Journal of Medical Ethics 30:198–203
Godard B et al. (2004) Strategies for consulting with the community: the cases of four large-scale genetic databases. Science and Engineering Ethics 10:457–477
Gottweis H, Petersen AR (2008) Biobanks: Governance in Comparative Perspective. Routledge, Abingdon, Oxon
Hansson MO (1998) Balancing the quality of consent. Journal of Medical Ethics 24:182–187
Hansson MG et al. (2006) Should donors be allowed to give broad consent to future biobank research? Lancet Oncology 7:266–269
Hansson MG, Levin M (Eds.) (2003) Biobanks as Resources for Health. Uppsala Universitet, Uppsala
Harris J (2005) Scientific research is a moral duty. Journal of Medical Ethics 31:242–248
Harris J, Woods S (2001) Rights and responsibility of individual participating in medical research. In: Doyal L, Tobias JS (Eds.) Informed Consent in Medical Research. BMJ Books, London, pp 276–282
Häyry M et al. (Eds.) (2007) The Ethics and Governance of Human Genetic Databases: European Perspectives. Cambridge University Press, Cambridge
Helgesson G, Eriksson S (2008) Against the principle that the individual shall have priority over science. Journal of Medical Ethics 34:54–56
Herrera CD (2003) Universal compulsory service in medical research. Theoretical Medicine 24:215–231
Hoedemaekers R et al. (2007) Solidarity and justice as guiding principles in genomic research. Bioethics 21:342–350
Holmen J et al. (2003) The Nord-Trøndelag Health Study 1995–97 (HUNT 2): objectives, contents, methods and participation. Norsk Epidemiologi 13:19–32
Kaye J (2004) Abandoning informed consent. In: Tutton R and Corrigan O (Eds.) Genetic Databases: Socio-Ethical Issues in the Collection and Use of DNA. Routledge, London, pp 117–138
Knoppers BM, Chadwick R (2005) Science and society – human genetic research: emerging trends in ethics. Nature Reviews Genetics 6:75–79
Kymlicka W (2002) Contemporary Political Philosophy. Oxford University Press, Oxford
Mackenbach JP et al. (2008) Socioeconomic inequalities in health in 22 European countries. New England Journal of Medicine 358:2468–2481
Marchand S et al. (1998) Class, health, and justice. Milbank Quarterly 76:449–467
Marmot MG (2004) Social causes of social inequalities in health. In: Anand S, Peter F, Sen A (Eds.) Public Health, Ethics, and Equity. Oxford University Press, Oxford, pp 37–61
Marmot MG et al. (1978) Employment grade and coronary heart disease in British civil servants. Journal of Epidemiology and Community Health 32:244–299
Mill JS (1859/2004) On Liberty and Other Writings. Cambridge University Press, Cambridge
Orentlicher D (2005) Making research a requirement of treatment: why we should sometimes let doctors pressure patients to participate in research. Hastings Center Report 35:20–28
Pogge TW (1989) Realizing Rawls. Cornell University Press, Ithaca

Rawls J (1999) A Theory of Justice, revised edition. Oxford University Press, Oxford
Rawls J (2001) Justice as Fairness: A Restatement. Kelly E (Ed.). Harvard University Press, Cambridge, MA
Rawls J (2005) Political Liberalism. Columbia University Press, New York
Rhodes R (2005) Rethinking research ethics. American Journal of Bioethics 5:7–28
Rhodes R (2008) In defense of the duty to participate in biomedical research. American Journal of Bioethics 8:37–44
Sandel MJ (1998) Liberalism and the Limits of Justice, 2nd edn. Cambridge University Press, Cambridge
Sen A (1992) Inequality Reexamined. Clarendon Press, Oxford
Sen A (2004) Why health equity? In: Anand S, Peter F, Sen A (Eds.) Public Health, Ethics, and Equity. Oxford University Press, Oxford, pp 21–33
Shapiro D (2007) Is the Welfare State Justified? Cambridge University Press, New York
Sutrop M (Ed.) (2004) Human genetic databases: ethical, legal and social issues (special issue). TRAMES 8(1/2)
Taylor C (1985a) What is human agency? In: Taylor C (Ed.) Philosophical Papers I. Cambridge University Press, Cambridge, pp 15–44
Taylor C (1985b) Atomism. In: Taylor C (Ed.) Philosophical Papers II. Cambridge University Press, Cambridge, pp 187–210
Taylor C (1985c) What's wrong with negative liberty. In: Taylor C (Ed.) Philosophical Papers II. Cambridge University Press, Cambridge, pp 211–229
Taylor C (1998) Sources of the Self. Cambridge University Press, Cambridge
Tutton R, Corrigan O (Eds.) (2004) Genetic Databases: Socio-Ethical Issues in the Collection and Use of DNA. Routledge, London
Weldon S (2004) Public consent or scientific citizenship. In: Tutton R, Corrigan O (Eds.) Genetic Databases: Socio-Ethical Issues in the Collection and Use of DNA. Routledge, London
Wendler D (2002) What research with stored samples teaches us about research with human subjects. Bioethics 16:33–54
Wikler D (2004) Personal and social responsibility for health. In: Sudhir A, Fabienne P, Sen A (Eds.) Public Health, Ethics, and Equity. Oxford University Press, Oxford, pp 109–134
Williams G (2005) Bioethics and large-scale biobanking: individualistic ethics and collective projects. Genomics, Society and Policy 1:50–66
Williams G, Schroeder D (2004) Human genetic banking: altruism, benefit and consent. New Genetics and Society 23:89–103
Winickoff DE, Winickoff RN (2003) The charitable trust as model for genomics biobanks. New England Journal of Medicine 349:1180–1184
World Health Organization (2003) Genetic Databases: Assessing the Benefits and the Impact on Human and Patient Rights. WHO, Geneva

Trust, Distrust and Co-production: The Relationship Between Research Biobanks and Donors

Pascal Ducournau and Roger Strand

Abstract This chapter addresses one so-called ethical aspect of biobanking, namely the relationship between biobanks for research and donors of human biological samples and personal health information. Central to bioethical theory and practice is the institution of informed consent and its potential to create trust. We present results from an observational study of the consent process during the recruitment to a local population DNA bank in Southern France as well as subsequent interviews with donors. Three types of donors were identified: (1) Persons holding a "natural trust" and who were quite uninterested in the information and consent procedure; (2) persons who expressed distrust, but nevertheless participated as donors; and (3) persons who appreciated the consent procedure as a sign of a well-organised institution. Although informed consent may appear partly irrelevant to the issue of trust for a large group of donors, we proceed to discuss the status and desirability of a strong focus on donors' trust in biobank experts. Indeed, more symmetry and distrust may be a creative potential in the co-production of science and society in the biobank era.

Introduction

The construction of biobanks for research – collections of human biological samples or data emerging from such samples – has become a central element in contemporary medical population-based research. Different human entities, be it individuals, institutions or other forms of collectives, play different roles in biobanking. For instance, researchers and research institutions contribute with their skilful work and research infrastructures. Investors and funding agencies provide the financial basis of the frequently expensive research activities. Last, but not least, donors contribute with samples of their own body and often with personal health information

P. Ducournau (✉)
CUFR JF Champollion - Albi/Unit 558, INSERM - National Institute of Health and Medical Research, Toulouse, France
e-mail: ducourna@cict.fr

J.H. Solbakk, et al. (eds.), *The Ethics of Research Biobanking*,
DOI: 10.1007/978-0-387-93872-1_9,

as well. All three parties – researchers, investors and donors – are easily seen to be indispensable to biobanking. To some degree, their motivation for participating in the biobank venture may also have common features, such as the belief in the benefit for the community and for mankind that the biobanks are expected to generate through scientific and medical progress. On the other hand, there are obvious differences between the parties when viewed as stakeholders. Researchers may be employees, paid for their labour, or may hold other vested interests in the enterprise. Non-profit investors may have a defined mission to facilitate medical research on behalf of society, while commercial actors depend upon long-term possibilities of profitable products and services, mainly within medical technology. Donors, on the other hand, are not expected to receive other than symbolic economic benefits from their participation; indeed, in many countries the contrary would amount to illegal commerce with one's own body. In some cases, the donors are patients who could hope to benefit on an individual basis or collective level from the expected scientific advances in the treatment of their disease. Often, however, citizens will have no easily identified interest in being a biobank donor except for the altruistic motive. Accordingly, from the perspective of those who desire the construction of large biobanks for research (for instance governments and health authorities in many countries) a particular challenge is to ensure and uphold the citizens' willingness to act as donors and allow research on their donated biological samples and personal health information.

The Discourse of Public Trust

Frequently, this challenge of recruitment is framed as a challenge of ensuring public *trust* in biobanks and population-based medical research, and donors' trust in particular (Hansson 2005; Tutton et al. 2004; Williams and Schroeder 2004). Furthermore, in the bioethics literature as well as in existing regulations, the information and consent procedures preceding the act of donation are seen as devices that may contribute to instil such trust. As such, this particular challenge may be seen in the light of a more general preoccupation in contemporary Western societies, namely the so-called crisis of public trust in science and expertise (as well as in the political elite). Episodes such as the so-called BSE scandal, the Bristol affair and the recurrent topic of the risk management of nuclear waste in the UK, and transfusions with HIV-infected blood in France, have played a crucial role in provoking such a crisis. Explanations of the crisis have been sought in institutional and moral deficiencies on the level of experts and authorities (such as in the mentioned "scandals") but also in public uneasiness about the production of risks (Beck 1992) and ethical problems in late-modern technosociety. A prime example of such uneasiness is seen in the European resistance towards genetically modified foods (Frewer 2003). Accordingly, both in biomedical research as well as in European science-and-society (FP6) and science-in-society (FP7) policy, a lot of attention is given to ensure accountability, transparency, ethical standards and proper risk management, with the aim of creating public trust, compliance and cooperation. In biomedical research, a primary strategy for achieving these objectives has been to implement informed consent practices and procedures.

In this chapter, we address two fundamental assumptions embedded in the argument about the importance of trust and trust-enhancing procedures in the context of biobank sample donation. First, we present empirical results that lead us to a further problematisation of the hypothesis that information and consent procedures build trust. This finding has normative consequences for the ethical regulation and design of biobanks: if the donors' trust is not obtained by consent and information procedures, such procedures are not necessarily sufficient to prevent irresponsible uses of data biobanks. Second, we challenge the appropriateness of framing the relationship between biobanks and their donors as a matter of ensuring trust. Indeed, we argue that such a framing reproduces asymmetrical relationships with respect to knowledge and power, relegating the donor to the role of the unknowledgeable and disempowered. To begin the work to overcome this deep asymmetry so characteristic of modern societies, the diverse research field dedicated to study the relationships between science, technology and society (STS) has moved towards a shared understanding of the necessity to study the three fields – science, technology and society – all as emerging out of the same large set of entangled social processes (Pickering 1995). In this sense, science and society are co-produced. This implies, however, that both scientists and citizens take part – directly or indirectly – in co-producing both science and society. Although scholars will disagree on the descriptive details and normative implications, the co-production perspective has proved fruitful in its different forms (Callon et al. 2001; Funtowicz and Ravetz 1993). In the final parts of this chapter, we propose some initial steps on the way from a framework of trust to a broader perspective upon biobanking, taking into account the lessons learnt from STS studies.

What is Trust?

First, however, we shall need to provide some clarification of the concept of trust. In English, the noun *trust* has several usages. One dimension of trust is the nature of the relationship to its object, which may range from a reliance on moral or other *qualities* of a person or thing ("faith") to a pragmatic expectation of success or reward ("hope"). A second, highly important distinction is that between trust as a psychological state and trust as an organisational factor of human action. In both cases, trust can be described as an assured reliance on something or somebody. In trust as an individual psychological state, the reliance consists in a belief or a sensation. Trust as an organisational factor of human action, on the other hand, consists in a certain consistency of action, for example in the voluntary abstention from own critical judgement. Often, one would like to think that the consistency of action is caused by the corresponding belief, such as when the patient will follow the doctor's orders because he believes the doctor to know better than himself. Nevertheless, the relation between psychological and behavioural reliance is neither necessary nor necessarily simple. The consistency of action may be entirely due to strategic motives or even of a contractual nature, exemplified by the institutions of *trust* in the world of commerce. There is also the phenomenon that Wynne (1996) called *virtual trust*, in which there may be no spontaneous belief in the person or object to be trusted, but the alternative to trust appears psychologically or pragmatically intolerable: If there is only one doctor in the village, you have no other tolerable choice than to trust him. According to Wynne, sudden loss of public trust may be due to the prevalence of virtual over "real" trust.

When a person trusts another person or institution, it appears that we may identify a certain asymmetry of knowledge. Thus, on the psychological level, the trusting person is in an intermediary state between being knowledgeable and ignorant. If somebody knows everything about another person, he does not have to trust him, but if he does not know anything about the other, he has no reason to trust him (Simmel 1991). Likewise, there is an element of power structure in trust, as trust as an organisational factor of action means to partially abstain from exercising critical, autonomous judgements.

Natural and Obvious Donor Trust in Biobanks and Biobankers

Biobanks for research pose a fundamental problem to the logic of individual autonomy and full information. It is simply impossible to predict the ultimate consequences of the research that will be performed on the data and the samples that have been collected. Research goals will change and surprises will appear, and uncertainty is a necessary dimension of biobanking as indeed of all research. Apparently, when the public give their samples and information, they have to base their decision on trust: they need to have hope in the biobanks and faith in the biobankers, much in the same way that we have faith in our banker's skills and honesty, and hope in the interest rate that the bank will be able to offer us.

It is in this context that the practical question emerges for those who operate the biobank: If donors' willingness to participate depends upon their trust, how are the biobankers to recruit and treat the donors in a trustworthy manner? A possible way to address the question of trust is to consider the procedure of informed consent. Indeed, in the literature informed consent is presented as a "ritual of trust" between biomedical actors and research participants or patients (Wolpe 1998).

One of us (Pascal Ducournau) has had the opportunity to conduct an empirical study of donors during the creation of a local population DNA bank in Southern France. Results from this study are presented and discussed in this chapter as well as in chapter "Users and Uses of the Biopolitics of Consent: A Study of DNA Banks". Methods and design have been described in detail (Ducournau 2005, 2007). Of special interest in this chapter is how a biobank donor group is formed (and later, how we were able to approach donors as informants to our study). Before turning to key findings, let us first review these recruitment processes.

In brief, donors were recruited by drawing of lots from electoral lists to build a control group study (but not only), in exchange for a cardiovascular health assessment. Another bank of the same type, not of the general population but of cardiovascular hospital patients, was also established and added to the first bank, to build a case group study. A total of 1,800 adult males with age in a predefined high-risk interval were solicited in either case, of which around 40% accepted to participate. The objective of the biobank project was to study interactions between genes and environment in the occurrence of cardiovascular disease and to investigate possibilities for new genetic tests allowing better predictions and prevention.

To obtain their consent, the recruited persons were asked to read an information sheet about the study, and if necessary, to ask questions, and then to sign a one-page information and consent form. By giving their consent, they were assumed to confirm that they had evaluated and accepted the consequences of participating in the study. The aim of the consent procedure can accordingly be described as an incentive to act, making a rational assessment of the participation, including possibilities to talk with the physician recruiter followed by the construction of a deliberate individual choice. This procedure has been characterised as putting the person in a new role with respect to medical solicitation: a self-overseeing actor of the uses of the body and its elements (Memmi 2001).[1] In the information sheet it was stated that the collected data would be used only for the objectives presented (the study of cardiovascular pathologies) and that the results of the genetic analysis would not be broken down to the level of individual DNA samples. Consequently, it was emphasized that the participants would not obtain individual results about their genome. Finally, in the information sheet it was stated that participation in the study would not entail any risk for the participants, except for the negligible risk connected to the letting of the blood sample: "Your participation in the study does not involve any particular risk. The blood test corresponds only to the taking of standard blood test".

Guarantees of anonymity and data confidentiality were given in the information sheet, and the potential participants were also informed that the study had been approved by the relevant ethics committees.

One of us (Ducournau) had the opportunity, following negotiation with the research team and the local ethics committees, to observe the signing of the consent form and then to have an interview with 61 participants in the days following their entry into the study. The semistructured interviews were designed to allow participants to talk about their motives for participation and to give their account and point of view on the form of the consent device and procedure (the signature ritual in particular) and also on the content of the information presented to them. The interview material was interpreted in terms of the degree and nature of the endorsement.

[1] In order to ensure that the participant gives his consent following a consideration of the objectives and consequences of the study for which he has been solicited, it is written at the end of the form: "After having read the information sheet, having talked about it and obtained answers to my questions, I freely and voluntarily accept to participate in the study on cardiovascular illnesses, in the creation of a DNA bank, and in sociological interviews" (The participant has the choice to tick "yes" or "no"). At the end, he or she writes "read and approved" before signing and dating. The doctor also signs the document "for the investigator" in a place reserved for that, located before the signature of the participant.

For instance, whether the participant read or not the information sheet, or asked questions, was considered interesting information in this respect.

A major finding was that approximately a third of the participants took on a strongly delegative role, leaving the question of the decision and of the understanding of the objectives of the study in the hands of the doctors and researchers. Thus, when presented with the informed consent device, supposed to give the individual the role of an "informed rational decision-maker", the participants responded with a logic of cooperation. This manifested itself in two types of attitudes: (1) They explained their own decision as ensuing from the advice of doctors and researchers, and (2) they conceived of the consequences of the action as beyond their knowledge and understanding. Hence, one participant explained:

> "I haven't read it (the information sheet) because if you considered it useful to go through, there was no problem, I agreed to it [...]. When she [the doctor] told me to participate in it, I did so [...] and I said to myself, if she does this, it is because she has a good reason [...] what sense will it make for me to know what it will be used for [the DNA]; I thought it could be useful to help research, to improve the treatment possibilities of coming generations of doctors, [...]. If they do this, they must have a reason. I think for them the reason for performing this study must be that it is cardiovascular disease". Question: "And you didn't try to get more information?" Answer: "When you choose to trust somebody, you simply do so. If you get killed, that's a pity. Anyway, you have to trust if you want to be treated".

In the discursive exposition of these delegative attitudes concerning the knowledge and the decision that catch the consent device on the wrong foot, we identified conceptions of the production activity of the medical and scientific knowledge. These conceptions were strongly marked by ideas of division of labour and an expert/layman asymmetry that is particularly visible in the lacking desire to be explained the aims of the research. There is, on the one hand, what is "useful" to know for the lay participant, and on the other hand, what is "useful" for the biomedical actors. The "not-trying-to-know" attitude can also be interpreted as a real disinterest. Then, a feeling of trust toward "medicine", "public research" (as opposed to private research) or researchers can be mobilised in this non-consideration of the medical and scientific objectives of the study for which the persons have been solicited. Indeed, trust allows for a kind of "cognitive economy" (Grossetti 2004) since it leads the doctors and researchers not to linger on the consequences of the action, not to "lose oneself" in various speculations; it allows the decision to be built on uncertainty. The participant does not know the objectives of the research but has *faith* in the researchers:

> Not knowing what this research is for, I don't really care [The information sheet], it's too long to read. [...]. For sure, I have asked if, for me, it could lead to any sort of discomfort... That's my concern. If they had asked me to take medication and then come back every month, I think I would not have agreed [...] I will not go to the bottom of things to find out whether what I think of genetics is right or entirely wrong. There are doctors, they are here for that, that's what I think. Thus you trust them or not. [...]. Trying to be enlightened is a waste of time. It doesn't lead to anything. [...] I'm a good participant. I don't ask questions [...] I have done mine, after that, the doctors, the researchers have to do the rest.

Our findings about natural trust are consistent with the extensive study of donors in the Swedish Umeå Genomics biobank (Hoeyer 2003; Hoeyer et al. 2004, 2005;

Hoeyer 2006),[2] in which the donors were portrayed as not much interested in the consent form and the specific information they were given, nor at all knowledgeable of the biobank and its research after donation. Indeed, their attitude was summarised in the title of Hoeyer (2003): "Science is really needed – that's all I know", and, we may add, all they needed to know to make their decision to participate. The findings led us to the concept of donors' relatively *wilful ignorance* (Michael 1996) of the objectives of the study. Wilful ignorance appears generally not to be considered in biomedical ethics where the degree of understanding of the protocols by the lay persons rather appears to be interpreted in terms of a more or less implicit "deficit model", envisaging the need for increased "enlightenment" of participants by educational means (Annas 2001; Moutel et al. 2001).

When Natural Trust Is Missing: Distrust and Virtual Trust

In contrast to the participants who did not take into account the medical and scientific objectives of the research, we found a different group of participants (less than 25% of the participants) who questioned the purpose of genetic research in general as well as the objectives of the study for which they had been recruited in particular. They had various reactions regarding the consent form and procedures, and notably also with regard to the signature procedure of the consent form. Far from seeing this as a ritual of autonomy or freedom of choice, they claimed to reveal in the consent form and the implicated procedures a way of eventually constraining the participant, in the sense of them having "no possible resort afterwards" if things "turned out badly", or a way for the biomedical agents to "unload themselves of all sort of consequences" by saying "nobody forced you" (Ducournau 2005, 2006). The information given on the aims and objectives of the study did not satisfy these participants' expectations. This group of participants did not share the conception of the laymen/expert relationship presented above, that is, in terms of fundamental asymmetry and division of work. They said for example that they had taken the opportunity to participate in genetic research to involve themselves in a domain that they did not want to leave to the specialists:

> "I said to myself: if I don't participate, I do not give my consent, in saying what I have to say, my thoughts will not be translated" (A participant who commented on the risks and uncertainties of genetics to the recruiting doctor).

These participants may remind the reader of Ulrich Beck's concept of the "modern distressed consciousness". They develop their own thinking about the risks of technoscience. The doubts and mistrust of these participants are articulated around issues such as respect for the confidentiality of the data stored in the bank, the "filing information" and the genetic manipulations, the cloning, the eugenics and the use of research – the aim of which could be a commercial one. In these ways of questioning, the rational assessment of the consequences of the action gets concerned

[2] For this, see also chapter "Embodied Gifting: Reflections on the Role of Information in Biobank Recruitment".

with the medical and scientific rationality itself. Rationality thus enters a "speculative era" on risks, on what cannot be seen, an era in which the modes of thought and representation of their connection in the visible world get loose (Beck 1992).

Illustrative of such a view is the following account by one of the participants of the moment when he received the information note:

> One understands while reading the letter that as a matter of fact, there's a taking (of blood) and that a bank of genes will be set up. However, the presentation in the letter is quite unclear and the explanations given by the doctor present are unclear as well, which gives reasons for doubt: how will this sample be used? It's a bit difficult to understand how the DNA will be used: whether it is for long-term use or, well, what is the intention behind storing it... One gets the impression that it is a bit like science-fiction; [...] one wonders what can be done with this material, what could be imagined... And imagination runs easily... since we have these stories of cloning, manipulation and so on.

Furthermore:

> They cover themselves by saying I gave the authorisation. I am talking about medicine, and science... they are safe since I accepted. But I have accepted only on one condition: that it does not leave its medical environment. But here, I have in fact no proof [...] of the travelling of my DNA.

One noteworthy fact is that these participants, in spite of their doubts and mixed feelings, did not refuse to give their consent. This finding is similar to Hoeyer's (2006) report of donor worries. Different reasons might explain this phenomenon. First, for some of these participants, facing the perceived impossibility to understand fully the aims and workings of the study, there was no other choice than to trust the biobankers. Trust represents then the only possible alternative to refusal (which they did not choose), but the kind of trust involved appears to be reminiscent of Wynne's conception of *virtual trust*. Second, and possibly complementarily, a refusal to participate could lead to a stop of the research itself, with an implied risk of generally stopping progress. Compliance appears here also as the preferred option, not the least for the persons who need new medical therapies (in the case group study for example). This situation is reminiscent of what Ravetz (2001) denoted as a "safety paradox": The research might be unsafe or have unknown and unwanted consequences or aspects, but the alternative might be even worse.

The Group Between

Between the first group of participants who were cooperating in a delegative way and the second group of relatively distrusting participants, a third group (approximately 1/3 of the interviewed participants) actually appeared to use the informed consent procedure as a "ritual of trust". The information provided in the information sheet, the consent form and the dialogues with the physician were utilised by these participants to explore the aims and workings of the study and were found sufficient to allow them to cooperate without any distress. One might say that they used the consent "device" as expected by legislators and ethicists: reading the information sheet, listening to the oral explanations and asking questions. For some informants, the information and consent procedures were described as a "contract" process producing trust with a procedure of signature that presents finally the rights and duties of each party involved in the project. However, their trust in the biobank project was not only generated by the procedure of consent. Trust was perceived as the result of a combination of different elements. A reference to Durkheimian sociology may help to explain this point: "All is not contractual in the contract" (Durkheim 1893/1930: 189). By this sentence Durkheim wanted to underline that a contract must be accompanied by moral involvement to be really effective. If the contractors do not respect the duties entailed in a contract, the contract cannot organise social relationships. All the contracts could be broken and any suitable relationship could be generated by a contract frame. A necessary condition for a successful outcome of a contractual relationship is located outside the contract, in the moral involvements of the parties to respect the contractual clauses. This is why a contract can work, and why it can constitute a basis for some social relationships. A parallel can here be drawn between Durkheim's analysis and the construction of trust by the procedure of consent. If the consent procedure is necessary to create trust between the parties, it is not a sufficient condition. Macrosocial conditions as trust in political, medical and scientific institutions are needed for making the interaction around the consent procedure work. The ritual of trust works if the donors have trust in the institution that is soliciting them:

> They make us sign this paper for our consent for the DNA research, yes! Even so it is quite confidential, DNA, isn't it? Well ... They penetrate the core of your intimacy since one knows who you really are ... So I suppose that is why they make us sign this paper [...] Personally, I had no need to sign this paper; they could take the DNA, as much as they wanted. But well, since they ask you, you see that the procedure is a serious one: well, they take precautions. It confirms the idea of the seriousness of the examination.

So, why did this participant say: "It confirms the idea of the seriousness of the examination"? In another part of the interview he explains that it is needed "to make trust in medicine" and that the public research institution involved in the project is a serious one. Obviously, trust is the result of the combination of different elements where the consent procedure is not the only one, and where there exist structural factors of trust as the trust in the experts of medical research and their institutions.

In another interview, the participant underlined the contractual dimension of the consent device and its power to reduce uncertainties in the relationship:

> I signed an agreement to do certain things. This agreement was using some pieces of information that I read in the mail notice I received, the agreement for the DNA sample etc, all the things that a citizen can legally verify or refuse concerning the exams that were made. There is no mystery; [...] It is a normal action to sign a form. This can be used as a discharge for them if there is a protest. This is a reciprocal contract. [...] This is a contract that says to me: we do not have to communicate the results of the DNA tests, and if you want you can ask the destruction of your DNA samples.

We can observe here that trust in the biobank project is also constructed in relation to a more general trust in the juridical and political system of respecting the rights of the "citizens".

In the three described groups of participants, trust appears then as a fundamental element in the construction of their co-operation. But the origin of this trust cannot be located in the consent and information procedures of the biobank recruitment. For the first group, there was too little attention paid to the information sheet and consent form to see the trust as derived from them. Trust pre-exists. The particular "research contract" presented to these participants gained its very importance through the "general contract" they mobilized with medicine, institutions and society: they believed in medical and scientific progress, in the responsibility and the expertise of public medical institutions. For the second group (the distrusting donors), trust did not appear in the same way, as resulting from the procedures of information or consent. On the contrary, these procedures were perceived as creating conditions of mistrust, and virtual trust appeared and was enforced upon them because of a lack of viable options. For the third group, i.e. the group in-between, trust was perceived more as constructed by the information and consent procedures provided, but it could not be reduced to these procedures, because macroelements of trust played a role as well, such as trust in the expert and political institutions.

Trust, Distrust or Co-Production: Towards an Alternative Design of Biobanks

The interview study presented above leads us to question the standard way of designing biobanks as organisations, and provides us also with some normative ideas for alternative designs. As explained in the introduction to this chapter, from the perspective of those who encourage, facilitate or perform the creation of large biobanks, it may seem important to ensure and uphold public trust – in particular donor trust – in the biobanks and the research they are meant to serve. From such a perspective, there is little doubt that biobanks are considered to represent something *good*, in ethical as well as political terms, but the inherent uncertainty in scientific research makes this goodness difficult or impossible to demonstrate beforehand. To the extent that the donors are not familiar with the workings of science, they accordingly have to trust the experts.

In the empirical part of this study, a majority of our informants indeed expressed such a trust. For a third of the informants the quality of the trust is what we have called "natural": It is a general trust in public medical research and the medical profession, and, as pointed out in Hoeyer et al.'s studies of Umeå Genomics, perhaps also a trust in local health/research personnel and local control and oversight. This trust appears to be in place before the information and consent procedures start. Indeed, in our data there are clues to support the speculation that the information procedure may decrease or at least disturb this natural trust by implicitly communicating that the donor ought to assimilate and critically assess technical information concerning the biobank. It should be remembered, though, that it is difficult to know how the donors would have reacted to the absence of any information and consenting procedures. The mere fact of its presence may give assurance, irrespective of how uninterested the donor may be in the specific details. Nevertheless, from our data we find it difficult to describe the information and consent procedures as a "ritual of trust" as Wolpe (1998) did. Rather, the information and consent procedures could be called a ritual of autonomy and contract, imposing a contractual relationship between biobank and donor.

One of the normative key questions addressed by the donors' natural trust in the organisational design of research biobanks is the ability of biobanks to deal responsibly with this kind of trust. In this case, consent and information procedures do not provide ethical guidance because trust is given a priori.[3] In the imagined case that the goals of the biobank project are ethically irresponsible or contestable, the naturally trusting part of the public might not pay any attention, not even if the irresponsibility of the goals is revealed in the information sheet. At best, biobankers may be sufficiently responsible to take account of this natural trust by assuming the moral duties of a trustee, though this is not obvious, as discussed by Williams and Schroeder (2004). At worst, they can use the consent and information procedure as a way to dissipate their sense of responsibility (Hoeyer 2006). Equally challenging, the biobankers may be unaware of ethical aspects that are contested in society in general or in academic discourse, aspects which in that case will be considered neither by biobankers nor by donors.

Efficiency or Empowered Citizenship

Several normative interpretations may be made from the possibility that the information and consenting procedures erode natural trust. In the official policies of many, if not most, countries research biobanks and biomedical research on the whole are seen as beneficial and desirable on a societal level. From this perspective, one might conclude that even though autonomy on behalf of the donors should be secured through informed consent procedures, natural trust is highly desirable in order to achieve efficient enrolment of donors. Indeed, for those who believe strongly in the value of large biobanks and at the same time believe themselves to be experts who ought to be trusted, one might fear that the loss of natural trust might cause lay people to be "led astray" into irrational forms of distrust. On the other hand, in particular in the European Union there is an increased attention to the development of democracy and empowered citizenship in a knowledge-based society. From this perspective, one might see natural trust as naïve and undesirable, as something that reproduces the asymmetrical relationship between lay and expert and should therefore be overcome. Indeed, in the current policy of the European Union (FP7 Science-in-Society), the co-production perspective is to some extent present. This

[3] Different ethical guidelines for biobank projects underline the importance to clearly expose the aims and workings of the studies in the information and consent sheets to prevent, e.g., misuse of genetic data (see Advise no. 77 of The National Consultative Ethics Committee in France: Ethical issues raised by collections of biological material and associated information data: "biobanks", "biolibraries").

document recognises that long-term scientific advance is unlikely without the support of knowledgeable, empowered citizens, and perhaps also undesirable without this support. Accordingly, one might emphasize the value of enlightened democracy and the importance of the citizens themselves in defining progress and societal benefit, at the cost of smooth and efficient enrolment of donors in the short-term perspective.

Switching from Trust via Virtual Trust to Distrust

Part of the nature of natural trust, however, is that it may enter into dialectic with powerful changes between trust and distrust. Such episodes are known not only from recent scandals mentioned above involving scientific experts, but more generally from our political history, such as the collapse of the former Soviet Union. We may hypothesize that such switches may happen more dramatically in cases of virtual trust or other forms of strongly asymmetrical trust relationships. Thus, even from a strategic pro-biobank perspective it may appear appealing to avoid a strong reliance on natural trust, which might flip or disappear in an unpredictable manner.

Distrust may be grounded in many ways. Our empirical data share the methodological weakness of many studies that our informants have accepted to participate in our study; and in our case, they actually also accepted to donate samples to the biobank. What we observe, however, is the apparent unproductiveness of the information and consent procedures in dissolving distrust. Rather, the procedure is taken to be a part of the system that is the object of distrust, a system much more broadly defined than the "biobank" or the particular research institution or team of researchers. As noted by Frewer (2003), although criticism of biotechnology may be presented in terms of risks and ethical challenges, its foundation can be a difference in value systems and fundamental forms of political disagreement. Critics of S&T development may see it as part of an accelerated, unsustainable capitalism and a run-away train, possibly producing monstrosities and misery in the long-term perspective. In this perspective, the information and consent procedures are offering anything but assurance. On the contrary, they may be seen as a confirmation of researchers and the "system" having a too narrow scope in what they consider relevant information, not seeing, or not wanting to see, that they play a part in the global problems.

Hence, the participants are led to give their approval inside a predefined scope which has never been the subject of a previous discussion. As Cresson (2000) recalls, in the much more general scope of the medical contract it is the doctors who generally give the definition while no attention is given to definition that the patients would give. The analogy to our case of DNA banks is clear. If the ethical committees approve of the protocols and the consent procedures used, and consent in this way constitutes a mediation between researchers and donors, the latter have little opportunity to intervene themselves in the definition of the scope in which they are asked to consent. Furthermore, by signing the consent form the donor has forfeited his possibility to oppose the framing of the problem. Still, our informants signed the form, which indicates the complexity of beliefs and sentiments that the individual may hold: trust, distrust, sense of duty, curiosity, or perhaps, for some of them, all at the same time.

From Trust to Co-production

We have argued that a model of biobank–donor relationships based on trust may in part reproduce undesirable power and knowledge asymmetries, and in part may be quite unpredictable as trust may lead to distrust. It remains to be investigated whether trust partially or wholly could be replaced by types of relationships in which the donors delegate less and participate more, and to what extent this would be desirable.

A first observation in this respect is that one should not speak in an unqualified manner about what is "desirable" without also asking "desirable for whom?" The authors of this study should not be viewed as advocates of "the citizens" or "the donors". As has already been argued, they are a heterogenous group with different opinions on biobanks. Rather, one should situate the perspective of the present analysis as de facto remaining in the academic discourse surrounding the regulatory issues of research biobanks. In this respect, "desirable" means "desirable in terms of the regulatory perspective". However, in this analysis, it is important to maintain an agnostic and critical distance to unconditional beliefs in the benefits of research biobanks and the importance of the efficient enrolment of donors to them. These beliefs, however predominant in current health policy discourses in Europe, do not form a necessary part of the regulatory perspective. Rather, they depend upon more general ideological beliefs in progress and science. Such beliefs need to be problematised both for intellectual and pragmatic reasons (Nordmann 2004).

To maintain the required analytical distance, we propose to consider a transition from the analytical framework of trust to that of co-production. The analytical framework of trust tends to frame thought in the categories of the experts and the lay people, the knowledgeable and the unknowledgeable, the informed and the uninformed, etc. A discourse of trust appears to assume a belief in the possibility of knowing in principle what is correct and what is good at the time of making the decision. This is less relevant, we believe, in contexts characterised by essential unpredictability and uncertainty, something which, in our view, is the case with biomedical research in general and research biobanks in particular. It is impossible to be "informed" in the classical sense because the future of the research is not determined at the time the material is collected. It is the future that determines its use.

The analytical concept of co-production (as originally introduced by Jasanoff and developed by Latour and the tradition of actor-network theory) was designed to be less myopic with respect to the study of science, society and the interface between them. A discourse of co-production allows us to consider a broader range of relationships between biobanks and donors. As explained above, such a consideration will not necessarily be seen as desirable, interesting or even relevant by all actors in the discourses of the regulatory issues of research biobanks. We think, however, that two (interrelated) issues may profit from an analysis from a perspective of co-production. First, there is the possibility that trust is ineffective and insufficient, and collapsing into distrust, as explained above. Second, there is a need for paying more attention to the aspect of unpredictability of the future of research and its implications for society. We shall argue that both issues may call for co-responsibility if broader attention is given to the actual and possible roles of donors and citizens in general in science, as well as the actual and possible roles of scientists in society, along the lines of analyses into social robustness of knowledge (Nowotny et al. 2001), post-normal science (Funtowicz and Ravetz 1993) and more generally the literature on governance of science and technology.

Callon and Licoppe (2000) argue that when scientific actors are on an equal footing with the public actors, trust is not necessary because the different persons who are involved in the interactions are not in asymmetric positions. Groups of HIV patients, persons with myopathies, are in France involved in this new kind of scientific and biomedical production of knowledge. They participate actively in the definitions of goals and modalities of research. For Callon and Licoppe, there is not in our contemporary society a crisis of public trust in science but an evolution of the organisation of research, that is including more and more "expert lay people".

A large and unsolved question is how, in practical terms, the more participatory approach could be achieved in the case of research biobanking. The few empirical studies of such attempts appear to conclude pessimistically about the prospects of real participation with a less than strongly asymmetric expert–donor relationship (Reardon 2001; Tupasela 2007; Tutton 2007). A possible beginning would be to invite participants to cooperate with biobankers in the design of the information and consent procedures, allowing new devices, possibly on the group or community level, for the debate of broad-scoped issues, including the societal and technological significance of the biobank. This beginning would not be enough, however, because the broader debate would probably identify a number of unsolved issues regarding the uncertainties of the implications of the research and regarding values and stakes in conflict. For instance, potential donors may disagree with research priorities or with the anticipated model of technology transfer in the case of commercially viable results. They might even disagree with the methodological design, perhaps arguing that *more* (or perhaps less) psychosocial variables should be integrated with the biological information. One might imagine donors to organize and negotiate their priorities with the research institution and the investors; and one could imagine scientific boards in which donor representatives act to promote the translation of their ethical and societal priorities into methodological design and the day-to-day research practice. This would amount to a case of what has been called "post-normal science" (*ciencia con la gente*) (Funtowicz and Ravetz 1993). Such an empowerment of the public would almost necessarily imply a reduction of the power of investors, central authorities and the research sector as they appear today, that is, often in mutual agreement and without too much intervention by the people. We say "almost necessarily", because if trust turns into distrust, the resulting lack of cooperation might disempower all parties.

The second large question is what society would gain from such an approach, and if and when it would be desirable. Indeed, it is hard to see that a democratization of science could be defended purely on its own terms and in conflict with scientific efficiency and quality. This is where the intellectual strong points of the co-production perspective need to be called upon, because it explicitly sees science and other parts of society as inextricably intertwined. Central in the theory of post-normal science is the demand that participation should be a means to heighten the *quality* of research, above all by increasing its relevance. We may interpret the case of myopathy and HIV patients in this way: Research and hence knowledge is improved by opening up methodological discourses in order to increase the relevance. In that case, it would be a matter of tailoring the objectives and hence the design of research to the needs of the patient groups. More generally, there is also a need to improve our civilisatory abilities to avoid the unintended and harmful effects of scientific research and knowledge, resulting from scientific research being essentially unpredictable and open-ended (Pickering 1995). In the terminology of Nowotny et al. (2001), knowledge needs to become socially robust.

There is at present no methodology to predict the unpredictable, and possibly there will never be. This does not mean that society cannot develop strategies by which we attend the unpredictabilities and uncertainties of research in a more responsible way. In such strategies, imaginative and critical thinking might play a key role. Indeed, with more attention given to "creative sceptics" some of the adverse effects of science and technology might have been avoided or limited (Harremoës et al. 2001). It strikes us that the group of informants characterised by virtual trust and distrust display imaginative critical thinking. Hence, without imposing the model of co-responsibility on each and every donor, from a regulatory perspective one may see sceptical and distrusting citizens as a particular intellectual resource in the struggle to elaborate good pathways of the co-production of science and society.

References

Annas GJ (2001) Reforming informed consent to genetic research. Journal of the American Medical Association 286:2326–2328

Beck U (1992) Risk Society. Towards a New Modernity. Translated from German by Mark Ritter, and with an Introduction by Scott Lash and Brian Wynne. Sage Publications, London

Callon M, Licoppe C (2000) La confiance et ses régimes: quelques leçons tirées de l'histoire des sciences. In: Laufer R, Orillard M (Eds.) La confiance en question. L'harmattan, Paris, pp 133–154

Callon M et al. (2001) Agir dans un monde incertain. Essai sur la démocratie technique. Seuil, Paris

Cresson G (2000) La confiance dans la relation médecin–patient. In: Cresson G, Schweyer FX (Eds.) Les usagers du système de soins. Editions de l'Ecole Nationale de Santé Publique, Rennes, pp 333–350

Ducournau P (2005) Le consentement à la recherche en épidémiologie génétique: un rituel de confiance en question. Sciences Sociales et Santé 23:5–36

Ducournau P (2006) Au-delà des textes et des procédures juridiques, qu'en est-il des droits de l'usager?. Une étude de cas sur le consentement éclairé et sa procédure de recueil, EMPAN, n° 64, p. 107–112
Ducournau P (2007) The viewpoint of DNA donors on the consent procedure. New Genetics and Society 26:105–116
Durkheim E (1930) La division du travail social. Presses Universitaires de France, Paris
Frewer L (2003) Societal issues and public attitudes towards genetically modified foods. Trends in Food Science and Technology 14:319–332
Funtowicz S, Ravetz J R (1993) Science for the post-normal age. Futures 25:739–755
Grossetti M (2004) Sociologie de l'imprévisible. Dynamiques de l'activité et des formes sociales. Presses Universitaires de France, Paris
Hansson MG (2005) Building on relationships of trust in biobank research. Journal of Medical Ethics 31:415–418
Harremoës P et al. (2001) Late lessons from early warnings: the precautionary principle 1896–2000. Environmental issue report no. 22. European Environment Agency, Copenhagen
Hoeyer K (2003) "Science is really needed – that's all I know": informed consent and the non-verbal practices of collecting blood for genetic research in northern Sweden. New Genetics and Society 22:229–244
Hoeyer K (2006) The power of ethics: a case study from Sweden on the social life of moral concerns in policy processes. Sociology of Health and Illness 28:785–801
Hoeyer K et al. (2004) Informed consent and biobanks: a population-based study of attitudes towards tissue donation for genetic research. Scandinavian Journal of Public Health 32:224–229
Hoeyer K et al. (2005) The ethics of research using biobanks – reason to question the importance attributed to informed consent. Archives of Internal Medicine 165:97–100
Memmi D (2001) Surveiller les corps aujourd'hui: un dispositif en mutation? In: Sfez L (Ed.) L'utopie de la santé parfaite. Presses Universitaires de France, Paris, pp 173–184
Michael M (1996) Ignoring Science: discourses of ignorance in the public understanding of science. In: Wynne B, Irwin A (Eds.) Misunderstanding Science? The Public Reconstruction of Science and Technology. Cambridge University Press, Cambridge, pp 107–125
Moutel G et al. (2001) Bio-libraries and DNA storage: assessment of patient perception of information. Medicine and Law 20:193–204
Nordmann A (2004) Converging technologies – shaping the future of European societies. Report for the European Commission via an expert group on foresighting the new technology wave. European Commission, Brussels
Nowotny H et al. (2001) Rethinking Science. Polity, Cambridge
Pickering A (1995) The Mangle of Practice. The University of Chicago Press, Chicago
Ravetz J R (2001) Safety paradoxes. Journal of Hazardous Materials 86:1–16
Reardon J (2001) The Humane Genome Diversity Project: A case study in coproduction. Social Studies of Science 31:357–388
Simmel G (1991) Secret et sociétés secrètes. Circé, Paris
Tupasela A (2007) Re-examining medical modernization: framing the public in Finnish biomedical research policy. Public Understanding of Science 16:63–78
Tutton R (2007) Constructing participation in genetic databases – Citizenship, governance, and ambivalence. Science, Technology and Human Values 32:172–195
Tutton R et al. (2004) Governing UK biobank: the importance of ensuring public trust. Trends in Biotechnology 22:284–285
Williams G, Schroeder D (2004) Human genetic banking: altruism, benefit and consent. New Genetics and Society 23:89–103
Wynne B et al. (1996) *Misunderstanding Science? The public Reconstruction of Science and Technology*, Cambridge, Cambridge University Press
Wolpe PR (1998) The triumph of autonomy in American bioethic: a sociological view. In: DeVries R, Sudebi J (Eds.) Bioethics and Society. Prentice Hall, Upper Saddle River, pp 38–59

Scientific Citizenship, Benefit, and Protection in Population-Based Research

Vilhjálmur Árnason

Abstract In the discussion of ethical issues concerning databases as resources for population research, two main positions have been predominant. On the one hand, a major emphasis has been on protecting the participants from being discriminated against or having their privacy violated. The other main emphasis has been on the substantial benefits that can be reaped from the research. I show how these often conflicting positions share an important underlying and hidden presumption, implying a too narrow vision of the citizen as a passive participant. I argue that it is important to explore alternative visions of the citizens in relation to population database research. For this purpose, I ask whether recent ideas of deliberative democracy and scientific citizenship provide us with a viable guiding vision of how to facilitate a more active and informed public engagement in database research society. I flesh out my ideas in terms of the debate about consent for participation in database research and show how different models of consent imply different visions of the citizen. I argue that a dynamic authorization model with an opt-out clause could contribute to conditions for more informed, active and critically aware citizens.

Introduction

Biopolitics and bioethical discourse implicitly reflect visions of the citizens that play a major role in policies about scientific research and biotechnology. These views resonate with general positions about the major functions of democracy and about the nature of citizenship. In this chapter, I first describe two typical views or ideal types of the citizen that I take to be prevailing in social and theoretical discourse about bioethical issues. I call them the protective view and the benefit view. I then explore an additional vision of the citizen which has been largely ignored but has more

V. Árnason
Department of Philosophy and Centre for Ethics, University of Iceland, Reykjavik, Iceland
e-mail: vilhjarn@hi.is

J.H. Solbakk, et al. (eds.), *The Ethics of Research Biobanking*,
DOI: 10.1007/978-0-387-93872-1_10, © Springer Science+Business Media, LLC 2009

active democratic features than the other two. I flesh out these views by showing how they have been made manifest in ideas about consent for population databases and biobank research. I also aim to show what implications they have for a broader social debate about biopolicy.

Since my reflection on these issues has mainly been inspired by the Icelandic experience I will draw my examples from there. This experience has been widely discussed (e.g. Greely 2000; Rose 2001; Árnason 2004). The dominating emphasis in the Icelandic discussion concerning participants' interests has been on privacy. The focus was mainly on two kinds of technical issues, a legal technicality about personal identifiability and coding techniques for storing the information (Gulcher and Stefánsson 2000). Privacy protection is certainly important from a moral point of view since personal data should not get into the hands of insurance companies, employers or others that could be motivated to use them for discriminatory purposes. But it is very limited to evaluate the interests of people mainly, not to say exclusively, from this perspective. There is a reason to believe that the extensive discussion about these matters precluded reflection on issues relating to active human agency, for example of consent and informed public debate.

Protecting Participants

The strong emphasis on security in the Icelandic discussion about the Health Sector database exemplifies what I call the *protective view* towards citizens. The view draws its name from the fact that either explicitly or implicitly the ethical regulation of biotechnology and research on humans emphasize above all the *protection* of people. Some of the major moral objectives in research ethics are protection of privacy, protection against risks (participants' welfare) and protection of vulnerable research subjects (a major requirement of justice). In all these cases, measures are to be taken that safeguard research participants and citizens in general from the possible hazards of biotechnology and misuse of information. Protection of autonomy is a more complicated matter but as it is usually fleshed out in the requirement of informed consent it tends to be reduced to a formal procedure which poses little or no challenge to the participant as an active, reflective agent. Moreover, such a narrow notion of autonomy can serve a questionable legitimizing function far beyond its scope (Árnason and Hjörleifsson 2007).

If we relate this to ideas of democracy, we see that these requirements of security and protection fit well with the function of liberal democracy to protect citizens against the misuse of both state power and marketing forces. This most often translates into the right of the citizens whose "private domain" needs to be protected against the invasion of powers that can manipulate people and make use of personal information in a discriminating way. This also squares well with the corresponding notion of citizenship which is seen in terms of the citizen as a bearer of basic civil rights (Marshall 1950). In the national debate about the Icelandic Health Sector Database case, an organization was formed with the particular objective of protecting the rights of citizens against both the misuse of public authority and a powerful private company, largely driven by market forces. Appropriately, the association is called "Mannvernd" in Icelandic, literally "human protection" which resonates well with "Persónuvernd", the name of the Icelandic Data Protection Agency.

It is instructive to see how the protective view made itself manifest in the debate about what kind of consent should be required for the Icelandic Health Sector Database. Some scholars criticized the plan for "lack of informed consent" (Greely 2000). Spokesmen of Mannvernd were among the hardest critics of the deCODE project and demanded informed individual consent. As can be seen from the homepage of the association, www.mannvernd.is/english/, this demand has often been supported by appealing to basic principles of research ethics or has even been put forth as a human right. However, this appeal to individual rights has also been used strategically by activist opponents of commercialized database research.

Specific consent implies that participants will be informed prior to donating their samples or data for research about its objectives, risks, benefits and other traditional ingredients of informed consent. A major problem with specific informed consent in this context is that it is unsuitable for multi-disease research on genetic collections. If data are collected for a particular research and no research can be carried out on the data that was not specified in the consent form, then any research with new questions requires re-contact with the participants. Participants find such continuous re-contact annoying and experience has shown that they are willing to give a wider consent and leave it up to the researchers and the regulatory committees to ensure that they are used fairly for the benefit of science and society (Hoeyer et al. 2004).

It could be argued that one of the important functions of informed consent is to protect people's well-being through their increased awareness of risk and their control of their research partiticipation. However, in the context of biobank research these important benefits will be better secured by other means than specific informed consent. Such consent requires detailed descriptions in scientific protocols which tend to overwhelm participants' intellectual capacity. The paradox is that the more information is provided, the less understanding is obtained, and the consent procedure becomes a mere formality. Instead of maximizing options for individual deliberation and control, opportunities for deliberation are actually lost in this way and the interest in human agency is not well served.

Benefiting Participants

The protective view, especially in relation to the issue of consent, has been partly in tension with a position which I label the benefit view. The reason for using this label is that here the emphasis is on the *benefits* that can be reaped from biotechnology and genetic research. These benefits can be either health-related, such as drug development, more effective predictive and preventive medicine, or benefits unrelated to health, such as increased employment opportunities for young scientists and other social and economic advantages that may flow from having thriving research companies. This medical and social utility position has been prevailing in political and economic discourse about biotechnology.

It is understandable that researchers reject specific individual consent for database and biobank research and prefer a version of an open consent. By an open consent is meant here that participants agree that their data will be used for any future scientific research permitted by the regulatory institutions. The main emphasis is thus laid upon institutional trust, such as trust in research ethics committees which would evaluate the participants' interests and act (as surrogates) on their behalf. This is indicative of the trend to regard genetic data collections as major resources to be mined for the benefit of society without the interference of the participating individuals who should simply trust regulating institutions to take care of their interests. In this way the benefit view lends itself to an open consent.

The emphasis on open consent has both communitarian and utilitarian flavours, depending on how the arguments are formulated. A utilitarian argument could be that public interests are best served by mining the data resource in an efficient way for drug development and other medical benefits. The Icelandic parliamentary discussion clearly had such a utilitarian tone which was increased by reference to additional advantages, such as increased employment opportunities for young scientists. There was also a strong appeal in the discussion to the national genome and medical records as social resources that should be exploited for the common good. In communitarian language, these can be called goods that we can only create in common and not in atomistic isolation (Sandel 1982). From this viewpoint, the emphasis should be on the duties of participants to contribute to progress in medicine and science no less than on having their privacy rights protected.

It can be argued that the benefit view towards biotechnology relates to the function of welfare democracy of ensuring that decisions are made in the collective interests and to further the common good. It also relates to the view on citizenship which emphasizes social and economic rights to security. In fact, this argument from collective interests is often used to accuse the protective position of emphasizing individual rights at the expense of social goods. This argument can be met from two angles. First, many of the social benefits that are promised by genetic population research are both debatable and uncertain. Moreover, even though they would bring benefits to the wealthier part of the global population, they contribute nothing to the most pressing task of improving basic health care in poor countries.

Second, arguments concerning the importance of individual consent are often met with statements to the effect that they put private interests above public interests and surely this is sometimes the case. However, there are important public interests at stake as well in maintaining the ethos of voluntary consent to participation in database research. Neglecting it may weaken a democratic society in the long run. A policy of open consent could also be detrimental to the public trust in science and thus destroy a major social asset. An open consent of this kind does not provide participants with the information necessary for them to make a meaningful choice, i.e. to act in a voluntary way on a basic understanding of the matter. It transfers the reflection on population research from the participants to regulatory institutions (saying in effect "leave the thinking to us"). Thus, motivations for scientific literacy and awareness of the public would be reduced, something which is not in the public interest. The benefit view as I have described it thus ignores important benefits related to human agency.

In light of this it can be a misleading description of the protective position to say that it focuses on individual interests that are contrary to general social goods, unless we have a very narrow understanding of such goods. Effective regulation of genetic population research which protects people against undue risk, hinders discrimination and manipulation of individuals, is in the public interest in the long run, even though it limits the leeway of researchers. From this perspective, the sharp distinction between individual and collective interests is misleading. Providing options for participants' deliberation and preserving other conditions for human agency and reflection are not mere private interests. However, these objectives are not best served by obtaining specific informed consent from otherwise passive participants.

In the context of my discussion of the relationship of these views to democracy and citizenship, their common shortcomings and limits have become more conspicuous than their differences. While the protective position puts security of individuals above other considerations, the benefit view regards the population as a collective resource for biotechnology. It is not surprising, therefore, that a prevailing position in biopolitics is a *combination* of these two views and their major limitations. This combination takes on the following form, for example in discussion about population biobank research: In order to mine the population for maximum benefits, privacy protection needs to be extraordinarily strong. In this way, strong data security becomes one of the very preconditions of the utility view. This combination characterized the database affair in Iceland.

In this combination, the otherwise contrary positions regarding database consent disclose an important underlying and hidden presumption concerning the scientific citizenry that is being created. Positions, which place the main emphasis either on protecting the participants' private domain from illegitimate interference or on providing them with material benefits, see people primarily in a passive role. They do not provide reasons for implementing policies that facilitate actions of the citizens in the public sphere. In this way they are part of a research culture which contributes to scientific illiteracy and disregards the active elements of human agency which are crucial for the democratic citizen.

This is not surprising because these two visions of the citizen tend to complement each other in contemporary society. These visions emphasize, on the one hand, the person in the domestic private sphere where the safeguarding of freedom from illegitimate interference is of primary importance. On the other hand, the citizen is seen as a consumer and worker in the economic sphere, contributing to the economic prosperity of society, upheld largely by high standards of health in the population. In this way, both the protective and the benefit positions relate more to people as private persons, consumers, workers and patients than as democratic citizens.

However, as I have indicated, both positions harbour elements that could be developed in directions which are more conducive to a reflexive democratic culture. This is more obvious in the protective position which aims to safeguard important conditions of human agency. By insisting on specific informed consent for participation in biobank research, opportunities can be created for people to reflect on their participation but at the cost of making biobank research practically impossible. The benefit view, on the other hand, justifies open consent by reference to the material benefits to be reaped from biobank research but at the cost of losing the important social benefits related to active human agency and deliberation. These positions need, therefore, to be complemented with emphasis on factors that can increase public awareness of population research and strengthen the conditions for their decisions and responsibility for participation in the research.

Engaging Citizens

The third view that I want to discuss draws upon ideas of the active citizen which has roots in republican ideas of citizenship and deliberative democracy (Benhabib 1996; Cohen 1997). This view does not reject the moral elements of the protective and the benefiting positions but seeks to overcome their shortcomings by taking other considerations into account. Clearly, one should not be forced to choose between either protecting individual privacy and contributing to social benefits or increasing the awareness of the citizenry about science and biotechnology. It is necessary to protect the citizens against the misuse of both private and public power in a democratic society, but this is a very limited view on the citizens' interests. The benefit view also harbours important considerations but the promised benefits can be questionable. This is especially the case when the biobank research is conducted by a private company, as in Iceland, because the mutuality of benefits that is secured in a social system of health care is absent. I am not saying that there are no public benefits to be reaped from commercial biobank research but that an appeal to them is not a sufficient justification of open consent.

Before considering the general implications it would have for biopolitics to take these elements of the active citizen more into account, I will consider its impact on the example of consent for biobank research. It is understandable that the question of consent for participation in database research has been in the limelight of discussions about genetic data collections and biobanks. Population data collections are *resources* for genetic research and it is impossible to describe in detail the research that will be performed on the data at the time of collection. This can lead to the following dilemma: *Either* data will be collected with specific informed consent which emphasizes interests of individual participants but radically diminishes the flexibility of researchers and the possible benefits of the research *or* data will be collected with open consent which maximizes research flexibility but can undermine the option for reflection and other conditions for moral agency of the participants. The challenge is to show how this dilemma can be dealt with without risking either the possible human welfare benefits or the moral agency interests at stake.

In order to avoid the pitfalls of the specific and the open consent, alternatives that are intended to strike a balance between the researchers' need for flexibility and the ethical demand for protection of participants' interests have been proposed (Greely 1999; Caulfield et al. 2003; Árnason 2004; Kaye 2004).[1] The main thrust of these proposals, which have different emphasis, is that participants should be asked to authorize the use of their data for described health care research. They would be informed about the conditions for use of the data, such as how the research will be regulated, how they will be connected to other data, who will have access to the information and how privacy will be secured, and that they will only be used for described health care purposes. Most importantly, participants would be told that they and/or their proxies will be regularly informed about the research practice and that they can at any time withdraw from particular research projects.

Such an authorization or permission would both allow participants "to meaningfully act on their continuing interests in their health information" (Caulfield et al. 2003) and provide science ethics committees with a meaningful ground for determining further use of the information. Such further use can be restricted to comparable research where members of research ethics committees can reasonably argue that the additional research would not have affected the participants' initial decision to participate. Such a policy could maintain the motivation for participants to reflect on their participation in research and to stay informed about how their data are used and for what purposes. An authorization policy might thus contribute to informed, reflective and responsible research participation that can underpin public trust in research practices. None of these would flow from an open consent policy for database research.

These considerations are relevant for avoiding two of the most serious dangers of scientific research on humans, those of deception and coercion. The authorization proposal implies that individuals are offered "simple and realistic ways of checking that what they consent to is indeed what happens and what they do not consent to does not happen" (O'Neill 2001). If the latter happens, they can opt out. In addition to strengthening the basis for non-deception, this last point aims at securing the purpose of non-coercion, since it implies that participants need not continue research against their will (Kristinsson and Árnason 2007). In this way interests associated with moral agency and the moral purpose of informed consent may be best secured and that, in the last analysis, is crucial in any evaluation of advantages to human society.

It is integral to the authorization model that participants will be encouraged by regulatory institutions to follow the research practices. This provides conditions for an active opt-out clause which is likely to create more informed and critically aware citizens and is also conducive to *informed* trust. This position thus enables active *scientific citizenship* because it emphasizes the creation of conditions or opportunities for citizens to reflect on their participation in scientific research. Contrary to the protective policy of specific informed consent, these conditions for participants'

[1] An interesting solution of the Icelandic National Bioethics Committee is to provide a "menu" of three types of consent which the participants themselves can choose between. Most opt for the widest one, which permits use of samples for other research than covered by the initial consent, provided that the National Bioethics Committee and The Protection of Privacy institute have approved the research.

deliberation do not come at the cost of a flexible biobank research. There is no requirement of a continuous re-consent in order to meet formal procedures, but a dynamic interchange which has the primary aim of keeping participants informed and aware. Such scientific citizenship need not thwart the possibilities of reaping the benefits of biobank research; it refuses, however, to reduce participants to being merely a passive part of a resource.

The notion of scientific citizenship is used here in a normative, critical way and not only as a descriptive term, where all kinds of reactions of the citizens to the new genetics are regarded as examples of biological citizenship or "different citizenship practices" in response to "new technologies which intervene on the body" (Rose and Novas 2005). This normative use of scientific citizenship can be criticized from the liberal viewpoint of "neutrality of rationale" for scientific policies (Kristinsson 2006). However, it must be emphasized that the idea is mainly *to offer participants the chance* to be active and reflective and not to *require* that they be so. It is an important tenet of liberalism that people are not passively subjected to policies and that they are provided with the *opportunity* to exercise their status as free and responsible agents. The conditions for this must not be reduced to information initially provided when consent is obtained but need to be seen in terms of options for a dynamic interchange with participants.

Guiding Vision

The objective is to create more informed or educated citizens who do not have to rely exclusively on expert knowledge but can use it in their deliberations about research participation. This, of course, is not something that can be easily realized but it is an important *vision to guide* our attempts in shaping citizens' awareness in society where biological research and biotechnology play an increasing role. This objective obviously requires that different biopolicies need to be introduced. I will only mention here two preconditions for such a biopolitics which takes the vision of the democratic citizen seriously: improved scientific education, preparing people for active participation in a society, and increased public deliberation about biopolitical matters.

Visions of the citizen are important in school curricula and education which increases scientific literacy that may contribute to more biopolitical awareness and thus create preconditions for policies which facilitate more public engagement. This effort must be aimed at the citizenry at large, at the "maxi public", so to speak, not only the "mini public" which is created in particular deliberative events. Such exercises in deliberative democracy can obviously be valuable but they need to build upon a comprehensive deliberative education of the citizens about biopolitical matters.[2]

[2] For examples of particular deliberative events, see the homepage of The W. Maurice Young Centre for Applied Ethics, University of British Columbia: http://ethics.ubc.ca/index.php?p=misc&id=5. For material aimed to educate the young in biopolitical matters, see for example the Danish

Another precondition for more democratic biopolitics is strengthened professional media and science reporting which provides the citizens with reliable information, critical analysis and creative scenarios about the socio-political implications of biotechnology. This calls for professional science journalists with insight into scientific discourse and ability to present it to the public. Improved scientific education and media can jointly facilitate informed public deliberation about biopolitical matters. This, however, will not do unless forums for public dialogue are created in society and the spaces of action and reflection open to citizens are expanded. This requires, in fact, that bioethics is not sharply distinguished from biopolitics (Hoeyer and Tutton 2005; Árnason and Hjörleifsson 2008).

The idea is clear although the task is certainly not easy. One thing to avoid, for example, is that public consultation be designed mainly as strategic means to ensure more public acceptance and institutional trust. This could result mainly in more docile public, more willingness to abide by the biopolicies that are shaped by the authorities. It is an important objective to increase trustworthiness of public policies but it is not the objective of democratic policies to construct citizens who are "vehicles" of a comprehensive biopower and "mechanisms of domination" over which they have no control (Foucault 1980). From this cynical angle it may not matter much how biopolicies are formed because the choice is merely between a vertical or horizontal exercise of power.

The Foucauldian perspective is of great heuristic value in the analysis of biopolitics but it provides limited guidance for the task of framing more constructive democratic biopolicies. As part of that task, it is necessary to create opportunities for citizens to develop their thinking and increase their understanding of science and impact on biopolicies. This vision implies a belief in the intrinsic value of consultation and public dialogue, more in the spirit of democratic deliberation. However, isolated deliberative events can be used simply to solicit citizens' values or preferences without engaging them in critical deliberation which requires that the participants adopt a civic standpoint.

Although the guiding vision is important, we need to, as Alan Irwin suggests, "move beyond general exhortation alone over such matters and instead explore the social processes, underlying assumptions and operational principles through which scientific citizenship is constructed in particular settings" (Irwin 2001: 15). It is important to move the discussion of public participation "from the level of sloganizing to an important focus for both social scientific and practical investigation and experimentation" (Irwin 2001: 16). Among the complexities involved in the shaping of more democratic biopolicies are questions like the following: What information is provided and how it is provided to the public? How are issues to be framed for public debate? How is public consultation to be institutionally located? No doubt, there will be a constant tension between science, politics and the public will, and this tension will take on various forms which depend on the subject matter. The

Ethics Council: www.etikoglivet.dk/sw11738.asp. Similar educational projects for the young are sponsored by the Norwegian Biotechnology Advisory Board and the Swedish National Council on Medical Ethics.

challenge is to transform this tension into a creative power for innovative policy making.

It is in the nature of creative democratic politics that it is in constant search for more efficient channels for people to be informed about decision-making and to increase their impact on policy making (Arblaster 1987). There is no universal solution to how this is to be done. The important thing is the willingness and effort to look for the appropriate approach in each case. In this chapter I have argued that one way to approach this task is by a policy of a dynamic authorization for the conditions of use of data in population databank research. It is clear, however, that much more thinking is needed in the context of research biobanking if we find it at all important to have a vision of an educated and engaged citizen. The exercise in deliberative democracy in relation to biobank research is now in its starting phase and there are interesting times ahead. It is of crucial importance that deliberative democracy is not used to facilitate the benefit view by making the "mini public" more accepting of the current practices and thus seeking to acquire a premature democratic legitimization.

A democratic legitimization in the spirit of deliberative democracy can only be reached by a preceding critical discussion in the public sphere, the outcome of which is translated into political will formation. As Joshua Cohen writes in the spirit of Habermas, "free deliberation among equals is the basis of legitimacy" (Cohen 1997: 72). This should not be understood as a realistic aim as much as a *critical idea* which can help identify the role of power, coercion and ignorance in social decision-making. This critical idea can be used for example to distinguish claims based on narrow self-interests from those conducive to the general public interests. This critical idea of freedom in public deliberation needs to be taken more into account in the exercise of deliberative democracy if it is to contribute to overcoming the limits of the protecting and the benefit positions with respect to research biobanking and create conditions for more informed and engaged citizens.[3]

References

Arblaster A (1987) Democracy. University of Minnesota Press, Minneapolis

Árnason V (2004) Coding and consent: moral challenges of the database project in Iceland. Bioethics 18:27–49

Árnason V, Hjörleifsson S (2007) Geneticization and bioethics: advancing debate and research. Med Health Care Philos 10:417–431

Árnason V, Hjörleifsson S (2008) Population databanks and democracy in light of the Icelandic experience. In: Launis V, Räikkä J (Eds.) Genetic Democracy. Springer Verlag, Heidelberg, pp 93–104

Benhabib S (1996) Toward a deliberative model of democratic legitimacy. In: Benhabib S (Ed.) Democracy and Difference: Contesting the Boundaries of the Political. Princeton University Press, Princeton, pp 67–94

Caulfield T et al. (2003) DNA databanks and consent: a suggested policy option involving an authorization model. BMC Med Ethics 4:1

[3] Thanks to the editors for helpful comments on this chapter.

Cohen J (1997) Deliberation and democratic legitimacy. In: Bohman J, Rehg W (Eds.) Deliberative Democracy: Essays on Reason and Politics. The MIT press, Cambridge, MA, pp 67–91
Foucault M (1980) Two lectures. Lecture two. In: Gordon C (Ed.) Power/Knowledge. The Harvester Press, Brighton, pp 92–108
Greely H (1999) Breaking the stalemate: a prospective regulatory framework for unforeseen research uses of human tissue samples and health information. Wake Forest Law Rev 34:737–766
Greely H (2000) Iceland's plan for genomics research: facts and implications. Jurimetrics 40:153–191
Gulcher J, Stefánsson K (2000) The Icelandic healthcare database and informed consent. N Engl J Med 342:1827–1830
Hoeyer K et al. (2004) Informed consent and biobanks: a population-based study of attitudes towards tissue donation for genetic research. Scand J Public Health 32:224–229
Hoeyer KL, Tutton R (2005) 'Ethics was here': studying the language-games of ethics in the case of UK biobank. Crit Public Health 15:385–397
Irwin A (2001) Constructing the scientific citizen: science and democracy in the biosciences. Public Underst Sci 10:1–18
Kaye J (2004) Broad consent — the only option for population genetic databases. In: Árnason G et al. (Eds.) Blood and Data: Ethical, Legal and Social Aspects of Human Genetic Databases. University of Iceland Press, Reykjavík, pp 103–109
Kristinsson S (2006) Autonomy and informed consent: a mistaken association? Med Health Care Philos 10:253–264
Kristinsson S, Árnason V (2007) Informed consent and human genetic database research. In: Häyry M et al. (Eds.) The Ethics and Governance of Human Genetic Databases: European Perspectives. Cambridge University Press, Cambridge, pp 199–216
Marshall TH (1950) Citizenship and Social Class and Other Essays. Cambridge University Press, Cambridge
O'Neill O (2001) Informed consent and genetic information. Stud Hist Philos Biol Biomed Sci 32:689–704
Rose H (2001) The Commodification of Bioinformation: The Icelandic Health Sector Database. The Wellcome Trust.
Rose N, Novas C (2005) Biological citizenship. In: Ong A, Collier SJ (Eds.) Global Assemblages: Technology, Politics, and Ethics as Anthropological Problems. Blackwell Publishing, Oxford, pp 439–463
Sandel M (1982) Liberalism and the Limits of Justice. Cambridge University Press, Cambridge

Part II
Research Biobanking: Towards a New Conceptual Approach

Mapping the Language of Research Biobanking: An Analogical Approach[1]

Bjørn Hofmann, Jan Helge Solbakk, and Søren Holm

Abstract New medical technologies provide us with new possibilities in health care and health care research. Depending on their degree of novelty, they may as well present us with a whole range of unforeseen normative challenges. Partly, this is due to a lack of appropriate norms to perceive and handle new technologies. This chapter investigates our ways of establishing such norms. We argue that in this respect analogies have at least two normative functions: they inform both our understanding and our conduct. Furthermore, as these functions are intertwined and can blur moral debates, a functional investigation of analogies can be a fruitful part of ethical analysis. We argue that although analogies can be conservative, they are nevertheless useful because they bring old concepts to bear upon new ones. We also argue that there are at least three ways in which analogies can be used in a creative manner. First, understandings of new technologies are quite different from the analogies that established them, and come to be analogies themselves. That is, the concepts may turn out to be quite different from the analogies that established them. Second, analogies transpose similarities from one area into another, where they previously had no bearing. Third, analogies tend to have a figurative function, bringing in something new and different from the content of the analogies.

[1] This chapter is based on Hofmann et al. 2006a, and is printed here with permission by Springer.

B. Hofmann (✉)
Section for Radiography and Health Technology, Department of Health, Care and Nursing, University College of Gjøvik, e-mail: bjoern.hofmann@hig.no

and

Section for Medical Ethics, Faculty of Medicine, University of Oslo, Oslo, Norway
e-mail: b.m.hofmann@medisin.uio.no

J.H. Solbakk, et al. (eds.), *The Ethics of Research Biobanking*,
DOI: 10.1007/978-0-387-93872-1_11, © Springer Science+Business Media, LLC 2009

Introduction

New medical technology often produces heated moral debates and creates work for an army of verbose bioethicists.[2] One of the reasons for this is that new technology is extremely productive, normatively speaking: it urges usto find norms of comprehension (what the technology is) and norms of conduct (how we should handle it). A crucial point in the formation of norms is the emergence of a new technology. What happens when a technology is being established? What begets and nourishes the normative processes? How do we come to understand and cope with the new and the unknown? One answer to this is: through analogies. Analogies tend to help us to constitute our understanding of the new phenomenon and guide us in our attempt at coping with it. As Roland Barthes declared: "no sooner is a form seen than it must resemble something: humanity seems doomed to analogy".[3] When we are faced with a new and unknown phenomenon, we tend to apply analogies in order to understand and cope with it. Moreover, analogies are at the basis of our reasoning (Lakoff and Johnson 1980).[4] This is particularly so with new technologies (Latour 1986: 173–183).

The objective of this chapter is to investigate the role of analogies in the formation of norms at the point of emergence of new technologies. In order to do so, we will use research biobanking as an example, i.e. the procurement, storage and use of biological material (and data) for research purposes. The reason for this is that it is a technology in emergence, a technology where the norms have not yet settled. Furthermore, it is an area that is rich in analogies. It will be argued that analogies have a double normative function in relation to new technologies:

- They shape our perceptions and conceptualizations – and thereby our comprehension – of phenomena
- They guide us in our handling of phenomena

For the first function we suggest the label "epistemic normativity", while "moral normativity" seems to be an appropriate label for the second function. Furthermore, the analysis reveals that analogies can be used to classify a phenomenon (classificatory), to predict phenomena (inductive) and to persuade of a certain conduct or regulation (persuasive). Moreover, analogies can be conservative, e.g. when stemming from existing and relatively fixed areas of life, or they can be creative, e.g. when they come from quite different areas of life and are used in untraditional ways. Analysing the analogies applied with respect to emerging technologies can be of help in clarifying the normative debate.

---

[2] Including, of course, the authors of this chapter.

[3] Roland Barthes is here cited from Silverman and Torode 1980: 248.

[4] We here use the term "analogy" synonymously with "metaphor" in cognitive linguistics. See also note 6 below. It is also argued that analogy is "the core of cognition" (Gentner et al. 2001).

Analogies

Analogies are used in a wide range of ways in relation to new and emerging technologies in general and in relation to research biobanking in particular. One prominent example of explicit and implicit uses of analogies as analytical tools to clarify the ethical and regulatory challenges raised by biobanking is a seminal article by George Annas (Annas 1999). Here Annas uses organ transplantation, blood transfusion and foetal tissue donation as analogies to both explicitly explore what placental blood biobanking is, and, implicitly, to argue for a particular way of conceiving of it and handling it. This double use of analogies corresponds well with general theories on analogies (Govier 2005) where analogies have both an explanatory and an argumentative function. However, the article does not use the term "analogy". Rather, Annas talks about "models". In another article using analogies in the field of medicine (Strand et al. 2004/2005), terms such as "metaphors" and "models" are used interchangeably for the same concept.[5] Hence, before we set out on our analogical endeavour, we have to define what we mean by analogy.

"Analogy" has its root in the Greek word *analogia* meaning "proportion", "correspondence" and "resemblance", and is defined as similarity in some respects between things that are otherwise dissimilar or a comparison based on such similarity. In "analogy" there is an aspectual comparison meaning that X resembles Y in certain aspects, and that there is a chance that other similarities will also be found. For example, the black and white photograph film has been used as an analogy to X-ray films. Thus, the key function of an analogy is the transfer of meaning from the analogue to the target. In other words: "An analogy establishes an interrelation between two different spheres or domains. It enables us to see aspects of the domain in question in the light of another domain" (Leuken 1997: 219).[6]

The point is not to claim that there is only one correct definition of "analogy", or even claim that "analogy" is the only acceptable term (Black 1962; Childress 2004). It is rather to suggest that applying comparisons from other areas to new, emerging ones is of particular interest to the ethics that is concerned with new technologies.

Epistemic Normativity of Analogies

Hence, although Annas uses the term "model", he quite clearly applies analogies (according to our terminology) to establish a concept of what a certain phenomenon is. The analogies of organ transplantation, blood transfusion and foetal tissue donation are applied to explore what becomes of biological material as a result of new technologies. These are not the only analogies that are used to understand biological

[5] Even scholars tend to use the terms "metaphors" and "analogies" interchangeably, e.g. Latour 1986: 246–247.

[6] As indicated in note 4, we apply the term "analogy" in accordance to ordinary language, and synonymous to the conceptual metaphor in cognitive linguistics (http://en.wikipedia.org/wiki/Analogy). Accessed on May 11, 2009.

material: waste, natural resources, organ donation, gift, commodity, stock market and recycling are prominent in the literature as well (Hofmann et al. 2006b).[7]

In the same manner, as analogies are used to understand biological material (as a result of new technology), analogies tend to be used to understand new technology in general. Analogies play a primary role in exploring and conceptualizing new technologies. The point is to find ways to reason about unfamiliar cases on the background of familiar ones. Explorative analogies can be classificatory: they can be used to classify certain phenomena. If an analogue has characteristics x, y and z, and the phenomenon in question, e.g. biological material, also has the characteristics x, y and z, it can be argued that the phenomenon should be classified in the same way as its analogue. Hence, if umbilical cord blood cells in a biobank have all the features of stored donated blood, the bank could be classified as a blood bank. This classificatory function of analogies is a priori in that the analogy can be made on the basis of reflection alone (Govier 2005: 1521–1524). The main point is consistency; similar cases have to be classified in the same way, because similar cases have to be treated similarly.[8]

A different kind of explorative application of analogies is inductive, where the analogies are well-known real cases that are used to predict the features of new phenomena, such as emergent technologies. We can use blood-bank blood as an analogy to predict characteristics of umbilical cord blood, including its social characteristics. However, how relevant and reasonable these analogies are, we do not know. The future value of umbilical cord blood is unknown, and analogies trying to help us predict an answer are speculative. The point is that the phenomenon in question shares some characteristics with its analogue, which makes us stipulate or predict that other characteristics will be shared as well.

In this manner analogies guide us in establishing our understanding of new technologies and phenomena, such as biological material, in at least two ways: by classification and by induction.[9]

Moral Normativity of Analogies

Correspondingly, analogies appear to play a central role in establishing certain modes of conduct related to new technologies. Analogies are conceived of as a device in argument, and are used to promote certain moral norms. If one can convince someone that biological material is waste, issues of property rights and remuneration are settled. The point is to bring undisputed cases to bear on an unsettled or disputed case.

The precedent system of law is based on this use of analogies, presuming a principle of consistency; equal cases must be treated equally. The point is exactly the

[7] For this see also chapter called "The Use of Analogical Reasoning in Umbilical Cord Blood Biobanking".

[8] This also entails that the pointing out of dis-analogies can have an important function.

[9] For other relevant examples see Ratto 2006; Maasen and Weingart 2000.

same in law; analogies are used to fill in "holes" that are not explicitly covered by law, by showing that certain cases resemble others that clearly fall under the law.

Analogies can have a classificatory function in argumentation in the same manner as in exploration. Analogies may be used to support a thesis that a thing may have a certain property (Whaley 1998) or that a certain case falls under a particular analogy, and therefore has a certain solution. For example, arguing that biological material has a series of properties in common with waste can be used as an argument that the "donor" (or more accurately in this case "discarder") has no property rights.[10]

In addition to the classificatory function of argumentative analogies, they also have a persuasive function (Whaley 1998; McCroskey and Combs 1969; Yanov 1996). Firstly, an analogy may be used to support one's argument or offer counter-arguments or refutations (Baaske 1991). For example, the gift analogy may be applied when arguing against reimbursement of the procurement of biological material, while the commercial bank analogy may be used to argue in favour of reimbursement. Secondly, analogies may be used (persuasively) as an influence device (McCroskey and Combs 1969: 333–339) promoting one's credibility or assaulting the opposition's character or competence.

Although we like to think that the supportive/refutative function of analogies is the most prominent in argumentation, analogies appear to be used extensively to modify credibility as well. One example is the famous Moore case[11] where the analogy of modification and manufacturing was applied to the work of the scientists to undermine Moore's claim that the cells of the cell line established from his removed spleen were "his cells". All persuasive uses of analogies may be effective, but are also subject to a series of fallacies (Govier 2005: 1521–1524).[12]

The point is not to give an exhaustive account of the argumentative function of analogies, but only to indicate that analogies do serve a variety of functions in argumentation, and that many of these functions may be at play in establishing moral norms for handling new technology in general, and with respect to biobanking in particular.[13] In sum, one can say that the explorative function of

[10] It is worth noting that in establishing new concepts due to the introduction of new technology other concepts may change as well. X-ray apparently changed our idea of "private parts" and of privacy as such, see for example Kevles 1997. In the same manner, research biobanking may change our concept of privacy and remuneration in health care. Hence, analogies and the concepts they establish may be morally normative in other areas than only with regards to a particular technology.

[11] Moore v. Regents of the University of California, 793 P.2d 479, (Cal. 1990).

[12] Moreover, analogies have other normative functions as well, e.g. in casuistry, where they are applied in order to make analogical inferences from related examples in order to reach conclusions in difficult cases and to set paradigm cases, see for example Jonsen and Toulmin 1988. Additionally, analogies are applied to analyse and develop ethics in itself. Examples like the survival lottery case (where organs are taken from one person in order to save the life of several persons), the trolley case (where a runaway trolley is proceeding down a track towards five workmen, but there exists a possibility of branching off the trolley to a track where there is only one workman), and other extreme examples have been used to explore moral intuitions and to refine and develop moral philosophy, see for example Thomson 1990; Kamm 2003.

[13] It is argued (White 2006) that analogies lack persuasive power, and that we need ethos, pathos and logos as prescribed by Aristotle in order to make analogies normative. We do think that the literature on analogies (and metaphors) and the examples given here and elsewhere (Neal 2006;

Table 1 Various roles of analogies

	Epistemic normativity	Moral normativity
Role of analogy	Explorative	Argumentative
Kinds of functions	Classificatory	Classificatory
	Inductive	Persuasive
		(a) support an argument
		(b) refute an argument
		(c) modify credibility

analogies can be employed in a classificatory and an inductive manner, and that the argumentative function of analogies can be used in a classificatory and a persuasive way (Table 1).[14]

Analogical Analysis

It is interesting to note the close relationship between the explorative and the argumentative function of analogies, i.e. between their epistemic and moral normativities. At the same time, as one is arguing for a certain concept of biological material in terms of analogies, one is promoting a certain conduct with regard to it. One of the reasons for this close relationship between the explorative and argumentative function of analogies may be that it is difficult to establish a practice with respect to a new technology if we do not know what it is. In order to conceptualize the new technology, we use analogies and it should not be surprising that the same analogies may have a morally normative function as well.

This relationship can itself be used argumentatively; under cover of pretending to investigate different understandings of biological material, the analogy can be used covertly in an argumentative way.[15] Conversely, in a moral debate, the analogies used argumentatively can turn out to have explorative elements. Accordingly, analogies may be used to reveal the way we conceive of a certain issue. They may be used to frame a certain domain, and to show which ways of seeing things are underlying a particular issue (Leuken 1997). For example, the organ donation analogy may be used in the case of umbilical cord biobanking in order to support not only certain understandings of biological material, but also to display and question the framework underlying such understandings.[16]

Holland 2006) are convincing. Besides it seems that the resemblance with familiar things or experiences in life can stir our emotions (pathos), convince us of its truth (logos) and evince the credibility of the analogist (ethos).

[14] Arguments from analogies are arguments in informal logic, and as such are inductive and weak, see for example Salmon 1973. Nevertheless, analogical arguments are important in ordinary language and they are arguments by showing (in contrast to arguments by saying), and as such important rhetorically, see Lueken 1997: 218.

[15] It is also important to note that a persuasive analogy can be used to hide aspects of the new situation. The "war on terror" analogy does for instance (intentionally?) hide big differences between this kind of "war" and conventional war. Furthermore, analogies may be used as normative devices under cover of being explorative. The selection of explorative analogies is hardly neutral.

[16] Leuken refers to Wittgensteins use of analogies in PI (§18) to underscore this.

Hence, sorting out the explorative and the argumentative function of analogies, as well as their classificatory, inductive and persuasive uses, may be of great value when debating new technologies. Moreover, analysing analogies can have a clarifying and emancipatory function, thus increasing the transparency of conceptual and moral debates. This raises the question of how to assess analogies and their uses.

What Is a Good Analogy?

The answer to this question is strongly dependent on the purpose of an analogy. If we intend to explore a new field, the criteria for a good analogy are quite different from the ones used if we intend to promote a certain conduct. In the latter case, great similarity between the analogue and the target gives weight to the argument. However, great similarity is not necessarily a prerequisite for a good analogy if the intention is to explore a new phenomenon (e.g. a new technology). As will be discussed below, some distance may add further value to an analogy.

Further, the conceptual aspect of an analogy can be used to add to its argumentative force. In this case it seems that an increased epistemic similarity will strengthen the moral argument. Thus, for example, the more we can convince of the similarity between umbilical cord blood and biological waste, the more forceful the analogy also becomes at promoting a particular conduct, i.e. of using the contents of umbilical cord blood biobanks without remuneration.

Correspondingly, the similarity between the analogue and the target with respect to moral norms can be used as an argument for a particular understanding, e.g. that certain biological material should be classified in a certain manner. By emphasizing the special moral importance of genetic information, we may strengthen a claim that all tissue has to be classified in the same category and receive special protection. Accordingly, one might argue that analogies carry different weight if they are used to argue from classificatory analogue to a target in an inductive or persuasive manner or the other way around.

Hence, the value of an analogy depends on the purpose and the context. The point is that these purposes can be hidden, and that we may initially be carried away by the sheer rhetorical force or novelty of an analogy. A closer analysis of analogies is therefore almost always necessary to reveal any covert normative implications. Only by revealing the complete analogical function in a particular context can one discuss its success. We will return to the question of how we can use analogies, both explorative and normative, in bioethical debates below, but first we will address the question of how analogies work in practice.

The Analogy Is Dead: Long Live the Analogy

We use analogies to establish norms of comprehension and conduct with respect to a phenomenon. The phenomenon can be a technology, such as molecular analysis of cells, new phenomena that the technology provides (DNA), or known things where new technology forces us to establish new norms because it makes the old norms obsolete (stem cells).

However, the phenomena that are conceptualized by analogies can themselves become analogies and be used quite independently of the analogies that established them. When analogically established things or practices themselves become analogies, the originally employed analogies appear to loose their primacy; they are dead (Lakoff and Johnson 1980). So, the thing itself (original target) can be used as an analogy quite independent of the analogies that established its concept. For example, when a concept of biobank is established, the bank analogy no longer plays any role, and "biobank" can be used as an analogy for other phenomena without any reference to commercial banking. The reason for this may be that the explorative and argumentative force of an analogy vanishes when the phenomenon has become conceptualized.[17]

The analogies tend to stiffen or congeal after norms of comprehension and conduct have been established, and the new technology (or phenomenon) can then itself be used as an analogy. For example, the gift analogy has been used to establish organ and tissue donation, whereas organ and tissue donation subsequently has been used as an analogy to argue against unconsented caesarean section.[18]

Thus, it appears that when a concept is established, the establishing analogies become obsolete. They no longer have bearing. One consequence of this is that analogical analysis is most fruitful at the emergence of new technologies, and, at a certain point, the explorative and argumentative analogies lose their reflexive function. This independence of the analogies that establish a certain concept raises the question of how independent a concept actually can be of its formative analogies in general. How much do the analogies bear on the concept they create?

Old Analogies for New Technologies?

So far we have said that analogies are used in explorative and argumentative ways and that they are important parts of moral debates about technology, especially emerging technologies, and that an analysis of analogies is of value to ethical analysis. However, what does analogical reasoning actually mean with respect to our ability to address new technologies? For instance, if our concepts of new technologies are based on "old analogies" i.e. analogies of established practices, do analogies not restrict our conceptialisations of new phenomena? Are analogies conservative? Can they address things that are really "unique, different or simply not captured by the existing analogy" (Johnson and Burger 2006). That is, is Gerald

[17] Some scholars would prefer terms such as "framed" or" normalized" instead. Subtle distinctions in this field is not the point here, but rather that norms of conception are established (and become fixed).

[18] In re. A.C. 1990, 57B A.2d 1235 (D.C.App.). This is only one example. We do not say that the argument from the analogy is valid.

Dworkin right when he calls analogy and precedent "the weapons of conservatives" (Dworkin 1988: 37)?

One may argue, correctly we think, that using established analogies from closely related areas may lead to the preservation both of old norms of comprehension and conduct and of their related practices instead of developing new concepts to understand and cope with the new phenomenon or technology actually at hand. In other words, trying to make new technologies fit images of existing technologies may not only generate conservative practices, but may as well obstruct our understanding of emerging technologies.[19]

The way we choose and use analogies when faced with new technologies may vary, but in most cases our analogical behaviour is triggered by similarities with the new phenomena. For example, in the debate on how to handle biobank material, more specifically umbilical cord blood, established analogies, such as waste and blood donation, were applied due to their physical and practical similarities (Annas 1999).

From this seems to follow that old analogies cannot be used to (1) understand radically new technologies or (2) understand genuinely new aspects of existing technologies. Consequently, the way to proceed would be to search for "new" analogies in order to cope with technological novelties. Thus, if the purpose is to explore a new technology, i.e. its elaborate epistemic normativity, one should rather apply analogies from quite distant areas so as to develop appropriate forms of understanding, instead of using established analogies from the same area or from closely related areas. In other words, we should take advantage of the polysemic nature of analogies (López 2006). As such, this could also shed light on alternative ways of handling technologies (moral normativity) that otherwise would not have been discussed.

Exploration by Analogy

With regard to exploration of procurement, storage and use of biological material for research purposes, we therefore suggest investigating the conceptual potential of analogies from a range of areas outside medical research, where people transfer something to a common institution. Examples of such analogies are ordinary commercial banking, associations, clubs (e.g. book clubs) or unions, libraries, military conscription, taxation, and management of pieces of art (Solbakk et al. 2004).[20]

Membership-related analogies could be of help in highlighting mutual relationships and responsibility, whereas commercial banking analogies, such as bank accounts, could be used to explore aspects of ownership, loan, interests and remuneration. Finally, we suggest employing insurance analogies to analyse aspects

[19] For a substantiation of this claim see Hofmann et al. 2006b and the chapter called "The Use of Analogical Reasoning in Umbilical Cord Blood Biobanking".

[20] This last analogy plays a prominent role in the commercial world (some art pieces are considered "invaluable", are protected, and have a cultural and symbolic dimension etc.) For this analogy, see also the chapter called "The Art of Biocollections".

of self-interest and risk.[21] Thus, when analogies are imported from quite distant areas, they may serve as fertilizers or catalysts in the shaping of our norms of comprehension.

The point is that transposing analogies from other areas of life creates a freer and more innovative ground for establishing new norms with respect to new technologies than just applying the most obvious and clear-cut analogies. There is a diversity in life that can make such a transposition of analogies fruitful. Furthermore, applying several analogies, instead of relying on single analogies, facilitates a creative rather than a conservative application of analogies (Shelley 2003; Holyoak and Thagard 1996).

Moral Argument by Analogy

As indicated earlier, the application of distant analogies is not symmetric with respect to their epistemic and moral normativity. An analogy from a distant area may be fruitful in an explorative sense, but not very convincing argumentatively. One reason for this may be that whether arguments are convincing appears to depend on how well we recognize the examples, i.e. how congruent the analogies are with our own experience. To take an example to illustrate this point: if we compare the donation of biological material to a research biobank with voluntary communal work (which in Norwegian has a special word, "dugnad")[22] to argue that this should be considered a voluntary contribution everybody should make, it is not likely that we will make a good case except in societies where the tradition of voluntary communal work is alive and well acknowledged. This corresponds with evidence that too extensive uses of analogies can reduce credibility (Whaley 1998). Furthermore, it is clear that transposing analogies is inductive, as one applies similarities in some areas to have bearing on other areas (where we yet have no definitive knowledge).

Hence, transposition of analogies from areas quite distant from the field in question appears to be more fruitful in exploration than in argumentation. Nevertheless, analogies from distant fields may generate new ideas for the handling of new technologies, even though their argumentative force is weak.

Old Dogs and Old Tricks: Are Analogies Doomed to Be Conservative?

Although this alternative approach to, or form of, "analogical behaviour" may enable us to address challenges related to the understanding and regulation of new phenomena in a different way, it may still be the subject of the same objection of being conservative. Analogies stemming from other areas are still analogies

[21] See Hofmann et al. 2006b and the chapter called "The Use of Analogical Reasoning in Umbilical Cord Blood Biobanking".

[22] For the dugnad analogy, see the chapter called "The Use of Analogical Reasoning in Umbilical Cord Blood Biobanking".

from existing and established fields, and as such, they may infringe upon our open-mindedness to the genuinely new. Thus, it could be argued that the principal function of analogical behaviour is to confirm already-established modes of conceptualization and forms of regulatory conduct. Even if the analogy makes us able to "see aspects of the domain in question in the light of another domain", this is still a view within the horizon of an established domain.

The difficult question is of course whether we can transcend the familiar and known when we are confronted with new phenomena, or whether Roland Barthes is right. However, even if he is, there are several reasons to believe that there are ways to teach old dogs new tricks, i.e. to attain new concepts from existing analogies. First, there appears to be empirical evidence available to demonstrate that our concepts of new phenomena may differ substantially from already-existing concepts as well as from analogies that were applied during the establishment of the new concepts. For example, our concept of DNA is quite different from anything that we had conceptualized or known in advance. This example furthermore indicates that although some concept of DNA was established early on, this does not preclude later gradual changes in our understanding. Correspondingly, our concepts of biobanks are dissimilar from those of commercial banks (although in many ways similar to our concepts of blood banks).

Second, it seems that analogies can have catalysing or fertilizing functions that reach beyond their epistemic and moral normativity. It is argued that analogies have a figurative function that goes beyond the "literal similarities" (Hawkes 1972). It is worth noting that this figurative function of analogies is as relevant in science as in other fields (Campbell 1920: 129; Hesse 1966;1981; Pickering 1999; Shelley 2003). There appears to be some kind of dialectics between the analogue and the trace resulting in a synthesis which is distinguished from both of them.

Recycling and Reshaping Analogies

The point that has been made is that analogies are applied in order to explore and argue for certain concepts of new technologies, and phenomena that stem from them, such as biological material. These analogies tend to be normatively productive in two different ways:

- Epistemically normative, i.e. to explore potential comprehension (e.g. what biological material is)
- Morally normative, i.e. to argue how things should be (e.g. how we should handle biological material)

Hence, analogies are normative in two different ways: they shape our comprehension and our conduct. Furthermore, the explorative and argumentative functions of analogies are related. If biological material is waste, we should not look for reimbursement, and this can make the conceptual and moral debates blurred. Additionally, the value of an analogy varies according to the purpose and the context of its application. Hence, an analysis of analogies can add to the moral debate and be a fruitful part of the ethics of new and emerging technologies.

Furthermore, the new concepts that the analogies establish may themselves serve as analogies. Biobanks may become analogies for other technologically related phenomena. Even more, the concepts may become analogies for changing the analogies that established it. For example, the waste analogy may be important for understanding (umbilical cord) biobank material. The biobank material may consequently change our understanding of waste. Hence, analogies tend to have some kind of fertilizing or catalysing function. But they are often themselves "consumed" in the process. Analogies are used to establish norms of comprehension and action, but then become obsolete (as a remainder). Analogies give life to the source of new analogies.

Altogether we must agree with Roland Barthes' claim that we are doomed to analogy, and that this indicates that analogies are conservative. Analogies tend to be epistemically and morally more forceful if the similarities between analogue and target are many. However, as this chapter has tried to explore, it is not necessarily so. In our view, the justified critique of the inherent conservatism of analogies can be countered by employing three arguments:

1. Empirical: in which new concepts are quite different from analogies available. They themselves can become analogies for the concepts they drew upon.
2. Transposed analogies: in which transposed similarities from one area to another bring new perspectives into the field. Therefore it can be useful to apply analogies from quite different areas.
3. Figurative function of analogy: in which they tend to have important creative functions resulting in uniquely new concepts and with potentially new conduct. Consequently, "old" analogies may be used to give rise to new concepts of technologies, be they old or new. It is possible to teach old dogs new tricks! This can be of relevance to the ethics of technology in general, as well as in the field of research biobanking.

Acknowledgments We are grateful to Jennifer Harris, Roger Strand, Jan Reinert Karlsen, Anne Cambon Thompson, Paula Lobato de Faria, and Anne Maria Skrikerud for their valuable comments to an earlier draft of this manuscript. We also thank the anonymous referees of Theoretical Medicine and Bioethtics for insightful comments and constructive suggestions. The financial support of the project "Mapping the language of research biobanking" (Contract No. 159864/S10) of the Norwegian Research Council and the Norwegian Institute of Public Health and the project "Research biobanking and the ethics of transparent communication" (Contract No. 182269) of the Norwegian Research Council are greatly acknowledged.

References

Annas GJ (1999) Waste and longing – the legal status of placental-blood banking. New England Journal of Medicine 340:1521–1524

Baaske K (1991) Analogic argument in public discourse: a reconsideration of the nature and function of analogy. In: Eemerenvan FH et al. (Eds.) Proceedings of the Second International Conference on Argument. Sic Sat, Amsterdam, pp 411–415

Black M (1962) Metaphors. In: Black M (Ed.) Models and Metaphors: Studies in Language and Philosophy. Cornell University Press, Ithaca, pp 19–43

Campbell NR (1920) Physics: The Element. Cambridge University Press, Cambridge

Childress JF (2004) Metaphor and analogy. In: Reich WT (Ed.) Encyclopedia of Bioethics, revised edition. Simon and Schuster Macmillan, New York, pp 1834–1843

Court of Appeals, D.C. In re. A.C. (1990) Atlantic Reporter 573:1235–1264

Dworkin G (1988) The Theory and Practice of Autonomy. Cambridge University Press, New York

Gentner G et al. (2001) The Analogical Mind: Perspectives from Cognitive Science. The MIT Press, Cambridge MA

Govier T (2005) A Practical Study of Argument, sixth edition. Wadsworth Publishing, Belmont

Hawkes T (1972) Metaphor. Methuen, London

Hesse B (1981) The function of analogies in science. In: Tweney R (Ed.) On Scientific Thinking. Columbia University Press, New York, pp 345–348

Hesse MB (1966) Models and Analogies in Science. University of Notre Dame Press, Notre Dame

Hofmann B et al. (2006a) Teaching old dogs new tricks: the role of analogies in bioethical analysis and argumentation concerning new technologies. Theoretical Medicine and Bioethics 27:397–413

Hofmann et al. (2006b) Analogical reasoning and technology in biobanking – an umbilical perspective. The American Journal of Bioethics 6:49–57

Holland S (2006) It's not what we say, exactly... or is it? American Journal of Bioethics 6:65–66

Holyoak KJ, Thagard P (1996) Mental Leaps: Analogy in Creative Thought. The MIT Press, Cambridge MA

Johnson S, Burger I (2006) Limitations and justifications for analogical reasoning. American Journal of Bioethics 6:59–61

Jonsen AR, Toulmin S (1988) The Abuse of Casuistry. University of California Press, Berkley

Kamm FM (2003) Harming some to save others. In: Darwell S (Ed.) Deontology. Blackwell Publishing, Oxford, pp 162–193

Kevles B (1997) Naked to the Bone: Medical Imaging in the Twentieth Century. Rutgers University Press, New Brunswick

Lakoff G, Johnson M (1980) Metaphors We Live By. University of Chicago Press, Chicago

Latour B (1986) Laboratory Life: The Construction of Scientific Facts. Princeton University Press, Princeton

López JJ (2006) Mapping metaphors and analogies. American Journal of Bioethics 6:61–63

Lueken GL (1997) On showing in argumentation. Philosophical Investigations 20:205–223

Maasen S, Weingart P (2000) Metaphors and the Dynamics of Knowledge. Routledge, London

McCroskey JC, Combs WH (1969) The effects of the use of analogy on attitude change and source credibility. Journal of Communication 19:333–339

Neal KC (2006) Analogical trends in US umbilical cord blood legislation. American Journal of Bioethics 6:68–70

Pickering N (1999) Metaphors and models in medicine. Theoretical Medicine and Bioethics 20:361–375

Ratto M (2006) Foundations and profiles: splicing metaphors in genetic databases and biobanks. Public Understanding of Science 15:31–53

Salmon WC (1973) Logic. Englewood Cliffs NJ, Prentice-Hall

Shelley C (2003) Multiple Analogies in Science and Philosophy. John Benjamins Publishing Company, Amsterdam/Philadelphia

Silverman D, Torode B (1980) The Material Word: Some Theories of Language and Its Limits. Routledge & Kegan Paul, London

Solbakk JH et al. (2004) Mapping the language of research-biobanks and health registries: from traditional biobanking to research biobanking. In: Árnason G et al. (Eds.) Blood and Data: Ethical, Legal and Social Aspects of Human Genetic Databases. University of Iceland Press and Center for Ethics, Reykjavik, pp 299–305

Strand R et al. (2004/2005) Complex Systems and Human Complexity in Medicine Complexus 2:2–6

Thomson JJ (1990) The trolley problem. In: Thomson JJ (Ed.) The Realm of Rights. Harvard University Press, Cambridge MA, pp 176–204

Whaley BB (1998) Evaluations of rebuttal analogy users: ethical and competence considerations. Argumentation 12:351–365

White GB (2006) Analogical power and Aristotle's model of persuasion. American Journal of Bioethics 6:67–68

Yanov D (1996) Ecologies of technological metaphors and the theme of control. Society for Philosophy and Control 1:1–15 http://scholar.lib.vt.edu/ejournals/SPT/v1_n3n4/Yanow.html Accessed May 11 2008)

The Use of Analogical Reasoning in Umbilical Cord Blood Biobanking[1]

Bjørn Hofmann, Jan Helge Solbakk, and Søren Holm

Abstract In this chapter we investigate the roles that analogies play in the processes of understanding and managing umbilical cord blood biobanking. The objective is to unveil analogies' role as analytical devices in exploring the "being" of the new technology as well as their normative function in conceptualizing its characteristics and how it should be applied. We demonstrate how analogies have both explorative and argumentative functions, and how none of the analogies alone are able to address all the challenges raised by cord blood biobanking.

Introduction

Biological material has been used for medical diagnosis and biomedical research for a long time. However, the emergence of new technologies for analysing biological material to gain information for diagnosis and treatment choice, as well as methods generating new therapeutic products, has made such material much more salient within clinical practice and biomedical research. In addition, these technologies have made the commercial asset value of biological material much more visible.

New developments in biotechnology, such as therapeutic use of (pluripotent) stem cells has made the traditional distinction between organs, tissues and cells

---

[1] This chapter is based on Hofmann et al. 2006a and is printed here with the consent of Taylor & Francis.

B. Hofmann (✉)
Section for Radiography and Health Technology, Department of Health, Care and Nursing, University College of Gjøvik, e-mail: bjoern.hofmann@hig.no
and
Section for Medical Ethics, Faculty of Medicine, University of Oslo, Oslo, Norway
e-mail: b.m.hofmann@medisin.uio.no

J.H. Solbakk, et al. (eds.), *The Ethics of Research Biobanking*,
DOI: 10.1007/978-0-387-93872-1_12, © Springer Science+Business Media, LLC 2009

less relevant (Raymond et al. 2002: 257–265).[2] One area where such new technology has given rise to a series of new possibilities and corresponding challenges is umbilical cord blood biobanking. Some of these challenges will be addressed in what follows. Umbilical cord blood biobanking is an especially interesting case because the material in the bank is of potential future therapeutic value to both the donor and others while at the same time it is of potential value to science. Umbilical cord blood haematopoietic stem cells are used for treatment of a wide variety of diseases, and have become a viable alternative source of haematopoietic stem cells to bone marrow transplantation (Rocha et al. 2006). Cord blood can be used in autologous as well as in allogenic transplantations and has given rise to both private and public biobanks. The intention behind our exploration of umbilical cord biobanking is to uncover the way we try to handle challenges related to new technologies, and the prominent role which analogies play in particular. New technologies pose fundamental questions of what the technology *is*, its correct understanding, and how it should be applied. We aim to show that we use an assortment of analogies to address the complex ontological, epistemological and ethical questions surrounding biobanking in its modern and technology-driven form, and that no single analogy seems able to cover the whole field on its own. What then are the main questions posed by biobank technology?

Big Challenges with Small Amounts of Blood

Within umbilical cord blood biobanking, i.e. the procurement, storage and use of umbilical cord blood, we are faced with a series of pertinent normative questions. The following list of questions is not exhaustive, although it may be exhausting. However, it illustrates how broad and deep the challenges are. Some of these questions relate to the issue of what biological material is; e.g. is it part of a person, and who owns the blood, the child, the mother or the parents (Lind 1994; Sugarman et al. 1997b; Zilberstein et al. 1997; Munzer 1999; Kline 2001; Dame and Sugarman 2001)? If it is conceived of as leftover or byproduct, what kind of rights does the child and its parents have with respect to umbilical cord blood (Knoppers and Laberge 1995), and does this depend on our understanding of the production process of which it is a byproduct? If they do not have property rights, do they have other rights with respect to accessing this material (in terms of stem cells, other umbilical cord blood products or information derived from these products)?

Other issues are related to challenges of regulation and management. For example, if the cord blood has potential for commercialization and commodification, how should this be regulated? Should biobanks be governed by the invisible hand of the market or should there be equitable profit sharing (Merz et al. 2002) and just distribution of estimated or actual outcomes (Merz et al. 2002; Smaglik 2000)? May biobank material be sold (across national borders) or is commercialization of such material unacceptable in principle (Holm 2004)? Should there be control of downstream use and patenting (Merz et al. 2002), and how should one avoid exploitation

[2] It is interesting to note that blood itself poses some of the same challenges: is it an organ or a cluster of cells?

of persons in a vulnerable situation, e.g. persons contributing to research or persons belonging to disease-associated advocacy groups etc. (Merz et al. 2002)? Moreover, what is the proper form of advertisement, if any? Is it correct to call paying for storage of umbilical cord blood "biological insurance"? What actually is the relationship between the clinic with its personnel and the company (agent or contractor) storing and analysing the biological material? In the case of private umbilical cord blood biobanking, what happens to the blood if payments are not made? Should the material and the information derived from it be given back to the donors, be destroyed or become property of the company or of national health authorities or should it be given away freely to research and/or technological development?

Moreover, how should the relationship between a donor and a receiver be conceived of in cases of life-saving donations? If one gives away something that may be of vital value to oneself to someone who needs it desperately, should one then be entitled to know the receiver? Conversely, if one receives something of potentially vital value, why should one then not be allowed to know who the donor is (to express thanks)?

Furthermore, how should risks related to donation from biobanks or the use of products developed from biobanks be handled? In particular, how should risks to the recipient in cases of "donor" diseases (HIV, genetic disorders) or disclosure of unwanted knowledge be handled (Sugarman et al. 1997b)? How should one deal with situations of insufficient minority representation (Ballen et al. 2002) or genetic discrimination (e.g. if not everybody is allowed to store umbilical cord blood in the bank)? Additional issues concern how to regulate alternative uses of biobank material (e.g. for forensic purposes, or for the purpose of gaining genetic information in the context of insurance or employment).

Other issues more directly address moral values and principles, in particular respect for autonomy, but also privacy and confidentiality issues (Burgio and Locatelli 1997; Sugarman et al. 1997a,b; Burgio et al. 2003). Is it possible to obtain informed consent from persons donating material to an umbilical cord blood biobank (Vawter et al. 2002), and can the consent be adequate (Hoeyer et al. 2005; Sugarman et al. 2002)? How is it possible in advance to inform about and handle perspectives of cost-benefit, when we at present do not know the future utility-potential of the procured and stored material? How should one address the implications of information overload, e.g. the situation in which "if subjects are too well informed they'll be less likely to participate" (Merz et al. 2002: 172). And what about the problem of directiveness of information when private or public umbilical cord blood biobanks seek to enrol new "clients" (Sugarman et al. 2002)?

Furthermore, should withdrawal from the bank be conditional or unconditional? Can the biological material be kept if it is anonymized, or does it – after withdrawal of consent – have to be destroyed? What about results and information gained from the cord blood? Are contributions to such biobanks made voluntarily or are they subject to undue influence (Nelson and Merz 2002)? How effective is the protection of research subjects (Annas 2001)? Moreover, how should the connection between health registers, genealogical databases and biobanks be handled? How should one handle genetic information about a child and his/her family in order to protect privacy and confidentiality (Askari et al. 2002)?

As indicated, this list of questions is by no means exhaustive. The reason the list is made so extensive is to illustrate the complexity of the field and the diversity of challenges we face, and the variety of interconnections that exist between these questions. The cases make it clear that the ontological question of what biological material *is* and the epistemological question of what the status of the knowledge that stems from such material is, are related to moral issues, such as whether we have property rights, whether it should be shared with others, and whether we should be protected against misuse. Furthermore, as the questions are interrelated, so are their answers.

Correspondingly, some of the questions raise general issues in relation to blood cord biobanking, whereas others are typical to a particular use of biological material (e.g. to allogenic use of cord blood, to private use, or to research). Hence, particular uses of biological material call for special ways of understanding and managing biobank material.

Before we analyse which analogies that have been applied to address these challenges, we will make a short contextual remark. It can be that the "commercial banking analogy has already become indispensable" and that it overshadows other analogies (Burns 2006: 49). In Europe the final form of umbilical cord biobanking is not settled yet. It is also important to remember that blood banking in Europe is almost exclusively public and does not involve payment to donors. Although we acknowledge that the commercial banking analogy is a central player in the field of biobanking and that it may continue to play a crucial role in future attempts at investigating the ethical, legal and social challenges raised by biobanking in its different modern forms and formats, we want to show that other analogies still play important roles. Moreover analogies tend to be applied (normatively) in US cord blood legislation (Neal 2006).

Analogies Applied

One forceful analogy that has been applied to address some of these issues is that of waste (Gluckman 2000; Senior 2001; Harris 2005). More specifically it has been used to handle issues such as "left over", commodification and enrolment. If biological material is perceived as waste, it becomes easy to make people give it away (for research or therapeutic purposes). Others have used the analogy of waste that is transformed into gold (Annas 1999) to put emphasis on the changed status of the material due to the emerging technology. Moreover, umbilical cord blood has been compared to natural resources (Senior 2001) to further highlight certain aspects of the (economic) value of biomaterial, while implicitly de-emphasizing other aspects. This analogy covers both the issue of property and/or ownership, commodity and justice and to a certain extent also that of autonomy.

Others have applied the analogy of organ transplantation. One implication of this has been to put emphasis on the *history* of invasive procurements:

> Placental blood is described as useful for the transplantation of stem cells. This phrase implies that the model [analogy] of organ transplantation should be adopted for the collection of placental blood. This similarity is perhaps natural, because historically the transplantation of bone marrow (the chief source of stem cells) has itself been treated as analogous to organ transplantation (Annas 1999: 1521).

Another implication of the organ transplantation analogy has been an emphasis on *consent* issues and on *living* donor-safety issues: "Thus, if we adopt the transplantation model for placental blood, *we are likely to focus on the risks to the donor and forbid commerce and sales*" (Annas 1999: 1521, emphasis added). A third implication is that the use of this analogy has led to vilification of any *commerce* in relation to umbilical cord blood:

> ... we prohibit the purchase and sale of human organs because we think these practices put donors at risk from potentially coercive monetary inducements and also because we highly value the "gift relationship" in organ transplantation as a rare and praiseworthy example of altruism (Annas 1999: 1521).

Finally, the organ transplantation analogy has appeared to be useful in dealing with property issues as well as issues of commodification and of cost-benefit. Blood transfusion has been used as an analogy to highlight issues of risk, property and commodity. The implication of this analogy has been a down-scaling of the risk of donors as well emphasis on *product* safety issues. Besides, it has been argued that the use of this analogy may help to legitimize *some* commerce in placental blood (Annas 1999).

Other analogies have been applied to address other challenges; only some are discussed briefly in what follows. Commodity itself has been used as an analogy to address the issue of profit (Nelkin and Andrews 1998; Cohen 2000; Munzer 1999), and the stock market (Merz et al. 2002: 969) has been applied as an analogy to highlight many of the issues related to economic values, autonomy and risk. Fetal tissue donation, e.g. using aborted fetuses for research or therapy, has been used as an analogy to emphasize ownership and/or property issues, issues of consent and decision-making authority as well as safety issues with respect to potential receivers of umbilical cord blood (Annas 1999). The gift analogy has been used to emphasize that donation represents a "rare and praiseworthy example of altruism" (Annas 1999: 1521). Also the analogy of sponsoring (Merz et al. 2002) has been applied to put emphasis on altruism, as well as to address autonomy and property issues. The recycling analogy (Senior 2001), on the other hand, has been applied to address issues of property, cost-benefit and risk. Viewing human tissue as one's home, extending one's own identity, has been used to argue for strong protection against invasions of privacy, whereas analogies of a public office have been applied in order to argue that biobanks are jointly built institutions for co-operation and common use (Eriksson 2003: 183). Figure 1 gives a short overview of the analogies and some of the challenges they are made to address.

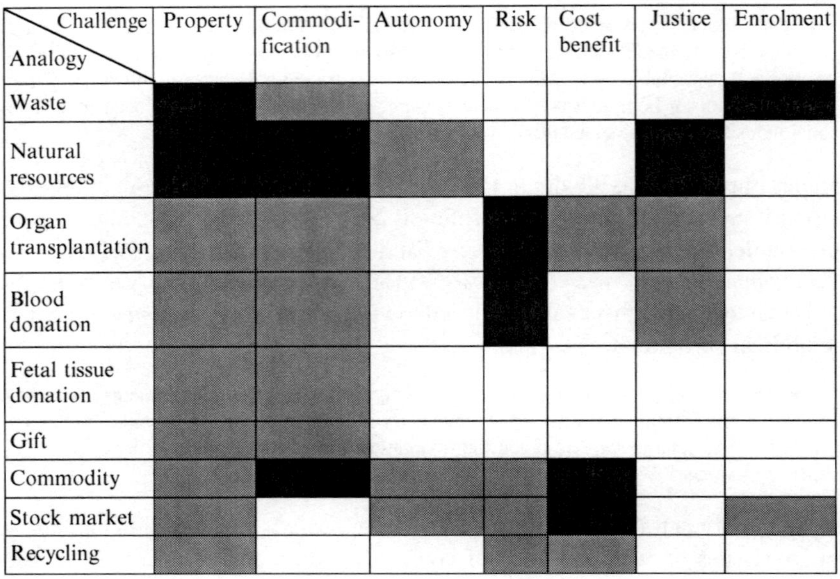

Challenge / Analogy	Property	Commodification	Autonomy	Risk	Cost benefit	Justice	Enrolment
Waste							
Natural resources							
Organ transplantation							
Blood donation							
Fetal tissue donation							
Gift							
Commodity							
Stock market							
Recycling							

Fig. 1 Some prominent analogies in the umbilical cord biobank debate (*rows*) and a tentative graphic outline of which challenges (*columns*) they address. *Black* indicates that the analogy addresses the challenge directly, *grey* that it addresses it more indirectly, whereas *white* indicates that the challenge is not addressed by the analogy (The graphic representation is inspired by Annas 1999.)

Analogical Reasoning

As was the case with the list of challenges raised by umbilical cord blood biobanking, the list of analogies presented above is by no means exhaustive. The overview of which challenges the various analogies have been used to address is at best cursory. Other analogies could have been applied for the same purposes, and the same analogies could probably have been used for other purposes as well. Nevertheless, we hope that our list and overview above have been able to demonstrate that analogies play an extensive role in the debate on how to appropriately understand and cope with umbilical cord blood biobanking and other forms of biobanking. From this at least five preliminary conclusions may be drawn.

First, not only do analogies change with new technologies and altered practices, but technologies and their practices change by dint of the use and promotion of analogies.

Second, analogies may not only serve as useful analytical tools in illuminating the normative terrain of research biobanking; their uses are in themselves normative. In this respect they may be said to have two functions, one *analytical* in the strict sense of the word and one *argumentative*. That is:

1. Analogies are applied in order to explore or analyse a certain issue, e.g. to sort out the ontological status of biological material, how to conceive of the knowledge

that stems from it and to map how we should actually act with respect to such material.
2. Additionally, analogies are applied in order to argue for certain conceptions and ways of handling the issue under scrutiny. That is, analogies are used to explore and map out the normative terrain of biobanking as well as to make normative claims.

Third, analogies are not exhaustive. That is, individually they are not able to deal with the full complexity of challenges that may emerge within a technology-driven field such as biobanking. Each analogy tends to have a restricted reach with respect to how many challenges they are able to address. Consequently, many analogies seem to be at play at the same time, even if one is to address one particular field, such as publicly funded umbilical cord blood research biobanks for allogenic use (Samuel et al. 2006). From this seems to follow that the function of analogies should be assessed from a variety of different angles and by the use of a variety of different parameters. Figure 1 gives but one account of such an assessment (with respect to the analytical function). However, it is quite clear from the analysis of analogies that only with great difficulty can such an assessment take place without becoming in itself argumentative (and normative).

Fourth, the relevance of analogies seems to be time-dependent, i.e. they tend to be more important at the emergence of a new technology. However, when the technology is conceptually settled, their relevance seems to decrease. In fact, the technology in question may itself become an analogy for other emerging technologies.[3] One of the reasons why we can still explore the analogies used in debates about umbilical cord blood biobanking is probably that this is a technology still in its infancy.

Fifth, part of the argumentative value of analogies is rhetorical. Many of the analogies used are value laden, and if I can convince my interlocutors to accept a particular analogy as a good analytic tool, I may also convince them to accept the valuation implicit in the analogy. A prime example is the analogy of "waste" which implies a low valuation of the material denoted as "waste" and a tendency to obscure the fact that some kinds of waste are extremely valuable even to the waste producer (e.g. the heat generated by the incineration of household waste is itself a kind of waste, but still valuable if used for central heating). That is, there is a relationship between the analytical and the argumentative roles of analogies, as noted previously. The acceptability and relevance of the explorative function of an analogy may increase its argumentative power.

Additionally, the selection and promotion of analogies has a normative function in and of itself. Because certain analogies are more suitable to emphasize particular aspects of a technology they will be more efficient in promoting certain conceptions and actions. Hence, an analysis of the application of analogies can be used to reveal and critically assess the vocabulary and the prevailing positions in the field of biobanking.

---

[3] For this, see also chapter "Mapping the Language of Research Biobanking: An Analogical Approach".

Analogies Explored

There is of course a wide variety of analogies not mentioned above that could be fruitful and interesting to apply – and which to date has not been applied – in discussions about umbilical cord blood biobanking. Here we suggest assessing the relevance of a set of self-interest-based and conscription-based analogies.

Commercial Banking

The first similarity between commercial banking and research biobanking relates to the notions of "input" and "output". One invests resources (biological material, money), and receives a return that is dependent on factors that are external to the investment. In commercial banking the output depends on the type of bank account and the interest rate, which in its turn depends on market factors.

In treatment biobanks, the output strongly depends on what kind of biobank holds the umbilical cord blood. If it is a private biobank where no other person other than the parents or the child can make use of the deposit, then the "interest" depends on the health of the holders of the "bank account". If they are healthy, there is no pay-back. However, if they in the future are afflicted by certain diseases, they may have an immense return. Therefore, it may be argued that a safe-deposit box – or maybe an insurance – analogy is more appropriate in this situation.

In contrast, if the blood enters a general pool where whoever fits certain criteria can receive it, and where one can get matching blood if one is in need, the situation complies very well with the commercial bank analogy. When we put money in the bank, the use of this money is beyond our control. However, the return is regulated. In the same way that it is not one's original money one gets back, it will normally not be one's own blood that one will receive. As when one puts money in the bank, one runs a risk (physical or social), and one may receive interest (profit).

In research biobanks for umbilical cord blood, the output strongly depends on the terms under which biological material is entered (e.g. whether the person is identifiable or if material is entered anonymously). In any case, the repayment is in terms of knowledge that can become of vital importance in the future or in terms of therapeutic uses. In the case of anonymous contributions, the repayment is knowledge and therapy in general. With identifiable contributions, the profit can be more particular in terms of knowledge or therapy that is of importance for the "account holder", e.g. by revealing important information about hidden diseases that can be treated or information about preventive treatment of dispositional conditions. One interesting issue is whether knowledge is the only currency for repayment. The banking analogy poses the question whether other kinds of repayment than money could be relevant. This leads to another aspect of the banking analogy: the practice of currency exchange.

The banking analogy can also be used to explore transactional issues. There may be many reasons for wanting to exchange biomaterial (and related information) within as well as across national borders. The pertinent question is how are we to conceive of and conduct such exchanges. One perspective would be to apply the analogy of currency exchange. If the biobank material has a certain value, one could argue that it could be converted into bio-currency, a notional token of its potential value. Correspondingly, one could argue for conversion of diagnostic bio-currency into research bio-currency and bio-currency to (analysed) data-currency.

This also opens up for the analogy of a savings account, where one could have an annual or monthly bank statement telling how much one's contribution is worth at present, or a tissue statement giving information on how many items are available, or what kind of knowledge has come out of the cord blood deposit so far. As many (grand)parents used to open a savings account for their newborn (grand)child, parents can now open a "research biobank account" for their newborn child. Accordingly, the money (i.e. the actual tokens) entered in a savings account would not be the same money one gets back. This would also be the situation in the case of a large common therapeutic cord blood biobank, i.e. one would get cord blood back on request, but not the same blood as entered.

In any case the analogy makes it clear that there is a repayment that depends on external factors as well as on the success of the bankers, that there can be an exchange of value, and that it is not the same item of value (money) that we deposit, which then later is returned to us. Additionally, the analogy highlights that there is a small risk related to depositing one's valuable items.

Insurance Analogy

Private companies for commercial umbilical cord biobanking have argued forcefully that what they offer to parents is "biological insurance". However, the explorative function of the insurance analogy has not been utilized at any depth. It is quite clear that the insurance analogy is relevant for the autologous therapeutic cord blood biobank. One deposits umbilical cord blood in the biobank and if something happens, e.g. if one gets a particular disease, one is entitled to a substantial "insurance payback". The payback is fixed according to available technology for diagnosis and treatment, and the benefit can be life saving. The insurance fees, which come in addition to the contribution of the cord blood, are regulated by the number of "insurers", as with many kinds of insurance in general. However, whereas insurance fees are usually risk dependent (e.g. young men pay more for motor insurance), the fee for cord blood biobanking is fixed.

With cord blood, the risk assessment is that about 1:20,000 healthy persons will gain from umbilical cord blood in the future (Brinch et al. 2004), but as with insurance in general, this is prospective and speculative. We still do not know much about how this blood will be used in the future.

Although this is the most obvious application of the insurance analogy, one may argue that although it highlights the risk aspect, it does not address the risk-spreading aspects of insurance. This aspect may be more relevant in research biobanking where the "insurance payback" is not the same cord blood, but knowledge that derives from research performed on such material, and that can become of vital value to each person contributing. If the contribution is anonymous (anonymized), the "insurance payback" is in terms of general knowledge. If the "insurance holder" is identifiable, the "insurance payback" can be more specific. The knowledge gained by the research project can be of vital value for him or her. However, the exact payback is not known when entering into the research, as it is the new knowledge (the "payback") that is the aim of the research. As with regular insurance, you do not know whether you get anything back, actually you hope that you will not (need to) get anything back, and if you get a payback, it may be quite different from what you put into it.

Moreover, it is worth noting that insurance is based on well-known probabilities, and that it is mono-axiological: it focuses only on economical value (including the economic value of uncertainty). This appears to be different with biobanking: the probabilities that you will need knowledge resulting from biobank research are not well known, and the values involved when you need it are certainly not only economic (or easily exchangeable to economic values). These aspects may make the gambling analogy more suitable.

Gambling Analogy

The gambling analogy highlights the aspect of uncertainty in the same way as the insurance analogy without, however, necessarily quantifying the uncertainty. A gambler enters the game because he wants the outcome (and for the fun of it) but seldom on the basis of risk-spreading calculations. Hence, one can enter cord blood into a biobank in order to hopefully win the big prize (life and health or important knowledge).

Correspondingly, a national lottery may be a relevant analogy for cord blood biobanking. In a national lottery you make an explicit contribution and whether you will win (or gain) is a matter of chance. If you win, the gain is substantial, but most people are aware of the chances being very small. In the Norwegian national LOTTO, the chance of winning any prize is 1:138 and the chance of winning the first prize is 1:5,400,000 and the maximum gain so far has been 1.3 million USD. Although the chance of winning (a high gain) is small, for most people the price of the lottery ticket is small as well. That is, the risk which they take is small, no matter how high the gain may be. Furthermore, some gamble in national lotteries not only because they hope to win, but also because they know that the earnings from the lotteries go to common causes (culture, education and sports in Norway). Hence, the gambling analogy also helps to unveil the role altruistic motivations may play in biobanking.

The national lottery analogy is also well suited to bring forth the uncertainty aspect of biobanking, as well as the low risk aspect and to a certain extent also the issue of enrolment. However, other challenges with umbilical cord blood biobanking are less well addressed.

Membership Analogies

Another analogy that may be relevant is the membership analogy. Although Merz et al. have addressed the issue of interest groups (advocacy groups) promoting recruitment for certain biobanks (Merz et al. 2002: 970), the analogy has not been fully explored. There appear to be other potentials embedded in this analogy. We can compare the participation in a research project conducted on umbilical cord blood to becoming a member of an association with a defined goal with which we identify. As a member of Amnesty International in a country with few political prisoners, one enters the organization motivated by ideas of freedom of speech and fair trial, and hopes that, as a member, one can contribute to these goals. As a member, one may receive information about the work of the organization and one may have some membership advantages. The membership (as well as the work) is voluntary and one has no property rights with respect to its results. The results may not be sold and enrolment is achieved by advancing information about the organization and by advertising.

Correspondingly, the enrolment in an umbilical cord blood research biobank may be conceived of as a membership, where one pays a membership fee, receives membership information and has certain membership advantages. One may not claim any property rights to the results of the research, or one may have a certain sharing in the results. Moreover, the membership analogy may be relevant to therapeutic blood cord biobanks as well. In this case, analogies of membership giving certain advantages may be more relevant than analogies of membership in ideal organizations. Several for profit clubs may be relevant, depending on whether the biobanks are for autologous or allogenic uses of cord blood.

Other relevant membership analogies are clubs (e.g. sports clubs, book clubs) and interest groups (e.g. environmental organizations). A particular kind of membership is related to organizations with compulsory membership, such as (in some countries) labour unions, health insurance organizations and military service organizations. Contribution to a common good is made compulsory in many aspects of life, and one could of course argue that donation of umbilical cord blood is an activity of this kind. Hence, donation of cord blood could be conceived of as a compulsory contribution to the common good. It is interesting to note that analogies of compulsory membership will presuppose that the gain from such a membership is substantial or that the cost is low, in order to legitimate its compulsion. The analogy of compulsory membership will therefore fit well with analogies such as waste, where the cost is low.

Another related analogy is the stewardship analogy. This analogy could be useful to highlight common ownership of umbilical cord blood, as it is not obvious whether it is the property of the child or the mother. A third party is entrusted the careful and responsible management of the biological material. The analogy of stewardship could be relevant both to research and therapeutic biobanks.

Thus, some membership analogies may be better for exploring research biobanking of cord blood (the first "common goal" analogies), whereas others may be

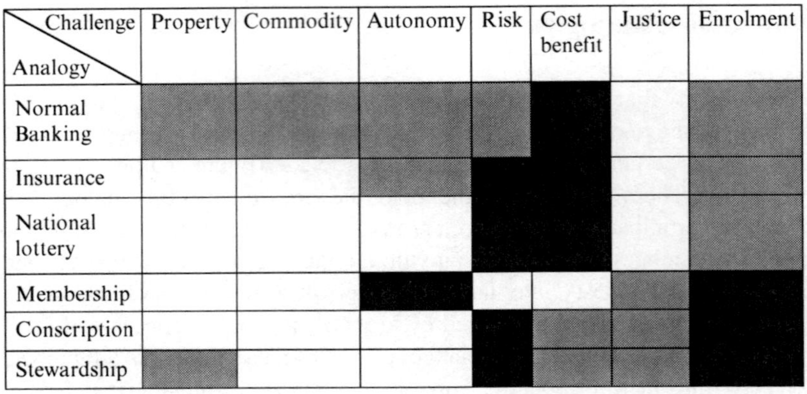

Challenge / Analogy	Property	Commodity	Autonomy	Risk	Cost benefit	Justice	Enrolment
Normal Banking							
Insurance							
National lottery							
Membership							
Conscription							
Stewardship							

Fig. 2 Self-interest analogies and conscription analogies and their ability to address the challenges with umbilical cord blood research biobanks. The assessment is by no means absolute, and is only used as an illustration. *Black* indicates that the analogy addresses the challenge directly, *grey* that it addresses it more indirectly, whereas *white* indicates that the challenge is not addressed by the analogy

better for exploring therapeutic biobanking (the latter "common activity" analogies). The features of new self-interest-based and conscription analogies are summarized in Fig. 2.

Concluding Remarks on Analogies to Analyse and to Argue

This chapter has reviewed a complex set of challenges related to umbilical cord blood biobanking and identified a whole range of analogies that may be used to address these challenges. Moreover, we have investigated a set of analogies that have been absent in the debate on umbilical cord blood biobanking, and indicated that these analogies may be fruitful for exploring (and arguing for) certain conceptions of biobanks so far ignored.

Our analysis has demonstrated that the analogies have both an analytical and an argumentative function. They are used to explore important issues in order to establish a conception of the biobank and to argue normatively in favour of particular conceptions and conducts with regard to banks of this kind.

Furthermore, analogies address ontological, epistemological as well as moral challenges raised by new technologies. Some analogies appear to be more appropriate for addressing and arguing for specific issues, e.g. regulation and management, moral issues and issues of ownership. Correspondingly, analogies are more or less appropriate for handling various aspects of umbilical cord blood biobanking, such as allogenic versus autologous uses, use for research or therapy, private or public uses etc.

Our analysis has shown that any single analogy is unable to address the complexity of challenges involved. On the other hand, analogies can fruitfully be combined in order to address various challenges (Hofmann et al. 2006b). And some analogies prove easier to combine than others, according to whether the issues they address are complementary; e.g. conscription analogies are easier to combine with the waste analogy than with commercial banking. This also shows that analogies that do not fit together because their conceptions or arguments conflict, and that analogies which do not address the issues that are conceived of as relevant and pressing, may be seen as less appropriate or be used as disanalogies.

Consequently, it seems justified to conclude that a variety of analogies are needed to cover the troublesome complexity of the field. Besides, restricting oneself to one analogy in order to understand and argue for a certain way of handling a technology may restrict our conceptions and actions. In order to thoroughly explore a technology, a variety of analogies should therefore be applied. This appears to be most important at the emergence of a technology, that is, before the technology in question has reached a state of conceptual saturation and fixture.

References

Annas G (1999) Waste and longing – the legal status of placental-blood banking. New England Journal of Medicine 340:1521–1524

Annas G (2001) Reforming informed consent to genetic research. Journal of the American Medical Association 286:2326–2328

Askari S et al. (2002) The role of the paternal health history in cord blood banking. Transfusion 42:1275–1278

Ballen KK et al. (2002) Racial and ethnic composition of volunteer cord blood donors: comparison with volunteer unrelated marrow donors. Transfusion 42:1246–1248

Brinch L et al. (2004) Therapeutic use of haematopoietic stem cells from cord blood: medical technology assessment based on a literature review. Oslo: The Norwegian Centre for Health Technology Assessment (SMM). Report 4/2004:60–78

Burgio GR, Locatelli F (1997) Transplant of bone marrow and cord blood hematopoietic stem cells in pediatric practice, revisited according to the fundamental principles of bioethics. Bone Marrow Transplantation 19:1163–1168

Burgio GR et al. (2003) Ethical reappraisal of 15 years of cord-blood transplantation. Lancet 361:250–252

Burns L (2006) Banking on the value of analogies in bioethics. American Journal of Bioethics 6:63–65

Cohen SB (2000) Blood money. Biologist 47:280

Dame L, Sugarman J (2001) Blood money: ethical and legal implications of treating cord blood as property. Journal of Pediatric Hematology/Oncology 23:409–410

Eriksson S (2003) Mapping the debate on informed consent. In: Hansson MG, Levin M (Eds.) Biobanks as Resources for Health. Uppsala, Research Program Ethics in Biomedicine, pp. 165–190

Gluckman E (2000) Ethical and legal aspects of placental/cord blood banking and transplant. Hematology Journal 1:67–69

Harris J (2005) On Cloning. Routledge, London

Hoeyer K et al. (2005) The ethics of research using biobanks: reason to question the importance attributed to informed consent. Archives of Internal Medicine 10:97–100

Hofmann B et al. (2006a) Analogical reasoning in handling emerging technologies: the case of umbilical cord blood biobanking. American Journal of Bioethics 6:49–57

Hofmann B (2006b) Analogy is Like Air-Invisible and Indispensable American Journal of Bioethics 6(6):W13–W14
Holm S (2004) Are all biobanks equal? An analysis of the right of remuneration of donors to tissue and stem cell banks. International Elsagen Conference: "Ethical, legal and social aspects of human genetic databases" & the XVIII European conference on philosophy of medicine and health care: "Genetics and Health Care" August 25–28, 2004, Reykjavik, Iceland. [Oral presentation]
Kline RM (2001) Whose blood is it, anyway? Scientific American 284:42–49
Knoppers BM, Laberge CM (1995) Research and stored tissues. Persons as sources, samples as persons? Journal of the American Medical Association 274:1806–1807
Lind SE (1994) Ethical considerations related to the collection and distribution of cord blood stem cells for transplantation to reconstitute hematopoietic function. Transfusion 34:828–834
Merz JF et al. (2002) Protecting subjects' interests in genetics research. American Journal of Human Genetics 70:965–971
Munzer SR (1999) The special case of property rights in umbilical cord blood for transplantation. Rutgers Law Review 51:493–568
Neal KC (2006) Analogical trends in US umbilical cord blood legislation. American Journal of Bioethics 6:68–70
Nelkin D, Andrews L (1998) Homo economicus: commercialization of body tissue in the age of biotechnology. Hastings Center Report 28:30–39
Nelson RM, Merz JF (2002) Voluntariness of consent for research: an empirical and conceptual review. Medical Care 40(9 Suppl):69–80
Raymond MA et al. (2002) Ethical, legal, economical issues raised by the use of human tissue in postgenomic research. Digestive Diseases 20:257–265
Rocha V, Gluckman E; Eurocord and European Blood and Marrow Transplant Group (2006) Clinical use of umbilical cord blood hematopoietic stem cells. Biology of Blood and Marrow Transplantation 12(1 Suppl 1):34–41
Samuel GN et al. (2006) Mixing metaphors in umbilical cord blood transplantation. American Journal of Bioethics 6:58–59
Senior K (2001) Cord blood: from waste product to valuable resource. Trends in Molecular Medicine 7:3–4
Smaglik P (2000) Tissue donors use their influence in deal over gene patent terms. Nature 407:821
Sugarman J et al. (1997a) Ethical issues in umbilical cord blood banking. Working Group on Ethical Issues in Umbilical Cord Blood Banking. Journal of the American Medical Association 278:938–943
Sugarman J et al. (1997b) Ethical aspects of banking placental blood for transplantation. Journal of the American Medical Association 274:1783–1785
Sugarman J et al. (2002) Optimization of informed consent for umbilical cord blood banking. American Journal of Obstetrics and Gynecology 187:1642–1646
Vawter DE et al. (2002) A phased consent policy for cord blood donation. Transfusion 42:1268–1274
Zilberstein M et al. (1997) Umbilical-cord-blood banking: lessons learned from gamete donation. Lancet 349:642–643

The Alexandria Plan: Creating Libraries for Human Tissue Research and Therapeutic Use

Laurie Zoloth

Abstract After the mapping of the theoretical human genome, and the discovery of the human stem cell, the task of collection begins, which requires the acquisition of large, representative and useful collections of human tissue. Such collection, storage and fair use immediately raise serious questions of justice: ownership, value and distribution. This chapter argues that the metaphor and praxis of "library" should be used rather than the marketplace nomenclature of "banking" for contemporary tissue collections. Thinking of such collections as the twenty-first century equivalent of the great libraries of antiquity will set in place rules that stress justice, access and a careful dignity for our collections. Unlike the term "bank" which set in place associations with markets, secrecy, competition and hierarchy, the "library" will allow us to understand why what we have in common must be held in common for us all. The chapter grounds practice of the creation of such libraries with the moral principle of "hospitality", with its ties to the treatment of the stranger and the nature of reciprocal need, and stresses the public nature of many libraries with a similar call for free and public access to scientific data.

Introduction

To know the truth of the world by scientific inquiry, in addition to being an activity of grace in its own right, is a public moral gesture of justice. To ask for such veracity implies both an ontology and a sociability – there is an otherness to the world that can be known only by observation, attention and listening. When the subject of the search for knowledge is *humanity*, in its multiplicity and its variety, one needs, is thus engaged by, and is in debt to this otherness. To all of the others and the otherness of the world that is the subject of the gaze – such is the reciprocity of research.

L. Zoloth
Northwestern University Center for Bioethics, Science and Society, 676 N. St. Clair St. Suite 1260, Chicago, IL 60611, USA, e-mail: lzoloth@northwestern.edu

J.H. Solbakk, et al. (eds.), *The Ethics of Research Biobanking*,
DOI: 10.1007/978-0-387-93872-1_13,

The speculations, hypotheses, observations and experimentations of the last 30 years have largely explored the molecular structure and function of cells, genes and proteins. Basic research on human genetics has led to a clearer understanding of the map of the world, to be sure, but of most immediate relevance, the research has led to new models of health and disease. It is this that re-links science to its first responsibility, even before we are seekers of truth; we are creatures in need of one another. To do any science, to make the claim of science, to conduct any search for truth, is to make a map of the world, indications of ownership and of permission. To map is to set boundaries and permissions. It is with these new models that understanding can lead from a description of things to a statement of intention of the possibility to change the facticity of suffering and befallen-ness in the human condition. This is the task behind all that is best about basic science research. I state this larger goal boldly at the outset of this chapter not in the service of science's promotion, but rather to achieve two ends. Since it is the claim that the nature, goal and purpose of basic research is not only an aesthetic pursuit, but a moral one, how can this claim be supported?

To do any mapping leads to the next logical step (for it is the structure of the imperial design of science): the collection and the "harvesting" of samples. This means samples of the human person. This act, this amassing of this human collection itself, is the subject of this chapter, for it is my argument that the collection of the collection, is akin to the amassing of the great catalogues of orders, genes and species in an Enlightenment that defined both "science" and "collections". The assemblage of knowledge in logical orders defined how truth was captured in the seventeenth century, and in the twenty-first century, the new genetic collections define our sense of what is true just as surely. It is my argument that the debates about how collections of knowledge would be accessed shaped the creation of the library systems, and the "publication" of knowledge in libraries was critical to the democratization and distribution of knowledge. In so doing, I will draw on three separate "books" or narratives of science to make my case that we are now at a critical moment in human history, in which the rules for collections of genetic material are being created. Thinking of such collections as the twenty-first century equivalent of great libraries will set in place rules that stress justice, access and a careful dignity for our collections. Unlike the term "bank" which set in place associations with markets, secrecy, competition and hierarchy, the "library" will allow us to understand why what we have in common must be held in common for us all.

First, I will argue that it is imperative that the *medical telos* be kept in mind whenever basic science is discussed – thus, it is important to remember that the whole point of human tissue libraries is this urgent medical telos. The tissues are from all, for all, in the case of public data collections, and from some to a private grouping, in private collections, but despite the distributive scheme, the telos is a constant. These collections are not like other sorts of collections of "stuff". Unlike other technologies, and unlike other collections, many of which are directed to the production of goods, or the creation of cultural artefacts, or the creation of wealth, medical science has another impulse at its heart, and it is one of hospitality, with all of its implications of debt and necessity. Science is a sort of knowledge,

I would note, that is "already spoken for", in the sense of Emmanuel Levinas. We care about basic research as moral philosophers not only because of its lovely and elegant beauty, or because we like explanations, but because we understand the world as fundamentally unfair. Science offers a chance for repair, and the field must be called to remember this task.

Second, I will argue that it is imperative that structures for justice, processes for deliberation, and systems for fair access to the social goods of science and the sacrifices it will call forth, must be set in place and fully disclosed long before the social goods are created. Unless both the systems of collection and the systems of distribution are understood, the potential for exploitation and exclusion in science as powerful as genomics and stem cell research is extreme. As these powerful new biotechnologies move towards actual results, and long before most have successfully been tested in clinical trials, the basic and translational research structures must be reflected upon, regulated fairly and made transparent. This research is distinctive in that it calls not only for reflection on how markets or states might structure access to therapies, and how clinical trials on human subjects might be regulated justly, but also that the basic research cannot proceed without the use of volunteer tissue donors. These donors include vulnerable populations such as the ill, and tissues with special meaning, such as gametes and DNA samples. Unlike in physics or nanotechnology, the human person is not only the subject of the research; the human body is the raw material, the reagent, for its enactment.

It is my contention that biotechnology research faces two sorts of crises. The first is a crisis of social forms, social perception and social support. For biotech to advance, new forms of funding, of information acquisition, and new technologies must be developed. New sorts of relationships and social perceptions must be engendered. These include relationships between moral actors who are strangers with considerable differences in power, authority and voice. The second is a crisis of language, in which the power of marketplace analogies and terms has nearly overtaken our views of the meaning of science. Thus, "banks" and "compensation", "competing entrepreneurs", "patents" and "privacy", are used for description of the most basic of processes that undergird science. In this chapter, I will describe a theory of justice and suggest a metaphor for this theory – the "library" as a place and process – for two fields, genomic research and human embryonic stem cell research using human gametes. It is my intent in this chapter not only to describe the history of the debate about justice in research, but to describe how the ethical issues that have emerged in the first decade of such research could be addressed if we could find common language for the sorts of processes needed in science. Here I will define some overt and covert moral arguments that continue to underwrite positions in the discourse and suggest that if we are to focus on justice, a core and shared value across many religions, then the creation and use of international cooperative libraries can allow an overlapping consensus on science policy. In particular, I will argue that the use of the term "library" is not only resonant with a shared political history and ethical values, it contrasts favorably with the current use of the term "banks" and is preferable to terms such as "museum" or "collection". Let me begin with two specific case descriptions of the problem of justice and human tissue exchanges to focus our inquiry.

Case 1: Human Embryonic Stem Cell Collections

Although only a few years have passed since their successful derivation was first published in scientific journals, research on human embryonic stem (hES) cells has received widespread attention both for the significant medical potential such research promises and for inspiring a number of significant ethical questions. A human embryo is destroyed in the process of obtaining stem cells from the inner cell mass of the five-day-old blastocyst and this fact has dominated civic discourse. It is because of this that the first years of ethical reflection on this research have centered on the question of the moral status of the human embryo and the warrant for its destruction in the course of even significant research. As the research has progressed, however, new scientific challenges and achievements have revealed new ethical concerns. One such challenge associated with the development and delivery of stem cell therapies into human subjects and later for use in patients, concerns immunological responses to transplanted cells, or the problem of histocompatibility. Unless the cellular therapy can be made histocompatible, its value would be compromised. Recent advances in possible approaches to this challenge, such as cloning or parthenogenesis, though representing only two of several options that might be pursued, raise the question of whether certain solutions to the problem of histocompatibility might be ethically more or less favorable than others. Hence, this author and other colleagues (Zoloth 2001; Faden et al. 2003) have turned our attention to assessing the emerging ethical issues accompanying new investigation into histocompatibility in human embryonic stem cell research.

The Ethics of Histocompatibility

The scientific and ethical challenges associated with cell line derivation represent only the first step regenerative medicine takes towards the realization of its full potential, for there are many other proposed uses for stem cell lines. In order for stem cell research to proceed successfully towards therapeutic fruitfulness in tissue replacement, however, stem cells must be usable in humans as transplantable tissue that will not be rejected by the transplant recipient. Though the use of stem cells for regenerative medicine via tissue transplantation is only one possible research end, it is, perhaps, the end most highly sought, and the end potentially offering the greatest medical value. Hence, the goal of histocompatibility will, to a large extent, inspire the next chapter of stem cell research. Insofar as the goal of stem cell research is therapeutic, a fundamental scientific challenge is the development of tissue that is not only organ specific, but that can be transplanted into patients without immune rejection.

Since it is understood that the cellular factors that trigger the cascade of rejection are mediated by proteins that are generated and controlled by genetic elements, one way that researchers have thought to resolve the vexing problem of histocompatibility is to develop and store multiple human cell lines or colonies that are derived from different embryos and that carry a wide variety of DNA. This idea is analogous to an existing solution to the problem of histocompatibility worked out by clinical investigators doing bone marrow transplantation (BMT). BMT is indicated for conditions in which replacement of bone marrow in the recipient makes possible a permanent cure from diseases that affect the bone marrow in which blood cells are created (such as leukaemias, some forms of sickle cell anemia, and other cancers). The cells of the recipient are ablated with high doses of radiation or chemotherapeutic agents, and new marrow cells, which are precursor cells to blood cells, are inserted from a donor. The best matches are with monozygotic twins, or other close family in many cases. However, for many, family does not provide a close match. Hence, there is now a Bone Marrow Registry in the United States, which allows a large set of donor HLA types to be collected and registered to allow for a wide set of possible matches from strangers in each particular case.

The promise of this solution is apparent, and has been used successfully with other sources of donated tissue. If pure lines of a wide HLA type variety could be normalized, it is theorized that many successful, reasonably matched donations could be made. If stem cell registries could be successfully established, offering a range of stem cell lines appropriate to the needs of a broad range of patients, this means of providing histocompatible stem cells for transplant offers several advantages. First, this approach to histocompatibility, though it offers only an imperfect typing of donors and recipients by comparison to nuclear transfer, could provide nearly universal access to stem cell therapies. While this imperfect match may result in a greater level of risk than nuclear transfer, it has the advantage of potentially being useful to all patients, including those for whom SCNT is not appropriate, feasible, or supportable in the current health care context, an advantage that the particularity of nuclear transfer lacks. Nuclear transfer, for example might be less useful for those patients suffering from inborn genetic illness. Stem cell collections by contrast could provide nearly universal access to stem cell therapies.

A second advantage of stem cell collections concerns the need for additional tissues. Unlike nuclear transfer which requires the ongoing need for large numbers of oocytes, collected cell lines, once established, would require no new egg donations from women. A significant feature of human embryonic stem cells is that they are "immortalized", that is, indefinitely, or at least, nearly so, and self-renewing. This ready-made scalability, built into the very structure of embryonic stem cells, provides the promise of a self-renewing tissue source for transplantation. The cells first developed by Thompson, for example, have remained stable, with little genetic variation, for the last decade.

At the present time, however, this promising idea faces several daunting obstacles, and hence, developing collections of cell lines remains an imperfect solution to the problem of histocompatibility. Unless there is a perfect match between donor and recipient (a rare occurrence), some form of chemical immunosuppression after the donation will still be required in some types (not all types) of tissue donations. These drugs have powerful and disturbing side-effects that are themselves physically challenging. They must be taken for the rest of the recipient's life to avoid the rejection of the graft by the host tissue (this disturbance in hospitality again).

Furthermore, the planning and selection of the lines presents an ethical and political problem. HLA types are not randomly distributed in human populations, but tend to be clustered in groups of ethnic-geographic origin (Klein and Sato 2000). At this time, nearly 2.4 million sequence variants (single nucleotide polymorphisms or SNPs) have been discovered in the human genome (Brooks 2001) which is likely only a first estimation of distinction. The SNPs that affect HLA tend to be found in patterns of population history. These include population size, bottlenecks in population growth, or a powerful genetic effect in one population group called a "hotspot", founder effects, isolation, admixture, or patterns of mate choice (Brooks 2001). In many cases, the SNPs that seem to determine histocompatibility are linked to ethnic and geographic populations. It has long been noted in the transplantation literature that these different ethnic groups have different responses to the problem of tissue and organ donation. Hence, within this existing BMT system, the needs of minority populations have been difficult to address adequately – for various social, cultural and religious reasons, Asian, African-American and Hispanic populations are significantly underrepresented in donor pools, and recipients in these minorities find it difficult to find BMT for transplant.

This issue has parallels in all forms of tissue transplantation, including organ donation, and has been particularly acute in terms of organs such as kidneys, where African-Americans have a higher rate of kidney disease, and a lower rate of kidney donation (Siminoff and Sturm 2000). This raises a complex issue of justice – both in terms of the responsibility of donors and the distribution of the available bone marrow types to the potential recipients. Despite public appeals for donation, these inequities persist. If we are to create a similar tissue collection of stem cells, a similar number would be necessary, and we anticipate even greater problems of donation. Further, if there is to be a large pool of cell lines for transplant that is available for all members of the human population, there are several daunting political and ethical problems. First is the problem of creating a systematic collection of cell types, and carefully including minority cell lines as well as ones that matched majorities in the population.

At the present time, tissue "banking" is a large national enterprise which requires a large system to support it, and not tangentially, substantial funding. Appeals are made to "target" populations to increase their rate of donation. Blood samples are drawn to seek volunteers for matches, and detailed records on the health status and whereabouts of donors are maintained and updated. Donation itself involves clinical or hospital stays, but is entirely based on the notion that one is donating a portion of one's tissue that is either renewable (as in BMT) or not needed (as in living donor organ donation). Donation of oocytes and embryos raises a different sort of concern.

There is the further challenge of developing a sufficient infrastructure for the recruitment of donors of ova, the collection of donor material, the storing of donor material, and the equitable distribution of donor material. This infrastructure does not now exist. It is beyond the capacity of a single company to bring such an infrastructure into being – such a structure will need careful oversight, regulation and public discussion. The problem of recruitment of minority HLA donors will tend to overlap ethnic groups and the sensitive and difficult issue of egg donation will have to be addressed. Furthermore, in hES research, no one knows how many cell lines will be needed to make a more or less exact HLA match feasible. This will vary by the level of homogeneity in the population. BMT radiation nullifies the initial immune response of the cells. The number may be in the hundreds or the thousands. The closer the matches, the better the chance of reducing the possibility of Graft v. Host disease (GvH).

How many cell lines will be needed to assure close matches so as to serve everyone within a diverse population? How will the collections be created to ensure that small HLA groups also have access to stem cell technology? The answer to such questions might require a widespread donation of "spare" embryos, representative of the population at large. This, in turn, will require a dramatic expansion of IVF – a technology driven by the needs of those with either the insurance or the private means to pay for this costly technology, and to pay for the "donation" of eggs. Hence if only existing embryos deemed "spare" in the current IVF system are used, it is likely that a limited pool of HLA types would be created, as only a limited demographic pool has access to the system in the first place, and since ethnicity and poverty plot closely in America, the pool is disproportionately White and of European ancestry. One could begin with a thorough-going testing of all current donors in IVF clinics. However, these tend to be drawn from two groups: women with the financial means to undergo IVF treatment and whose own eggs are used in the process and women who are recruited to donate eggs for others, who tend to be selected for phenotypic criteria, and who need the money that is paid for the "donation" exchange. Few Asian or African-American women are recruited, which would impact the creation and the ultimate justice of the stem cell collection. Thus, to create a justly constructed stem cell collection will require the creation of embryos to provide stem cells with no intention of the embryos even being used for reproduction. The very act of seeking matching donors implies that the donor is a woman who is willing to undergo a process of hyper-stimulation of her ovaries, and the extraction of her eggs, to create the embryos in question. This implies that far more women and couples would be required to participate in the enterprise than are now used for IVF, allowing embryos to be produced for research alone. For many citizens who understood stem cell research as only using embryos already destined for destruction in any case, the deliberate creation of embryos in order to destroy them for their stem cells is morally reprehensible. For others, including the many families with children or family members affected by disease, this presents no ethical problem. In fact couples, many of whom are past childbearing, have already queried researchers about how to donate eggs for research.

If there were a cell collection representative of only 70–80% of the population, this would constitute a reserve of cell lines potentially useful for a large number of people, but not all people. However, at the present time, even this hoped-for expansion seems unrealistic. At present, we have only sampled less than one percent of the population. We would need to do "ethnic profiling" in order to acquire more hES lines, a politically troubling prospect. Asking women to donate eggs for this purpose is problematic on several accounts. First, it would entail the deliberate creation of embryos for research, a step seen as distinctive and as ethically problematic by many who support the use of "spare embryos". There are also the risks of IVF to women – risks associated with the hormonal stimulus of hyper-ovulation. Hormonal stimulation is a process that takes place over months, affecting both physical and emotional responses, and the retrieval of eggs is highly invasive.

Clearly, even if the above problems were to have been addressed, much experimentation would be necessary to substantiate this approach. It may be possible to use a technique called the "mixed lymphocyte reaction test" that could screen prospective recipients to make the matching more accurate. But such a test may only be predictive of relative success. Much research will still be needed to understand why cells differ in their reactivity and in how the DC react to new settings. Cardiomyocytes and islet cells are very different, for example in their immune reactivity. Islet cells require a closer HLA match than cardiomyocytes and neurons may not require close matching. One would have to test HLA matching in different cell types to establish how close a match would be sufficient, which raises issues of how to structure clinical trials, how to fund such research, and how to protect the human subjects of the trials. This approach in any case would require a large number of embryos to be created and destroyed for both the serious research suggested above and for the stability of the enduring international stem cell collection. Developing more cell lines will require the destruction of more embryos – something that will clearly be strenuously opposed by those who believe that the endowment of personhood is coincidental with the generation of new human life. It is unclear, for example, how long tissues of stem cells can be maintained. However, the creation of a stable "library" of stem cells, if it could be achieved, and if the public accepts the justice considerations that would call for a population-wide set of donated eggs across the spectrum of human SNP variation, does offer the faint promise that if the collection of stem cells could be stabilized (as it is in mice stem cell research), then no further or far fewer embryos would be needed in the process.

Finally, there are political considerations. The idea of developing stem cell collections was contrary to the policy of the Bush administration, for it would require large numbers of cell lines, far more than the 72 presently authorized by the NIH compromise guidelines and surely more than the fewer than 20 actually useable ones. It is not only the United States, however, that has proven uneasy about stem cell research and egg exchange. Indeed, when the ISSCR met for a year to craft international guidelines for the research on stem cells, it was the issue of compensation for egg exchangers that was the most contentious part of the guidelines. In the fragile political environment, such a strategy would present significant difficulties until policy changes support such a proactive strategy. Let us now turn to a second case, in which genetic collections raise different ethical issues.

Case 2: NUGene and Genomic Population Studies

Northwestern University is the site of the second oldest phenotypic/genotypic database in North America, and one of its largest to date. The donors of genetic material are all patients who are members of the faculty physician group, and thus are closely monitored. The data is collected, the medical and genetic records linked and the entire set of data elaborately protected by secret codes, separate databases and computer interfaces. Yet, despite the goal of securing 100,000 participants, in the first six years of collection, fewer than 1,500 have decided to donate even a few cc's of blood, taken when annual blood tests are done.

NUgene is based in the same theory of all tissue "banks", that the collection of many examples of gene–body interaction would be critical in finding etiology and therapy. The goal is to allow genome wide hunts for common genetic diseases. The wider the participation, the better the possible data will be. Data is then sold to biotech or pharmaceutical companies, in addition to other universities, as a research tool. Like the stem cell collections, a diverse ethno-geographic population is needed – hence the appeal of collection in a region like Chicago, a gathering place of immigrants since the early 1800s.

Unlike stem cell banks, in which matching identities is key, the legal fiction of NUgene is total anonymity – a closed system so tightly guarded that the IRB overseers recommended to the project that even if a tangential finding is uncovered by researchers on a project, the patient ought not be identified and approached for therapeutic response (the signal case is one in which a BRCA gene is found, or a marker for Alzheimer's for example).

Why is Collection an Ethical Issue?

Collection of human tissue is said to be an ethically vexed proposition. Yet this is odd, for one of the first human activities of science was the organization of observations, and the collection of objects to observe first hand, often in captivity or dead, to better understand them. Opponents of collections raise a consistent set of concerns to what would otherwise be seen as the equivalent of the earliest projects of sixteenth century science – the amassing of large amounts of biological samples to create a categorical system. The fears that surround science now are considerable, but this is, at least in part, due to the idea that science and its regulation by civic law is the modern surrogate for morality and its regulation by religious communities and their norms. Such an observation has become a commonplace of bioethics. The biological sciences move from observation to creation, the scientists move from scholars to cultural mediators, and the philosophers become bioethicists.

In this way, some of the functions of religion are the "business" of bioethics, but all without the mediation of grace, or love or intrinsic community in religious traditions. The functions of religion are both intimately tied to the present (for example the rules for using the body of others, the place of the market in the life of a community, the uses of power, the proper regulations on sex, the limits on what can be made or known) and they are also always about the future (a world to come, a prophetic call, the consequences of sex, the consequences of justice or its absence). Yet a problem emerges: for the history of religious thought is constructed, in the Western traditions, with the idea of ensoulment. It is this idea that is the most troubling in modernity. In a way, the intellectual move from faith to science is a movement about what matters to us as the "real". This explains why the DNA molecule has assumed such significance – it seems to us to be the actual self.

DNA in this view is ipseity, the very core of the self, the stand in, in our times, for what medievals called The Soul. Thus, once a particular DNA molecular assemblage is established, to destroy it would amount to killing. Collections of tissues have become collections of DNA; hence these collections, once more akin to samples in the British Museum, have begun to take on the quality of the collection of souls, something rather more churchlike. Further, just as the regulation and protection of women's bodies is a concern for the religious, the issue of the vulnerable woman subject is also a concern for bioethicists. Yet, just as the assemblage of samples was a critical part of the early science of biology, the assemblage of genetic and cell samples is a critical part of modern biotechnology. If there is a positive obligation to the duty to heal, then we have duties to persons who are suffering, a religious duty. We have a duty to heal the suffering other, which comes from religion, moral philosophy or intrinsic altruism. If this is the case, then we should not allow persons to block that duty or moral action and thus it is not warranted to block the moral action of healing to avoid the destruction of a blastocyst, or to avoid fears of privacy violation. The collection is both/and – a collection of genetics identities that makes us wary, and a collection of tissues that, in their similarities and distinctiveness, may yet yield critical clue to the nature of disease.

Many discussions about tissue banking end after discussing this impasse. Yet, in the next case, I will argue that our basic research will uncover new problems, and as basic becomes translational research, the problem of identity only begins a larger discussion about the sort of future we wish to live in. Once the collections are assembled, how will issues of just distribution be organized in translational research and therapies?

Case 3: Ovarian Tissue: Disclosure, Fidelity, and Privacy

Northwestern University is also home to a new project that involves taking ovarian tissue from women and girls who are facing the impending loss of their fertility because they have cancer and are about to be treated. The treatments for most cancers (especially pediatric cancers of the most common varieties) are invasive, with significant side effects. While radiation and chemotherapies have changed cancer outcomes considerably, the survivors of cancer and its treatment are unlikely to remain fertile, for chemo and radiation operate by killing newly emerging cells. This largely destroys the capacity to make gametes. Sperm can be extracted from patients facing treatment prior to cancer therapy, and frozen, but extraction of eggs from a woman facing a cancer diagnosis is an entirely different matter, and not just because of the signal difficulty in obtaining egg in a mature state. Eggs, unlike sperm, have to be surgically extracted from a women's body, and moreover, while freezing eggs is possible, thawing and then fertilizing them is still largely experimental. For some, hormonal injections, in vitro fertilization and the creation of embryos is a possibility, but for many, it is not (and, of course, that has also created the issue of hundreds of thousands of stored embryos).

The idea in research ambiences is to find ways of cryopreserving ovarian follicles that can later be thawed and coaxed into creating fertile eggs for IVF. The strategy works in mice and in some primate models. At this stage, the researchers do not know if it will work in humans, but in advance of this knowledge, they are creating a tissue collection from women, both to study and potentially for use if the technique ever becomes efficient and safe. But what is the nature of storage itself? What is this collection, after all – a source for research, or a storage of valuable gametes? Who bears the cost, and for how long? The researchers face a complex dilemma. As the research was initiated, each patient was not only told about the procedure and about the experimental nature of basic science, but that they would be kept fully informed about their tissue samples. In the process of the research, 80% of their tissue is frozen directly after surgery to be available for use should the research experiment ever be successful, and 20% is donated for research. If eggs are obtained and are not used for immediate fertilization, they too are frozen.

Now there is a question of what to tell patients about the fate of the 20% of tissue that they donated to research as the research begins to yield results. Recently the protocol was rewritten, and now, like many protocols in genetic research, all tissue is de-identified. In genetic research, scientists are not specific about the fate or condition of one person's tissue, phenotype or genotype. In NUGene research, many samples are stored and archived along with physical histories, but elaborate codes are kept to maintain complete anonymity. There is no relationship, as is recommended in controlled clinical trials, between researcher and physician. In these cases, if a finding is uncovered that may impact a person's health, the plan is to give a generalized account of the research being done in the lab (perhaps in the form of a newsletter) that would alert all physicians and all subjects about the fact, with general admonitions to seek private, non-research testing of personal genomes at personal expense. This policy has recently been in dispute by some (Church 2008). However, the first set of patients was told that they would be kept abreast of their tissue status, in a personal and direct way. The procedures are not anonymous – they are done by a known physician; the tissue is well labeled. The research team does the informed consent, and each case is personal and contextual. Thus a great deal of information is known about each case. At stake is whether the language and initial promise should be changed. If so, and if the new consent form is used in the future, what is the duty towards the first patients?

Patients obviously have vested interest in the research and the tissue, for if the tissue is bad, meaning it does not mature into viable eggs, perhaps the other 80%

is flawed in the same way and it is not worth the financial strain to store it. It is important to learn exactly how much this storage would cost per year. Who should pay? Shouldn't this be covered by insurance? But if it is not, then patients need to decide if it is prudent or simply wasteful to freeze the 80% for years and years in the hope that: (a) they will be able to thaw successfully, (b) the research will work and (c) their eggs will fertilize. If this seems to be impossible (as was a case with an older woman with a higher FSH; and is the case of a 16-year-old girl whose tissue is bad), then it might be considered unfair to allow a hopeless hope to be maintained at the expense of the research subject/patient. But conversely, the team recently has removed tissue from a different 16-year-old who was only going to freeze tissue, but in that process found two or three mature eggs, and is freezing both. In this case, the subject/researcher might be told that her eggs seemed more viable to prevent her from becoming discouraged. In both cases, the vigor of the 20% of the tissue will affect how they feel about the 80%.

This raises a side issue – what does full disclosure mean? For example, the researchers have told a subject that her tissue was successfully taken but not that they also retrieved an egg and that they have been freezing both sets of tissue because "egg freezing falls into the category of tissue". This is troubling if you are fully disclosing, for you are saying that eggs do not freeze well as a part of why you want follicles, but also because it demonstrates the complexity of the level of disclosure. The team is divided on how to proceed. This is reasonable, and thus needs careful reflection and revisiting. The issue of disclosure may be one of the largest issues they face. There is a conflict between wanting to support full disclosure and a participatory model versus wanting to maintain the sort of anonymity that characterizes other genetic research collections. Patients want to know what's going on with what is an intimate part of their body, and upon which their future is engaged. In fact, the team will need to keep in contact with them for a long time to come and the question will inevitably be raised about whether it is realistic to hope for a genetic family. More issues will emerge, such as the problem of finding that she has genetic diseases which might affect fertility, or that may cause her not to reproduce with her own eggs, such as Tay Sachs, or onco-genes themselves. As more research uncovers more knowledge of genetic linkages, the genetic knowledge in the stored tissues will become more important. The tissue at issue is not the part that is saved for later use, but rather the tissue donated to research. The team does not wish to tell "good news" about the robustness of the tissue, and create what could be false hopes in research that cannot be replicated or developed into therapeutic use in the future. But is there a moral obligation to inform patients that their tissue is not responding correctly, or is "bad", so that they can release their other tissue and avoid years of payment for freezing and years of false hope about their tissue? What if five years hence a new breakthrough is made and the "bad" tissue is no longer "bad" for it could be stimulated in some different manner? Once you begin to speculate about the future, either good or bad, then you are engaging in speculation of the most ephemeral sort. The fact that the knowledge base is mutable on all fronts means that even characterization is difficult and potentially misleading. It is not as if there is any factual support or firm data in these cases.

Moral appeals are made from many directions in this case. The structure of genetic research to which this is being compared was based on considerations of privacy and objectivity. To avoid bias, clinical trials are largely de-indentified. In libraries of genetic material, the argument is made that the material is data points and should not be linked to persons lest that knowledge influence the science in some way. But this research cannot be so separated. In genetic research, if it is linked to identities, it would allow undue exposure and knowledge of risks that may adversely affect the person, since gene identity is linked to the self so profoundly. This is a reality in this case as well. Thus, there is precedent for de-linkage and information being made general to the research population as a whole. In some collections, the processes seemed flipped – thus, the more information was potentially available, the fewer chances for disclosure were offered. Patients were told only of a general advance in knowledge. In the NUGene project, when new genes and linkages were discovered, fear was such that patients would be upset if their DNA could be traced, and some researchers felt that this "upset" was more harmful than telling them that they were at risk of disease. Patients were "re-consented" with the new protocol via a mailed post card when the decision not to inform patients of genetic variants, even harmful ones, was made. But some can argue that the tissue once separated from the patient as "research material" is now to some extent in a separate category – it is more akin to any other tissue or material used in basic research, such as HELA cells. The knowledge base is mutable and speculative – the entire category of "good" and "bad" is subject to this, as is the ultimate outcome of the research. Perhaps actual informed consent would merely focus on this issue. People change their minds in research and can withdraw participation but this can be affected by their sense of outcomes, for unlike research in which the patient's body is affected, the body of the patient will change, age, etc., and the tissue also changes in that more will become known about this and the freezing process. Patients, research subjects and tissue donors are all promised different things and the relationship varies. Their issues ideally should not be confused. Yet, in this protocol, women are subjects, donors and patients – at different times in the course of the experiment. This raised the core ethical problem for this particular tissue collection. Is it just for Northwestern's scientists to act as the broker for families after fertility, or is it only using the tissue for a basic research trial? Are the doctors still the doctor of the patient? What sort of collection is the collection of ovarian tissue?

Endemic Problems: Money and the Secret Body

The fears about collection, and the cultural anxiety associated with this are centered on two issues: privacy and power. That the issue of privacy and its antidote, informed consent, have dominated the reasons to oppose collection in principle is a reflection of late modernity. This is true for two reasons. First, the idea of a private life is a late addition to human history. The very idea that one's illness could (or should) be a secret will, a commonplace in large anonymous urban populations, was not the case in antiquity. Second, because the idea of the state and secret policing of one's inner life is a resonant trope in modernity. The structure of consent gives a sense of control (although this may be an illusion, for evil doers likely to use knowledge for harm or control are unlikely to respect such a process). The second endemic issue is the issue of the marketplace: is it just to take material from human bodies and use it for profit? Is it ethical to create a free market solution for tissue extraction and exchange? How should persons exchange aspects of self and why? One could argue that in these instances the issue is not the nature of genetics, but the nature of secrets (and perhaps more to the point, secrets of sex and family); and the nature of property exchanges; and the limits of use of the body of the other; and the way one must treat befallen-ness; and that such matters have long been discussed as religious legal concerns, not genetic novelties. In bioethics, the strongest resistance to strong autonomy-based argument arises from moral philosophy and religion.

Each question has a justice problem at its heart; each is situated in an unfinished context; for health care, like other social goods, is unequally distributed. We are led to ask: what sort of an act is it to exchange parts of the self? This is an epistemic question. How we *know* the nature of the act is dependent on one's metaphor by which we name it. Each exchange had a particular history, which lead to certain justifying arguments, and most reflect tension between libertarian and egalitarian theories of justice. Such an act involves more than the act of collection, for once the data or tissue is collected, it is manipulated, and thus participation involves cascade of moral gestures. It is an act of moral citizenship. It is the creation of a context for rare and protected acts between strangers.

How should we create a theory of justice for such an exchange? What should we call the theory? What are the institutions called forth by such a theory? Arguments from different histories drive debates. For example: organ or blood donation, egg exchanges in IVF clinics, or research on human subjects are all likely examples. In organ donation, another new sort of relationship between strangers involving one's body parts, the marketplace, is carefully excluded to assure public support. Blood donation, once a marketplace exchange, was due to the possibilities of abuse rethought and changed to a gift relationship. In IVF, a new entity created a new sort of human embryo, but a human embryo nonetheless. We also create the disaggregation of the process of reproduction. Is our collection of tissue like experiments on healthy volunteers?

Moral Philosophy Means Many Arguments are Valid

In thinking about each of these cases, all of which involve some new aspects of tissue or genetic material collection, it is important that alternative arguments be voiced. This includes the voices of religions which focus on the duty which we owe the stranger whose proximity and vulnerability is the context of our possession and desire; the question of the just limits of the free sphere of exchanges in a world given to our care; the question about the worth and merit of relationships in light of essential social contracts we regard as decent and just. Hence, the moral question considers the act, the participants, the context, the alternatives, the history of such action, and the value of the exchange. For example, one could argue for a justice of the market.

The marketplace argument for how to handle the collection of human samples is coherent and dominates. However, those making it should then be consistent. Our "marketplace" takes place in a tragic world that must be negotiated with decency and justice. We create collections, but unlike earlier "explorers" we ask: "what sort of world do we create with our research?" It is in the name of this question that certain relationships or institutions may be blocked as exchanges. There is a role for blocked exchanges and blocked uses of power, especially in powerful new sciences.

The question of how we organize collections is linked to several deeper issues: if we believe that it is good to donate tissue for science, then how are we to motivate good acts? (Is the market the only way? Is everything for sale, or are there some acts outside the market?) Our duties to donate to collections arise largely from our duty to attend to suffering and happiness which are marks of our life in social world. Since Kant, and the Scottish Enlightenment, philosophy has concerned itself with the limits of the marketplace relative to the human body – what can be bought and sold with justice – since everything (of course) can have a price.

Collections are organized using the two solutions offered in positive law: to create the best standards in the market (labour laws, safety, unions, regulations, anti-trust) or to bracket some things as outside the constraints of the market. Collections of human material could be seen in different ways. In Greek philosophy's understanding, the three fields of human interaction are *gens, socius* and *universitas*. If we see collections as a matter of social or academic interest, it would allow for supererogatory acts (volunteers, exchange of organs and tissues, charity, libraries, education, and health care). Collections also raise implied issues of use. Once societies have all these social goods/things/products/treasures/knowledge in one place, how are they to be used and by whom? How do societies achieve justice in research?

This question opens into others and assumes several complex and different questions of justice: What research should be funded? How should lines be collected? How are patents and licenses established fairly? Or how would open access work? What is the best way to establish a system of distribution? How does a society decide what is just? In a world of scarcity, how ought a society to justly distribute scarce goods and services? In light of the particular and poignant crisis of health care what would be the language of such choices? How can states be accountable for justice? How can an international community reflect on justice? Should collections be regulated on standard candidates for material principles of distribution, such as numerical equality, need, individual effort, social contribution or merit?

Further, all our theories share in common the presuppositions of the liberal tradition. All rest on the assurance of the primacy of the individual person with liberty, rights, duties, and the ability to engage in voluntary consent. This existed prior to the social contract itself, the social contract that is entered into by rational free agents operating from an original position that was either historical or hypothetical, that created the liberal state. Many of our ideas about collections arise from our attachment to liberty, private property, and entitlement, which immediately raises the problem of ownership and the rights of each individual to own his or her own resources.

But in tissue collections, what are we to make of genetic difference and injustice? If we turn to egalitarian theories of justice then other issues are at stake. Thus, each of us has inescapable and essential rights and obligations towards one another that cannot be ignored and our rights, obligations, duties and needs arise from something we share as persons, which is common to all, and must be respected by all. Our commitment to equality is based on an ability to make rational choices that honour this equality at the heart of this theory of justice. First among these duties was the notion that justice was rooted in equality, equality due to the basis of shared human embodiment and participation in a mutually consensual human society.

While imperfect, liberal theory and its attendant reliance on the contract autonomy of consent and private decision-making were the basis for many health care dilemmas. Research on tissues followed the same norms. For example when a particular patient, Ms. Lane, had a particular avid cancer tumor whose cells grew rapidly and thus were useful for research, the cells were seen as a commodity that could be exchanged. HELA cells are an example of how pure research (*universitas*) distributes social goods. Ms. Lane's tumor was removed in cancer surgery, no consent or knowledge to the family was seen as needed, thus a pure transformation to commodity as a cell line was made, and there was free use in all labs. But as my examples illustrate, gene "banks" are far more complex, and the rise of biopharmaceutical industries have made such collections more necessary, more valuable, and more socially impactful. Along with the rise in need and interest (here that "bank" language assumes real potency), comes the consideration of justice.

Here is the statement of our problem: How can we set in place a fair and just system of access to the good ends of medicine and science implied by collections of human tissue, using a fair and just process that protects donors and recipients and aims for fair and just goals for humanity? Given that we understand that we live in an unjust world, whose problems cannot be wholly solved by this or any single institution, how are we to regard what may be our generation's most important science infrastructure project, the amassing of genetic samples and stem cell lines for research? It is with the premise of the next part of this chapter that we must begin, as do so many human projects, with naming correctly.

The Alexandria Solution: What We Have in Common. A Theory of Libraries as a Theory of Justice

A theory of justice could create an alternative to the market-based solutions on which the current collection of genetic material is made and organized. It is the premise of this chapter that common nomenclature is important, for it sets in place the possibilities for a common future. This section will argue that a justice-based solution would lead us towards such commonalities and not towards increasing rigid codes for how we contract between fearful strangers. Thus, let me argue that the best model for genetic "banks" are not banks at all, but libraries, long the way that human societies organized and shared knowledge.

The paradigm of exchange in the marketplace is always one in which gain and loss are calculations designed to be weighed and assessed. The library is both a place and an event – a venue that is not of the marketplace, nor entirely within the marketplace, as is a bank, but one that exists in societies in which marketplaces and libraries are both a part of well-regulated civilizations. In fact, many new uses of the Internet in science (PLOS, the turn to web-distributed journal, etc.) use developing models such as open source donations and withdrawal systems which are characteristics of library use. Let us state that one of the clear goals of genetic or stem cell libraries is preserving lines of human genetic material for universal research and perhaps therapeutic use. This is not an unfreighted claim, but let me state it plainly: while commerce is not an evil activity, the making of profit from genetic collections ought not to be the point of such research. Libraries are full of books that people were paid to make (write, bind, print, illustrate, etc.) but the premise of the library – as opposed to the book store – is that the objects enable the generation of ideas, things, and new research, but that the library itself is not the location of profit and loss. There are rules about the limits of use, fees and penalties for misuse, for unfair holding, or plagiarism, and these rules and norms exist for one purpose – to maximize circulation to the greatest possible number of citizens. Thus, libraries are tied to human liberty, creativity and citizenship. Libraries are a mixed economic model: authors are paid. Yet, authors write in part to create a map of the moral universe, authors like to think about narratives and find the work rewarding and socially useful. They are paid to publish books, and they have copyrights on the work and get paid if others wish to use them, but in a greater sense, the writer of any piece (including, for example, this very one) is also engaged in a moral activity, giving the text to the other, for the intellectual and moral possession of the other. Publication is, in the Hegelian sense of the word, "witnessed speech acts" that function as signs of identity, and the identity is as unique as one's physical DNA.

Libraries are public in another way. Like "big science" and large genetic banks, the enterprise is too large and complex to be private, and donations and taxes all are used to create them. Governments at different levels – federal (Library of Congress), state and city officers set up rules for use and sharing within the library, and every community has norms of conduct for use. But all share these qualities: one must set up a library based on internationally accepted rules for order (Dewy Decimal, or Library of Congress); they prohibit and punish malfeasance, and set fines; one must be a card-carrying member and be screened at some level; this membership can be withdrawn; the users of the actual product use it for free, and borrow books, learn from them, they tell their stories or create with the knowledge and thus, knowledge is "produced" based on this publicly shared set of commodities.

The library functions as a sign of other things: in a city, it is a neutral zone for public discourse, for events at the city library demonstrate a certain citizenship. It serves as a core moral teaching of honour and fidelity for there is an honour system for borrowing, use and return that cannot really be "enforced". It uses an open stack model to create a free association between the users and the knowledge. Thus, the ethical value offered by the library is not "autonomy"; it is "hospitality". Libraries are hospitable because we agree that justice and fair play are best in play when knowledge is openly shared and that fairness, citizenship, and public voice in democracy need such centers for the outworking of citizenship.

The Great Library in Alexandria

The Great Library of Alexandria, Egypt was one of antiquity's major intellectual projects and it still contends with any modern project of scholarship in its importance and vision. Ptolemy III of Egypt mandated that all visitors to the city were required to surrender all books and scrolls they possessed before they entered; the scrolls were then swiftly copied by official scribes:

> Sometimes the copies were so precise that the originals were put into the Library, and the copies were delivered to the unsuspecting previous owners. This process also helped to create a reservoir of books in the relatively new city. The Ptolemies also purchased additional materials from throughout the Mediterranean area, including from Rhodes and Athens. According to the earliest source of information, the pseudoepigraphic *Letter of Aristeas*, the Library was initially organized by Demetrius of Phaleron. Demetrius was a student of Aristotle. Initially the Library was closely linked to a 'museum', or research center, that seems to have focused primarily on editing texts. Libraries were important for textual research in the ancient world, since the same text often existed in several different versions of varying quality and veracity. The editors at the Library of Alexandria are especially well known for their work on Homeric texts. The more famous editors generally also held the title of head librarian. The geographical diversity of the scholars suggests that the Library was in fact a major center for research and learning. King Ptolemy II Philadelphus (309–246 BC) is said to have set 500,000 scrolls as an objective. Mark Antony was supposed to have given Cleopatra over 200,000 scrolls for the Library as a wedding gift. These scrolls were taken from the great Library of Pergamum, impoverishing its collection. Carl Sagan, in his series Cosmos, states that the Library contained nearly one million scrolls. (Wikipedia 2007).

The Carnegie Libraries

The Great Library was destroyed by fire, but the idea of a library as the collection of social resources was maintained throughout the classic and medieval periods with the Church acting as the repository of all manner of scrolls. The Vatican's collection, and increasingly the collection of royalty and wealthy families allowed all that was known of wisdom to be saved across generations, even at times when few could read. The amassing of huge personal and private libraries became a part of all wealthy homes and libraries as private collections flourished. The idea of the library as a democratic, public arena is linked closely to the American democratic movement. A new vision of the library emerged with the beginning of the new Republic, supported by Benjamin Franklin, who, as a printer, had access to most books published in the colonies and who advocated for the public library. What made the idea of a "free library" in every town possible were the reading habits of a highly literate immigrant class and the convictions of one man.

Andrew Carnegie was an impoverished Scottish teenager who fled a jail term for labour organizing, and arrived in the US aged 13. He worked as a labourer, then a telegrapher of unusual skill on the railroads. There he learned how the systems and schedules worked, and when the Union army needed to ship large numbers of men to the front lines quickly, he was able to understand the power of the new rail technology in war, and organized US troop transfers in the Civil War. He parlayed his savings into another new technology needed by the war and created and owned the first steel mills, which enabled the infrastructure and the guns needed to defeat the South. He retired at age 40, as one of the largest and most successful businessmen in America having made a huge fortune by a combination of grueling work and very good timing, and went back to England, this time to study at Oxford. Carnegie is thus famous, not only for this classic rags to riches narrative, and for his war time business profits but also for two tragedies in American history and for one deeply redeeming moral act. The first was the Johnstown flood (1889) in which 2,200 people died after their entire town was swept away when an earthen dam built to provide a private lake to Carnegie's hunting club gave way. The second was in 1892 when a 143-day strike at Carnegie's steel mill was crushed after a private army of Pinkerton guards and state troopers advanced with weapons on unarmed strikers, women and children. It was a record so hated in many small towns, that later, town councils would refuse his charity out of protest for these practices. Yet he wrote and published a widely read tract called "Wealth" in which he advocated for the duty of philanthropy. One of the many philanthropic projects was his pledge to create and fund a library in every small town in America:

> Of the 2,509 libraries funded between 1883 and 1929, 1,689 were built in the United States, 660 in Britain and Ireland, 156 in Canada, and others in Australia, New Zealand, the Caribbean, and Fiji. Very few towns that requested a grant and agreed to his terms were refused. When the last grant was made in 1919, there were 3,500 libraries in the United States, nearly half of them paid for by Carnegie without any public funds. In the early 20th century, a Carnegie library was the most imposing structure in hundreds of small American communities from Maine to California. Contrary to the belief of many people, most of the library buildings were unique, displaying a number of different Beaux-Arts and other architectural styles, including Italian Renaissance, Baroque, Classical Revival and Spanish Colonial. Each style was chosen by the community and was typically simple and formal, welcoming patrons to enter through a prominent doorway, nearly always accessed via a staircase. The staircase was intended to show that the person was elevating himself. Similarly, outside virtually every branch a lamppost or lantern symbolized enlightenment. (Wikipedia 2008).

Nearly all of Carnegie's libraries were built according to "The Carnegie Formula" which required the town that received the gift to demonstrate the need for a public library, provide the building site, and annually provide ten percent of the cost of the library's construction to support its operation. It was a massive gift, the equivalent of approximately $2 per person (Wikipedia 2008).

Using this history, and using the language of the "library" and not "bank" ties our contemporary effort to amass huge collections of our most valuable resources to our most serious traditions of democracy and civility. Libraries work, unlike the marketplace, because of the ethical norms of reciprocity, hospitality and sacrifice. It is, of course, sacrifice that is at the heart of donation of tissue in the first place, and this gift is obscured if it is cloaked and shadowed by a contract relationship as if it were a business deal and as if the main moral activity is of secrecy, privacy and non-disclosure. It is hospitality that is at the heart of long-term projects for the social good. Such theories of hospitality and sacrifice imply a long-range view of "the good", one that will take generations to see. We see the idea of intergenerational trust in the great libraries of civilization, all of which take generations to build, many of which are housed in building planned in one century and completed in one or two others. The telos of the project was the incentive, not immediacy of the utility value. There is, as well, a value in living in a just world with rational projects such as libraries and collections such as these.

A theory of the library enacts a theory of duty, with no "right" such as ones for privacy or profit, but rather the expectation of generosity. The term "library" implies a curious geography – when one sits in a library, say a small Carnegie library in a small town in the American plains, or a crowded street in an urban center, one is both "home" in one's local neighborhood and internationally connected to the life of the mind – everyone's intellectual neighbor. Is such a neighborhood possible? How will we be neighbors to one another? How is the next door of the future affected by the needs of the neighborhood? All of these are the questions in play just under the nomenclature.

A theory of the library does not solve every question: for example, who has access to the knowledge as it is still unfolding, and what the status of the participant is. For projects such as NUGene, Oncofertility, and stem cell research, the use of the term "library" suggests other ways of thinking about it, however.

Justice is the ground for the law, but hospitality, I would conclude, is the ground for medical research. Thus, I would strongly argue that an "International BioLibrary" should be our goal for the Great Collections of the twenty-first century, as ambitious as the organizers of the Alexandria library, whose insistence on participation marked the place of civic entrance, and as democratic as the builders of the Carnegie system, whose devotion to communities has created a venue for thoughtful engagement and a place of civic duty, that endures today.

References

Brooks LD (2001) Developing a Haplotype Map of the Human Genome for Finding Genes Related to Health and Disease. Background paper for the Haplotype Mapping Project, NIH, July 18–19, 2001

Church G (2008) From Genetic Privacy to Open Consent. Nature Reviews Genetics: Perspectives http://arep.med.harvard.edu/pdf/Lunshof08.pdf. Accessed Dec. 1, 2008

Faden RR et al. (2003) Public Stem Cell Banks: Considerations of Justice in Stem Cell Research and Therapy. The Hastings Center Report 33:13–27

Klein J, Sato A (2000) The HLA System – Second of Two Parts. New England Journal of Medicine 343:782–786

Siminoff L, Sturm CM (2000) African American Reluctance to Donate: Beliefs and Attitudes about Organ Donation and Implications for Policy. Kennedy Institute of Ethics Journal 10:59–74

Wikipedia. Library of Alexandria. Wikipedia 2007 http://en.wikipedia.org/wiki/Alexandria_library. Accessed Dec. 1, 2008

Wikipedia. Carnegie library. Wikipedia, 2008 http://en.wikipedia.org/wiki/Carnegie_libraries. Accessed Dec. 1, 2008

Zoloth L (2001) The Human Embryonic Stem Cell Debate: Science, Ethics and Public Policy. MIT Press, Boston

The Art of Biocollections

Anne Hambro Alnæs

Abstract This chapter examines and discusses certain similarities and differences between established national art collections and evolving public biobanks. Such a comparison has the merit of sharpening our awareness concerning the rights and duties pertaining between collectors and donors. Tracing the way in which some works of art have been acquired in the past, and considering more recent examples of bioprospecting, it becomes evident that collecting exists along a continuum from people's altruistic donations, via deposits, to commercial acquisitions, as well as illicit appropriations hardly discernable from confiscation and theft. Comparing collections of biologicals with art galleries shows that analogies are polysemic and depend on being interpreted in line with some, but not with other connotations, if they are to add to our understanding. Both national art galleries and depositories of biologicals represent iconic and indexical representations of considerable value for future scientific research and as archives for posterity. It is up to future researchers to unlock the as yet unknowable information embedded in present biological depositories. This chapter aims at shedding light on which rules for preserving, dissolving, selling, or abandoning different kinds of collections should prevail. Analogies have a didactic potential, which at the same time carry normative implications.

Introduction

The uniting theme of this book is how to develop analytical tools to understand, define, and discuss ways of conceiving and managing depositories of human biological material. As the Norwegian law on "biobanks" implies (Norway 2003), the potential of such material is threefold: to increase diagnostic competence, to develop innovative therapy, and to enhance research endeavours to further medical

A.H. Alnæs
Section for Medical Ethics, Faculty of Medicine, University of Oslo, Oslo, Norway
e-mail: ahambro@online.no

J.H. Solbakk, et al. (eds.), *The Ethics of Research Biobanking*,
DOI: 10.1007/978-0-387-93872-1_14, © Springer Science+Business Media, LLC 2009

knowledge for future generations. These three kinds of biological depositories, subsumed under the term "biobank", render it somewhat difficult to capture the essence of these fledgling institutions (i.e. their ontology) with a single expression or analogy, as there always will be features in what is sought compared, which draw in different directions. However, the aim of an analogy is *not* to demonstrate a homologous relationship between the known and the unknown, but rather to discover differences in likenesses, and similarities between what *appear* to be different features. If not, the analogy would cease to serve the purpose of functioning as a heuristic tool.

In the present chapter, I want to explore the connotations elicited by existing nomenclature, "biobank", and to compare these with the implications and undertones embedded in art collections. These may be, I suggest, better suited to convey the creative potential of the different *kinds* of value as well as a number of immaterial *values* that are at stake in the expanding "wealth" of stored biological material. Perhaps a term like "biogallery" could serve as a heuristic device to bring forth other practical as well as ethical dimensions in the wake of establishing biological depositories?

When discussing the merits and disadvantages of the term "biobank", we must of course realize that we have already long passed the time of its birth, and the joining of "bio" and "bank" cannot now easily be undone. The choice of inputs to this metaphor was, I will argue, non-arbitrary and inherently a priori value laden. However, the point is not to replace the common use of the term "biobank" which is well ingrained; rather to examine whether alternative terminology could sharpen our awareness about the non-obvious, and perhaps dubious, associations emanating from the term biobank (henceforth without inverted commas). By suggesting a notion like"bio-gallery" I intend to compare the repository of human biologicals[1] to public art collections. National Galleries are the property of a nation's citizens, which, according to most institutional statutes, cannot sell its pictures or sculptures, or exchange them for other art valuables in the open market. These are in a double sense price*less*, even though the cost of insuring them when they occasionally go on tour to other galleries temporarily necessitates evaluating individual works of art in economic terms.

People's concerns about biobanks are partly due to an uneasiness some experience about the possible commodification or misuse of tissues derived from our bodies. Biologicals convey information about our "inner" selves, which many do not wish to make available to others. The possibility of identifying genes predisposing individuals to a variety of diseases, and cross-linking this information with other data, such as health care files and insurance policies, causes anxiety and raises questions about consent, benefit sharing, the regulatory power and possible politicization of ethics boards and data protection. However, as Søren Holm (Holm 2007) provokingly points out, if doctors are allowed and expected to pass on information about a would-be insurer's health record to the company in which he wants to be

[1] A useful term to cover the plethora of biological materials, introduced in the early 1980s. See Landecker 1999: 204.

insured, there is no principled argument which should make it legally and ethically unacceptable to inform these companies of that person's genetic predisposition.

The use of analogies, metaphors and metonyms affords a way of probing into something unknown. This is a bold step to take for representatives of the exact sciences, such as medicine. Analogies are often used as rhetorical devices to reassure people that the "newness" of a concept is not at all threatening, but instead, rather close to something everyone is familiar with – only dressed up in new clothes. Hofmann et al. point out that, if the purpose of an analogy is to instill a certain conduct, the closer the two analogy components are, the more persuasive the comparison will be (Hofmann et al. 2006).[2] If we on the other hand prefer to use analogies as a way of exploring something unfamiliar, then a modicum of distance between the familiar and what is sought explained has greater potential. However, I will claim that analogies which make us see new connections are able to combine the two purposes just mentioned, even though analogies at times may be cognitively and emotionally demanding. Analogies, metaphors and models rarely involve a unidirectional transfer of meaning from source domain to target domain, but rather affect each other in a two-way, reciprocal manner. Metaphors that "work" consist of an imaginative bringing together of words and ideas stemming from different domains. As a result, and in hindsight, this subtly changes and expands their original meaning. Lakoff and Johnson describe the imaginative merging of words as "cross-space mapping" (Lakoff and Johnson 1980).

An expression such as the "living dead" can serve as an example: it is built on the Christian promise of eternal life after death in this world, a notion familiar to most people in the Western cultural sphere. It has since been transported into the secular realm to refer to patients who are neither alive, nor quite dead (according to traditional standards), i.e. those *liminal* persons evolving in organ donation situations, in the era of transplantation medicine.

For those involved in transplantation medicine, the "living dead" is a far less offensive term than "cadaver-donor" (commonly used in transplantation literature under the acronym CD); besides, it is better suited to capture what many still regard as the mysteries and miracles of transplantation medicine. The innovative effect of the living dead metaphor is that the decisive factor for being declared dead have changed (from cardiovascular to brain death), just as the criteria for being considered alive nowadays no longer depend exclusively on maintaining a heart beat, but rather on blood circulation to the brain.

If the analogy between national art collections and central depositories of biologicals is to have any leverage, we first need to agree whether possible similarities between works of art and human biological material[3] may have any argumentative

[2] For this, see also chapter "Mapping the Language of Research Biobanking: An Analogical Approach".

[3] The German pathologist Gunter von Hagen did something which can be compared to Duchamp's historical elevating of the urinal as a work of art. Through a plastinating preservation technique he developed, von Hagen exhibited a collection of dead persons' bodies in postures from everyday life, such as sitting at a desk, running or driving a motor bicycle. He also displayed aborted embryos.

hold. A patient's blood and tissue samples are clearly not the result of artistic visions. In the world of medical research, cells and tissues have to be extracted, isolated, preserved and developed through the re-working and analysis by laboratory technicians, pathologists or haematologists. It is only the reworking and interpreting of what was our bodies' physical materials which makes them scientifically valuable and hence, in a certain sense, comparable to works of art. In the hands of experts, otherwise perishable biological tissue is metamorphosed into specimens to be inspected and analysed for their morphological properties and biomedical patterns. It is the adding of fixants and transformation by various biomedical techniques, that prevents biologicals from degrading into non-informational waste.

The fact that the establishment of national galleries is generally linked to the era of nation building does not render the concept obsolete for comparison purposes. On the contrary, the shifting patterns of national boundaries, e.g. the rebirth of countries which until 1989 formed part of the Soviet Union, such as the Baltic countries, show how establishing national identity through carefully targeted media-footage, e.g. about the uniquely heterogenic or homogenous character of one's population, remain an inherent aspect of nation building. Collecting and organizing the genetic pool of one's countrymen (as took place in Estland and Iceland) may similarly be seen as capturing the vitality of a dormant, but nonetheless national, resource. Indeed, this can be understood as the post-modern equivalent of the way people in previous centuries regarded their painters', composers' or writers' works as symbols of national identity.

For the reasons mentioned so far, it is timely to imagine alternative scenarios for how biological collections should be conceived and managed, before the national, or transnational policies (such as EU-legislation) governing them become immutable and fixed in issues of national prestige. When I use the word "imagine" it is to emphasize the importance of sharpening one's antennae for what may lie around the next innovation bend, instead of deducing and solving problems solely on the basis of other institutions' experience. As a well-known phrase reminds us: all comparisons are odious, which means that we must be wary of exaggerating the similarities at the expense of important dissimilarities between the associative components in analogies and metaphors, such as *biobank* or national *bio-gallery*.

Overview

In the following I shall first present some arguments for why I consider biobank a somewhat unfortunate way of referring to collections of human biological material subsumed in the term biologicals. Second, I will point out some perhaps unobvious (a) similarities and (b) differences between the familiar (art collections), and what is sought explained (the depositories of biologicals). Third, I will probe into the various kinds of relationships that exist and are reproduced between valuable objects, their collectors and custodians, and the "consumers" of these objects. I refer to "consumers" in a double sense: (1) figuratively, as when people visit and view works of art, or art historians who research and write articles and books on the basis of works of art; or (2) in the case of biologicals, literally, in connection with biomedical researchers and the recipients of therapeutic substances such as e.g. blood or bone marrow. Fourth, before concluding, I shall consider various kinds of opposites to capture different paired ways of parting with and acquiring objects of value.

Problems Related to Terminology

First, why am I to a certain extent sceptic as regards the term biobank? Fundamentally, it is because we here are dealing with a neologism which masks its unacknowledged metaphoric construction and therefore risks narrowing our view. An indication of tropes' seductive and persuasive power can be seen in the way politicians use metaphors. On a more general level, when we hear speakers using metaphors we see the issue being discussed from the speaker's point of view. We are enticed into following his or her arguments and hence tend more easily to agree to conclusions drawn.

The term biobank was clearly not created ex nihilo. I presume that the concept was born with intended alliteration, like in *b*lood *b*ank, *b*ody *b*uilding and in *b*ed and *b*reakfast. Catchy coining makes new expressions easier to remember (compare biobank to e.g. "biogenetic storehouse", "biotank" (ref. "think tank") or "biotarium" [reminiscent of other well-known concepts like "planetarium" or "arboretum"]). Moreover, the coupled *b*s in blood bank and biobank facilitate the design of visibly pleasing and easy to remember logos, to be used in recruitment campaigns.

In a cautioning commentary to Hofmann et al.'s article about the analytic usefulness of analogies, López (2006) points out that the French Commité Consultatif National d'Éthique (CCNE 2003) was careful not to use the term biobank and instead studiously referred to "*collections* of biological material and associated information data" (my emphasis). By examining the construction behind taken-for-granted metaphors such as biobank, we can appreciate how politicians as well as representatives of biocapitalism – such as the pharmaceutical industry – use neologisms to subtly persuade the public into accepting *their* point of view (here: the primarily positive potentials of biobanks).

Neologisms, especially those created in the absence of existing forms, are often constructed *experimentally* and aim to capture what initially is seen as the essence of the new entity. However, joining the financial sphere of "banking" with the universal givens of biology, as in the prefix "bio", can be seen as a devious way of placing biologicals within the realm of market economics and imposing a profit-maximizing way of thinking. This association proximity risks foreclosing alternative ways of perceiving the ontology, goals, dangers, constraints and opportunities that lie in the accumulation and management of biological material and data. One of them is the important question of altruism which has little to do with the principles of banking.

Titmuss' celebrated work on blood donation (Titmuss 1970) illustrates the importance of counting on, rather than discounting the appeal that altruism still has for many people. Biotechnological development has of course 'since' vastly expanded the therapeutic and research promise contained in blood and blood products, and the realization of this potential has necessitated big economic investments which would not have occurred unless investors were given some control over investments and believed that these would generate profit. As a result, what started out as gifts has been reworked and developed into objects circulating as commodities, thus negating the sharp borderline which Titmuss argued existed between gifts and commodities. Describing this situation Kopytoff remarked that there has always existed a universal tug-of-war between the tendency of all economies to expand the jurisdiction of commoditization, and of all cultures to restrict it (Kopytoff 1986). The commoditization of human biologicals, e.g. blood, semen, ova and the recent possibility to "rent-a-womb", is no exception, and exemplifies how these issues are enmeshed in profound bioethical and moral webs of meaning.

Whereas altruism arguably is not a relevant concern for understanding financial banks, it is crucial for the constitution and management of collections of biologicals. These presuppose people's generosity and willingness to contribute without any other benefit sharing than a furthering of scientific insight and development of new medicines, achievements which possibly only future generations may "profit" from. Spreading doubts about people's altruism on a general level (e.g. as a Marxian form of false consciousness) risks, in my view, to backfire onto the more specific level of recruiting contributors to biobanks. This downplaying of any altruistic motivation reflects a Hobbesian belief in a "nasty and brutal world" in which people neither desire nor are able to behave unselfishly. Titmuss' analysis of blood-donation practices strongly suggests that an a priori negative view on people's altruism is mistaken. Indeed, his evidence showed that the number of voluntary blood donors in the UK actually decreased when blood donations were remunerated in cash. Titmuss' informants maintained that exchanging blood for money constituted a trivializing commodification, which to many donors acted as a disincentive. Applied to the discourse on biobanks, it is not inconceivable that contributors are similarly motivated by altruistic concerns in contributing to the general welfare of society, i.e. reflecting a communitarian approach. It should be pointed out though, that alternative definitions of altruism complicate the picture. According to some theoreticians, altruism is restricted to acts that involve placing the interests of others well ahead of one's own, i.e. excluding all acts in which self-interest is involved. In Titmuss' understanding, though, a degree of self-interest does not necessarily disqualify or preclude altruistic motivation. Donating a kidney to a close relation is to a certain extent *also* in the donor's self-interest, because the latter's wellbeing may depend on the ailing relative staying alive and being relieved from suffering.

In a more narrow understanding, altruism is motivated by a regard for the wellbeing of others for its own sake. However, according to a third point of view, self-interest and altruism are not necessarily always incompatible. Altruistic acts can for instance also be a way of increasing self-esteem, while at the same time intentionally benefiting others. This is perhaps not an uncommon mix of motivations among those who donate their art to public collections; for in addition to augmenting their own self-image, they may perhaps also enhance their public esteem and social status, which again can be used as an entrance ticket to a country's cultural and established élite. According to a narrow altruism concept, this kind of motivation conceivably renders such donation acts slightly less selfless.

My other misgiving about the term biobank has to do with the way it gerrymanders trust. Future contributors to biological depositories will – presumably unconsciously – transfer society's generally positive attitude towards blood donation to biobanks, thereby reducing and/or postponing possible resistance towards such depositories. To quote Pierre Bourdieu, by playing on the linguistic resemblance between "blood banks" and biobanks, the latter gain "symbolic capital" from the former (Bourdieu 1977). According to Bourdieu, such symbolic capital serves as the subtle but necessary means through which the production and reproduction of social institutions is achieved, smoothing over possible resistance.[4]

As a consequence of the scandals of HIV (contaminated blood which caused the death of many haemophiliacs), people's previous blanket trust in blood banks has to a certain extent been eroded. It turned out that blood supplies were not only composed of blood from healthy altruistic citizens, but that stores were pooled with blood *purchased* from donors recruited in countries lacking sufficient control to eliminate HIV contamination. As a result, even receiving blood transfusions in connection with surgery has become a health risk in several countries.

A third reason for my unease with the biobank term is due to its additional and unfortunate connotations. Casting biobank contributors in the role of financial depositors, for instance, risks overshadowing the role of *altruism* which was, and still is, a crucial criterion motivating donors of blood and bone marrow (as it certainly also is for many donors of paintings, sculptures and other works of art). The role of "depositors" is on the other hand more in line with prevailing ethical principles of autonomy, enabling contributors to be framed as "participants" in "partnerships", i.e. in what appears to be reciprocal relationships with the collectors.

When we add the many negative connotations that now also stick to commercial banks, such as exploitative interest rates, corruption, risky investment schemes and bankruptcies (sic), it would probably have been well advised to have had a deeper discussion of the neologism[5] biobank, before choosing terms for legislation.

Similarities Between Biobanks and Art Galleries

It may still seem facetious to draw an analogy between national art galleries and the contents of a country's biological depositories. For what does a Renoir painting have in common with biological specimens, genetic information twins, or statistics on blood types and HLA-matching?

[4] For more on symbolic capital, see below.

[5] Random House Dictionary, second edition, 1991 does not contain the word, nor does the Encyclopaedia Britannica of 2007.

First, both national art galleries and depositories of biologicals function are archives. They are, respectively, storehouses of our cultural heritage and storehouses of biomedical information about a country's citizens in need of experts' handling. They can be interpreted as testimonies of identity, on a national or personal level.

Second, such collections have a number of organizational features in common:

(a) Biobanks and art museums both collect their material according to established rules of inclusion and exclusion. Only pictures of acclaimed quality are deemed worthy of hanging on the walls of National Galleries. Art works are selected according to qualitative criteria such as skill, originality, and the artist's reputation and position in the history of art. The institutions containing biologicals are similarly run according to prescribed rules and regulations, whether they be collections of pathology slides from cancer patients, registries of people who have declared themselves willing to become bone marrow donors, storages of frozen blood products in blood banks, or institutions containing data on a given population's individual DNA profiles (such as those existing in Iceland and Estland).
(b) Both types of collections depend on recruiting voluntary donors. Affluent art collectors are courted by gallery directors and fund raisers, in the hope of acquiring their art treasures either as altruistic donations, or as bequeathals. In the case of biobanks, recruitment consists of finding people who will consent to the transfer of their bodily tissues to institutions dedicated to either diagnostic, therapeutic or research purposes. I intentionally use the perhaps fuzzy term "transfer", to indicate that the character of the exchange of biomedical entities remains unclear. Are they *gifts* with or without strings attached, *donations* given altruistically, temporary *deposits* which can be withdrawn at any time, or more or less voluntary *extractions* carried out at hospitals or in doctors' surgeries?
(c) Just as banks attract new (and keep old) customers by inferring reliability and confidentiality, so biobanks depend on the general public's long-term trust as regards these institutions' ethical soundness. Contributors expect openness about biobanks' motives and expected findings, as well as the risks and benefits befalling donors. In a similar way, the exhibition policies of art galleries are open and available to be seen and critiqued by reviewers and the general public.
(d) Art collections, diagnostic records, biological samples and genetic profiles all represent iconic and indexical representations of considerable value. Bone marrow "banks" are virtual depositories in that they contain indexical information about potential donors' immunological profile. The costs of cross-matching and transporting vials of haematopoietic tissue are covered by the national health care system of the recipient. In a similar way, when works of art are sent abroad or to other galleries in the country, on temporary loans, the Ministry of Culture finances insurance expenditure and transport, pending on formal request and approval.
(e) The most significant feature linking art collections and biobanks is that they both must be understood as systems of *communication*, and as encoding meaning. We read "meaning" into pictures and sculptures through training the eye to see beyond the colours, shapes and lines, and to recognize and interpret topics

and scenes through our cultural heritage. In a similar way pathology slides, MR-*imaging*, X-rays and PET *scans* are meaningless squiggles and clouded patterns to the medically uninitiated, but convey information about disease and irregularities to the specialist.

Some Differences Between "Biobanks" and Art Galleries

As institutions, biobanks and art galleries also differ in several important respects. First, National art collections are accessible to the general public, whereas health records, genetic profiles and other biologicals are encoded sources of information available only to accredited officials and researchers.

A second distinction has to do with the contributors' legal rights to destroy what they have stored, compared to art collectors' disposal rights. As already mentioned, contributors to biobanks can demand that their own biologicals be withdrawn and eliminated. Under certain circumstances, such as bankruptcy, the custodians of these depositories may be forced to demolish their entire collection, even without the contributors' permission, to prevent sensitive information from falling into wrong hands or being misused. Life insurance companies could for instance use knowledge about clients' genetic predispositions as a way of raising premiums.

National galleries on the other hand, are enjoined to at all costs preserve and maintain whatever is entrusted them. Thus, an artist who has deposited or donated a work of art to a National Gallery cannot, even if s/he thinks the work is out of date or badly executed, demand that the Gallery destroy it.

The rights of private art collectors as regards their acquisitions are perhaps less clear. Collectors who buy at auctions are customers who choose to invest their money in paintings, instead of in expensive buildings or other forms of property. The Japanese businessman and art collector Ryoei Saito, who bought van Gogh's "Portrait of Dr. Gachet" at the then record-price of $82 million, stipulated that when he died, the masterpiece was to be cremated with him. It is perhaps only thanks to the economic decline of Mr. Saito's paper manufacturing firm, which forced him to sell van Gogh's masterpiece (and Renoir's *At the Moulin de la Galette*), which prevented them from being irrevocably destroyed.

A third seminal difference has to do with the financial running of these two kinds of institutions. National Galleries are usually non-profit institutions aimed at benefiting the general public, educating children and students, and providing bases for art curators and researchers. National Galleries are primarily financed by governmental grants, which cover the costs of employees' wages, the acquisition of new works of art, expenditure and insurance premiums.

Depositories of biologicals, of which there may be several in a given country, can – but need not – be governed by commercial interests.[6] Research biobanks, which e.g. aim at isolating viruses and locating gene sequences that heighten

[6] According to current Norwegian legislation, those who seek permission to establish a biobank are duty bound to inform the Ministry about economic interests and possible profits.

targeted people's risk of acquiring various diseases, prepare the ground for the development of new medicines and vaccines. However, the process of getting new pharmaceuticals from the laboratory, through a period of trials, to hospitals and doctors' offices, is arduous, time consuming and costly and usually in need of commercial backers. Investors who place capital in such ventures are motivated by the possibility of patenting their products and reaping profit.

A fourth difference is of course that works in art galleries are valuable and appreciated also due to the deep *emotions* which they evoke about the human condition. We feel pity on seeing Munch's portraits of TB-infected children, horror at witnessing Goya's rendering of an execution scene, and delight at seeing Breughel's "Children's Games". Paintings kindle our interest in ways of living in the past and open our eyes to problems in contemporary society.

In contrast, depositories of biologicals are the object of medical professionals' enquiring and unemotional gaze, something to study (research institutes), or a source from which to provide therapy for patients in need (of e.g. bone marrow). However, breakthrough discoveries and intellectual insight stemming from research on biologicals may evoke scientists' pride and the general public's admiration, which shows that achievements in the field of biomedicine are not entirely without emotional value.

A fifth disparity lies in the use of money, when forming collections and depositories of biologicals. While works of art can be bought as part of a National Gallery's policy to complete or expand its collection – through Governmental grants or sponsorship – when and if desirable works become available on the market, in most jurisdictions it is illegal to pay for raw biologicals. Collections of biologicals depend on recruiting willing donors or by passing laws which make the collecting of various medical data mandatory.

The Relationship Between Collectors and Collectables

Collections of art and the way these are formed and run can perhaps provide a tool with which to scrutinize the different relationships that evolve between the differently positioned actors and the materials about which they compete.

Despite the laws that have been passed in several but not all countries, biobanks as yet mostly function somewhat ad hoc and according to trial and error, whereas National Art Galleries are well-established institutions. However, it is precisely such a diachronic perspective which may enable us to profit from the way art galleries over time have solved some of the problems that now beset contemporary collections of biologicals.

The contested ways in which much art has been collected across centuries, can now serve as a background against which to gauge the way contributors of biologicals should be recruited. National Galleries, on their side, might perhaps today profit from emulating biobank practice, i.e. by having an ethical committee which can be consulted, for instance as regards possible conflict of roles. Is it e.g. ethically and politically warranted that a National Gallery's Board-chairman functions as a private art collector at the same time s/he is in charge of a nation's main art collection, as was the case in Norway from 2003–2007? Or is a more stringent separation of roles called for so as to avoid any conflict of interest? All kinds of collecting implies a coveting eye, strategic conduct, and a set of motives, which are not necessarily always transparent to those whose valuables are sought (see for instance the Moore case, discussed below). An important issue is therefore to consider the relationship between collector and contributor, and to uncover whatever hidden purposes may accompany the collecting process both as regards art and biologicals.

As briefly mentioned, we have not yet satisfactorily settled what kind of transfer takes place when human biological material is removed from people's bodies and transferred to depositories. The role, rights and duties of the contributors and managers are also somewhat unclear. Are the contributors to biobanks *customers*, as in saving banks, or *citizens* expected to do their duty for the common weal, both, or something in between? And who are the possessors and true managers of biobanks: The Health Ministry, the officially appointed national bioethics committee or the combined forces of medical and biotechnical researches and the pharmaceutical industry? The manner in which these transfers take place has implicit consequences for the way depositories of biologicals are to be managed.

I want to draw attention to the contributors' understanding of the exchange taking place and the very different rationalities guiding collectors' practices. Contestable degrees of ownership are involved when valuables "move house" from their erstwhile owners, and are categorized and given institutional "labels". This applies equally to works of art and to biologicals, except in cases when the biologicals have been anonymized and therefore cannot be traced back to the donors. What happens to known donors' and depositors' rights when an institution changes its legal statutes (such as happened when Norway's National Gallery changed from being a national institution under Ministerial leadership, to a free-standing foundation)? Can contributors, whether in art galleries or in bio-depositories, be forced to accommodate to a given institution's new structure, or do new agreements and informed consent procedures need to be re-negotiated?

As will be seen, collecting exists along a continuum that stretches from people's altruistic donations, via deposits, to commercial acquisitions, to illicit appropriations, hardly discernable from confiscation and theft. Expanding our view to the way art collections have been formed in the past may sharpen our awareness about the overtones now present in the gathering of other valuables, such as biologicals.

In the subsections which follow, I start with the known (art galleries), in an attempt to shed light on the unknown (depositories of biologicals).

Altruistic Donations

On one side of the continuum we find the selfless givers of both art and biologicals who donate without strings attached. When Olaf Schou donated his collection of

116 paintings by Munch and other seminal Norwegian artists to the National Gallery in the early twentieth century, they were intended to be shown, or stored, entirely according to the director's discretion. Munch, who normally did not like to part with his paintings, was willing to sell his works to Schou, knowing that they eventually would be donated to the National Gallery.

The altruistic donations of both biologicals and art become complicated and emotionally taut when the person or institution destined to become the recipient, for various reasons, is prevented from accepting the proffered gifts.

An ethically delicate situation may for instance arise when an artist wishes to donate one or several works, and the intended recipients consider them to be of insufficient quality, or the artist poses conditions as regards the way the work(s) must be exhibited. Sometimes the artist wanting to give away a painting or sculpture does so for an ulterior motive, namely to enhance his/her CV by including a sentence such as "the artist is represented in the National Gallery". If the Gallery accepts the donation, the artist is bereft of the pecuniary income s/he might have derived from a sale on the open market, on the other hand it enhances his/her "symbolic capital" (Bourdieu 1977).

Within the field of biologicals, a pure, uncorrupted form of altruistic donation takes place when the next of kin of a person suffering sudden death, consent to letting the deceased's organs be used for transplantation purposes. In Norway, families know they will never receive any external form of gratitude or recognition. The only kind of verbal – and therefore symbolic – reciprocity lies in the brief thank you letter from the National Transplant Unit to the family who gave their consent. The next of kin's donation is therefore based on altruism,[7] although, if the dead person had pre-signed a donor card (which has only been possible in Norway since 2001), or in other ways made his/her positive attitude to donation known, it is the *deceased* who should rightfully be seen as the true altruist. However, compared to people who will their paintings or sculptures to a National Gallery instead of letting their family inherit the valuables, organ-donor cardholders do not in any way reduce the value of their estate. Their organs cannot be of any use to their survivors, nor can solid organs be preserved for future sale, which, besides, would be illegal.

Even if organs for transplantation are in short supply, legislation and ethical considerations sometimes make it necessary for the health care in charge of the donation to refuse the next of kin's "offer" of their deceased relative's organs. While conducting fieldwork at a hospital (Hambro Alnæs 2001) to study next of kin motivations in connection with organ donation, a doctor-informant told about a case in which the family had been adamant about their newly deceased's desire to have his organs used for transplantation purposes. The trouble was that the would-be donor did not qualify as brain dead, i.e. there still remained some blood circulation to the brain. For ethical reasons, the doctor was not willing to keep his patient on the mechanical ventilator, as it was uncertain how much time was needed to fulfil the legal requirements of brain death. In this case, the doctor felt compelled to decline the family's offer, a response which the family experienced as deeply humiliating. (Patients

[7] Strictly speaking, they consent to transferring objects to patients in need, which they do not, and have never owned. They (only) donate *on behalf of* their dead relative.

suffering from diseases such as diabetes or cancer are excluded from becoming donors [except in cases of totally encapsuled brain tumours], even if the deceased and/or his or her next of kin were in favour of organ donation).

Countries which include a protocol called "Donation After Cardiac Death (DCD)" (as opposed to donation after *brain* death), defend their practice as a way of satisfying the expressed wishes of the person about to die, as long as there also exists a foregone agreement with the next of kin. However, such DCD-protocols are contingent on expanding the medico-legal criteria for donation, an issue which remains ethically fraught for many health care employees. The donation of organs, in other words, involves not only givers and recipients, but also mediators (doctors, nurses, transplant coordinators) whose skills and communicatory competence are of the essence for the transfer of these highly sought after and valuable biological gifts.

A willingness to donate biologicals or works of art is, in other words, not necessarily contingent on any duty to receive what is offered. Artists and art collectors cannot count on having their works accepted. This, of course, goes inherently against the sociologist Marcel Mauss (1990) well-known analysis of the universal rules involved in the exchange of gifts which – as he observed among pre-modern Maori – consisted in the duty to give, to receive and to reciprocate.

Deposits

Deposits can be altruistic, such as when rare and extremely valuable musical instruments are given on *loan* by collectors to promising young artists who cannot afford purchasing them. Sometimes these loans are done anonymously, through an intermediary; sometimes the owners are well-known companies who wish to enhance their status by supporting and encouraging young artists' careers. These deposits can be likened to a right of use, or ususfructus, a form of temporary possession – which precludes the right to sell – as opposed to ownership.

Some painters similarly lend their works of art to galleries as deposits, an arrangement which can be seen as an alternative to costly storage with the added advantage that they will be seen by the public under the protection of guards and relatively safe from theft. Depending on agreement, the artist can in principle temporarily or permanently withdraw his/her work so as to take part in other exhibitions.

The borderline between donations and deposits is not always crystal clear, either in connection with works of art or biologicals. For the curators of National Galleries, the conditions attendant upon donations can even give cause to legal problems. One such incident occurred when the director of Norway's National Gallery wanted to re-hang its pictures as part of the new profile he wished to present to the public after the "new" National Museum[8] was established in 2003. This involved splitting the priceless Langaard collection, which had been donated on condition that the works

[8] Based on the amalgamation of four national art institutions.

of art be exhibited as a unified collection in a single room. The heirs were provoked, and threatened to withdraw the entire collection. Compared to Mr. Schou's donation, the strings attached to the Langaard collection make the latter seem less of an unconditional gift, more like a deposit.

As regards biologicals, the principle of donations under restricted conditions seems most in line with present Norwegian biobank legislation. Current law specifies that the contributors of biologicals can withdraw their material at any time, and without giving any reason, as long as the samples are not anonymized or already used in a publication which has appeared or is about to go into print. Thus, while the contributors of biologicals cannot exert traditional property rights over their material once it has been sampled, they have retained significant dispositional rights, an arrangement which empowers contributors in accordance with the principle of autonomy.

Contributors of biologicals need to be approached again if their material is to be used for research which differs from the project they consented to originally. By their withdrawal the contributors can demonstrate their disapproval, e.g. if they consider the new protocol unethical. An exception, however, is when the projects in question are part of a national overview such as e.g. Norway's personalized registry of all cancer diagnoses, or the national registry of all diagnosed causes of death. In these cases, the need to secure epidemiological data overrides the autonomy of the individuals who provide the bases for these registries.

If the donors of biologicals maintain the right to withdraw their material at any time, then surely their contributions resemble deposits more than gifts, and Langaard more than Schou. Biological contributors' rights at present seem to fit better within the more self-interested frame of "biobanks" than within the domain of altruistic gifts. If they haven't been anonymized, biological deposits appear to be reversible and retractable. This situation differs from the unconditional altruism evident in the donation of organs. The next of kin in organ donation situations are called on to act as communitarian-minded *citizens*, whereas the contributors of samples retain *customer*-rights.

Why the contributors of other kinds of tissues, including blood, should have different and stronger rights, needs further explaining. Reserving oneself against the use of one's donated blood in certain research projects corresponds to a view that a person's tissue – after it has been detached and treated with fixants – continues to represent a person's inalienable identity. This metonymic way of thinking, i.e. founded on association by contiguity, is the principle behind sympathetic magic,[9] which seems a far cry from the rationalities otherwise guiding biomedicine. But practice does not seem to be fully consistent. If blood donors learn that their donation might go to treat wounded soldiers in a war of which they disapprove, the donors can all the same not withdraw their blood. Bone marrow donors, on the other hand, can decide not to go ahead with their planned donation, even when the recipient has already started his/her often gruelling de-construction of immunological defences, in preparation of transplantation.

[9] See Jakobson 1956.

Power

Whereas the depositories of biologicals presuppose recruitment strategies, the job of leaders of National Galleries lies in encouraging private collectors to relinquish their art treasures to enhance the public weal. They are collectors *on behalf of* the general public. Although both, seemingly, act in the role of supplicants, they do so from positions of power. While power is notoriously difficult to define (Barth 1993), most people know how powerlessness is experienced.

From the point of view of the Greek nation, the acquisition of the Elgin Marbles was the result of an unethical exercise of asymmetric power, a point of view which can be said to resemble Mr. Moore's opinion on Dr. Golde's appropriation of his lymphokines (see below). For several decades now the Greek Government has argued that the Parthenon Marbles should be returned to their "homeland", as they are the most profound symbol of Greek history and identity. The British Museum argues that Thomas Bruce, the seventh Earl of Elgin and ambassador to the Ottoman Empire, had purchased the sculptures and frieze legally. The Greeks, however, claim that Elgin took advantage of the Ottoman occupation of Greece, by obtaining a vague and untraceable *firman* (license to purchase) from the Sultan. To begin with, the Sultan gave Lord Elgin permission to remove the freestanding sculptures; however, Elgin used the volatile political situation to help himself to the monumental frieze forming part of the main temple as well. The British Museum rejoinder has been that the marbles would have eroded and been lost for posterity if Lord Elgin had not purchased these monumental sculptures and brought them to London. According to the British Museum, the Marbles were legally bought, and now morally owned by the museum.

For the collectors of biologicals, this example of power wielding from the early nineteenth century serves as a reminder that informed consent consists of more than signing a document and that consent of this kind is only valid if the contributors fully understand and agree to what and why they give. From a different perspective, the Elgin Marbles also exemplify the importance of salvaging valuables from destruction. There is little evidence that the Greeks at the time of the purchase attached any significance to the ruins of the Parthenon. This and other examples from the art world may also serve as a cautionary note for collectors of biologicals to keep samples and data in professional storage, even when their immediate value is not evident. Neither the donors nor the collectors of biological specimens can foresee what scientific insight may evolve from such collections. Unless the storage of biologicals represents a public health hazard, the duty to responsibly preserve what can arguably be seen as a biomedical resource is not different from a national museum's duty to store and maintain the works of art in their collection, either within the Museum itself or in storehouses elsewhere. Donations of both kinds of valuables are per force based on the donors' implicit trust that they will not be misused or squandered and that custodians will ensure that the deposits are treated according to appropriate ethical rules and accountability.

Let us consider the power dimension in connection with the recruitment of contributors of biologicals. Such recruitment usually takes place at hospitals, blood banks, and/or in doctors' surgeries. Doctors' offices and hospitals are spaces in which the situation can be crudely defined as: patient in need of therapy or guidance seeks doctor's care and advice. When GPs or hospital doctors ask their patients to participate in research protocols, this approach has been criticized for being slipped in between other health concerns and as being insufficiently explained. In medical consultations, the asymmetric distribution of power stems partly from doctors' virtual monopoly in interpreting medical facts, risks and proposal of therapy. Access to medical information on the Internet has not changed the situation drastically: patients – according to some professional (personal communication) – still lack the education and insight to differentiate between robust arguments based on complex knowledge and more spurious, un-evidenced articles; so the "homework" the patient does ahead of his meeting with the doctor, is often riddled with misunderstandings in need of time-consuming clearing up.

As regards bone marrow donors, recruitment is layered. Blood banks carry out the initial, preliminary drafting, which is followed by a more comprehensive enrolment process, if the potential donor after a period of reflection is still willing to participate. Enrolment involves HLA-typing, medical examinations and a thorough explanation about what being a bone marrow donor entails for the donor, such as sometimes having to donate several times (e.g. if the first transplantation is not successful), the necessity of having injections to stimulate the production of stem cells, and the possibility of having to undergo general anaesthesia if peripheral blood donation is unsuccessful. Informing about bone marrow transplantation would be one-sided unless the donor also understands the life-saving benefits bone marrow transplantation represents for the recipient. However, such information can hardly fail to place an enormous responsibility and moral pressure on the potential donor, hence rendering it extremely difficult to cancel donations at the last minute. Although the two-step process may appear to be a way of diminishing the asymmetry between the "brokers" of bone marrow, i.e. the doctors and the donors, the situation can equally well be interpreted as exemplifying a form of symbolic violence (see below).

The exercise of power is particularly contestable when trust is assumed but violated. Since cell lines are such a contested issue and the primary goal of several research biobanks, I want to revert to two now famous instances.

The well-known HeLa cell line, named after an African-American Baltimore housewife, Henrietta Lachs, was developed on the basis of her cervical cancer tissue which had unusual cell division properties. Over the years, copies derived from this patented cell line were first given to researchers free of charge, later sold to scientists and laboratories for considerable amounts. Nobody at the time asked the patient for permission to use her tissue, and despite later requests from her husband and children, Lachs' survivors never received any economic remuneration.

The famous Mo-line (later renamed "RLC") was the result of what in hindsight came to be seen as the result of an ethically questionable relationship between patient and therapist. Mr. Moore suffered from hairy-cell leukaemia. During several months of treatment and before advising his patient to have his spleen surgically removed, Mr. Moore's doctor, David Golde, discovered that his patient produced unusually large quantities of lymphokines. These, he knew, could be used to develop a potentially profitable cell line. During eight months following splenectomy, Dr. Golde repeatedly called his patient in for follow-up consultations which he used as a pretext for taking further blood-, bone marrow-, skin- and sperm samples.

Mr. Moore's suspicions were raised when Dr. Golde requested him to sign a document in which he granted the University of California rights to any cell line or product made from his blood, a document he refused to sign. Mr. Moore instead hired a lawyer, and a ten-year Odyssey from court to court ensued. Mr. Moore claimed that Dr. Golde had taken tests on false premises, without first seeking his consent or informing him about the potential economic profit. Mr. Moore's repeated court appearances were by many understood as being primarily motivated by a desire to be awarded a share in the economic benefits ensuing from what he saw as "his" cell line. In addition, Mr. Moore claimed that Dr. Golde's manipulation of him had blocked his opportunity of donating tissue to "enable *other* researchers to make the most of these discoveries".

Seen from a power perspective, it can be argued that given his serious illness, Mr. Moore lacked adequate informed consent *competence*. As Dr. Golde had been silent on the question of possible economic gain, Mr. Moore was implicitly led to believe that the numerous tests he was required to take were to provide him with the best diagnosis and therapeutic possibilities. Dr. Golde's conduct can thus be seen as setting the stage for a *therapeutic misconception*, which occurs when a patient believes his tests are taken to enhance curative possibilities. As became clear, Mr. Moore's samples were part of a different agenda. Dr. Golde needed to secure sole access to, and control over, Mr. Moore's exceptional lymphokine-rich spleen, in order to file for patent rights and subsequently reap the economic profit from the cell line he planned to and succeeded in developing (in collaboration with his research assistant Shirley Quan).

Mr. Moore's well-organized effort to ensure some form of benefit sharing between the contributor and scientist came to nothing. The successive law suits focused on distinguishing between on the one hand the raw material (Mr. Moore's spleen), which would have been useless if it hadn't been developed as a research vehicle, and on the other hand the re-worked tissue which through Dr. Golde's intervention and scientific prowess resulted in an entirely different entity, of great biovalue. Some saw the courts' rulings as condoning confiscation, others as ensuring that research efforts should not be stymied by demands from the person whose biologicals are used. Denying Mr. Moore rights to the dividends of the Mo-cell line was seen as a way of empowering the research community, which in turn could benefit the general public. Mr. Moore's demands to exercise autonomy over his own tissue were seen by the court as a legally irrelevant topic of contention.

The conflict stood, and stands, between intellectual property right advocates who claim that the scientific endeavour depends on rewarding innovators and tempting investors; and those who are concerned about the need to protect the equivalent of a biomedical "commons"; the latter wish to protect certain kinds of human tissues and information about tissues and genes from commodification. The conflict in many ways mirrors the competitive struggle that took place in England from the twelfth to the nineteenth century, over the so-called enclosure, i.e. the land over which the community held disposition rights. The manorial lords gradually succeeded in privatizing the "commons" in order to increase their own amount of full-time pasturage, at the expense of grazing rights for villagers' livestock.

The power relations between collectors, donors and recipients are contextual, contingent on timing, context and the often unequal distribution of knowledge between the interactants about the objects' real value. After Mr. Moore unsuspectingly had let Dr. Golde take samples from his spleen, and agreed to splenectomy, he was de facto rendered powerless. Dr. Golde had the raw material from which to produce biovalue, exemplified in the ensuing Mo-cell line.

In comparison, doctors hold less power in necro-donation cases. However strongly the requesting doctor is motivated to secure organs for donation, the outcome in each case depends wholly on consent from the bereaved family and/or prior confirmed statement from the deceased patient. Potential bone marrow donors are in a similar position of power vis-à-vis the requesting doctor and the patient in need of such therapeutic HSCs. Donors can withdraw their offer to donate at any point in the process without incurring any negative repercussions from the medical establishment or doctors.

Art collectors' power sometimes resides in their superior knowledge compared to those who apparently voluntarily give, or sell their objects for a pittance. As mentioned above, the Parthenon Marbles were not seen by the Greeks as valuables at the time Lord Elgin negotiated his deal with the Sultan. Nevertheless, for the Greek nation, the frieze, particularly, represents a key symbol (Ortner 1973) of Greek identity and nationhood. After lengthy legal battles, the Paul Getty Museum has recently returned a substantial number of art treasures to Greece and Italy because of incriminating evidence about how these objects had been "found", acquired and exported to the US.

The exercise of power is thus closely linked to the unequal knowledge of positioned interactants. A way of reducing this gap in connection with the collection of biologicals is to empower the national bioethics boards, and for the government to support the spread of knowledge to the general public, even, or especially, when existent legislation is scrutinized and critiqued. Part of the officially appointed Norwegian Biotechnology Advisory Board's (*Bioteknologinemnda*) role in Norway is to increase the transparency of the biomedical community by e.g. arranging regular public meetings where national and international experts come together to discuss contentious issues within research and health politics.

Appropriations–Extractions–Confiscations

These are not synonyms to be used interchangeably, as their appropriate use depends on context and crucially on the differently positioned actors' knowledge and motivations. There exists a significant difference between coercing people into handing over objects which they themselves value and wish to keep, and the transferral of objects when one of the partners knows he is dealing with something of great potential value, whereas the other believes s/he is giving away trivialities, as the Moore case and the Elgin Marbles exemplify.

Dr. Golde's treatment of Mr. Moore constitutes an example of what has later been critiqued as "biopiracy" and "bioprospecting" (Nelkin and Andrews 1998).

In the world of art collecting, the Soviet Union's justification for appropriating the war booty stemming from World War II represents a noteworthy example of declared confiscation. The Soviet army seized 300,000 works of art, among them a rare Gutenberg Bible, paintings by Matisse, Renoir and Manet, and the Trojan gold treasure discovered by Heinrich Schliemann. Some 50 years after the peace agreement (1995), this trove of art resurfaced from hiding and was openly appropriated by the Russian State as *compensation* for the devastations caused by the German army, even though many of the works consisted of paintings and art objects stolen by the Nazis from innocent Jews. According to the *Duma's* ruling, only victims in countries who fought against the Germans had the right to claim restitution of cultural valuables.

In Western Europe, the Austrian government's treatment of the Rothschild art treasures, confiscated by the Gestapo and their Austrian accomplices after *Anschluss* in 1938, affords another example of extraction euphemized as "donations". In 1947, Louis Rothschild's niece Clarice was given custody over the crates of art treasures, which had been systematically categorized and stored in a salt mine outside Salzburg. This was not, however, the same as reclaiming the family's stolen goods. Even though the Rothschilds were war victims, the Austrian government decided to apply a law introduced after the World War I as a pretext for preventing Rothschild's private collection from leaving the country. After much legal haggling, export licenses were provided over a period from 1947 to 1950, but only in exchange for "donations" to Austrian museums and galleries. Ironically, the labels informing visitors about these exhibits now read: "dedicated by Clarice Rothschild, in memory of Alphonse Rothschild".

This first instance of art appropriation exemplifies an extreme case of cutting of strings between the rightful heirs on the one hand and the paintings and other art treasures now forming part of Russia's art collections. In the second case, the Austrian Government used an old law as a bargaining tactic to pry loose some of the Rothschild treasures from the heirs of their erstwhile owners. The cases illustrate what Callon has aptly called "disentangling" processes (Callon 1998).

Callon first employed the word "entangled" to describe the character of organs from brain-dead patients. These perishable valuables cannot be stored outside the body, and must be reattached to the blood vessels of the recipients within the course of a few hours. The transfer of organs is also restricted by their immunological profile, to prevent the graft from being rejected by the recipient. Consenting to donation presupposes relinquishing ownership and disposition rights, which in its turn depends on cutting the symbolic strings attached to these organs.

Callon's "entanglement" and "disentanglement" terms can be used to critique disposition rights over other kinds of biologicals. As just mentioned, the next of kin relinquish all rights when they consent to organ donation. In connection with other non-anonymized biologicals, however, the contributors maintain dispositional rights, for instance the possibility of withdrawing their material if they do not wish it to be used in a particular research protocol. This is noteworthy, considering that most contributors in donation situations regard their deceased's heart, lungs and cornea or other organs as significantly more inalienable than a vial of blood.

The doctors helping themselves to Mr. Moore's and Henriette Lachs' "raw material", without prior consent, can be seen as an illegitimate cutting of strings. This may not be quite the equivalent of appropriation, confiscation or extraction, since the question of ownership was debatable, even if it in practice amounted to the same. Disentanglement can perhaps be likened to the cutting of the umbilical cord. The birthing mother does not "own" her child, resembling the way contributors of samples do not "own" their samples either.

In UK legislation it is quite clear that a person cannot own his/her body. S/he has sole disposition rights over her/his body while alive, but only as against others' possible claims. Bodily gifts are irreversible, and the giver relinquishes all rights after the transferral, although even this unquestioned doctrine has recently been disputed.[10] In the Moore-case, the US Supreme Court judges' ruling established a fundamental distinction between undeveloped human biological materials, which normally would have been treated as "waste", and the biological entities resulting from *inventions* based on such material. In the judge's view, Dr. Golde's error consisted in a breach of fiduciary duty and a failure to inform his patient about the intentions of his research. But from Mr. Moore's perspective, the illicit extraction of his lymphokines amounted to appropriation and confiscation.

To briefly return to the question of power, the ways of appropriating, extracting, confiscating or surreptitiously stealing biologicals described earlier up, can all be seen as instances of Bourdieu's "symbolic violence".

Bourdieu coined the expression as an extension of his term "symbolic power". He urged social scientists to always be on the outlook for and identify power, particularly where it is least obvious. When power asymmetry is accepted and referred to as "natural" by the dominated, there is reason for others to reflect on the researcher's role and ulterior motives. As opposed to the overt, enforceable power embedded in legislation, Bourdieu defines symbolic violence as that invisible capacity:

> "to impose the means for comprehending and adapting to the social world by representing economic and political power in disguised, taken-for-granted forms", and "only through the complicity of those who do not want to know that they are subject to it or even that they themselves exercise it" (Bourdieu 1991: 164).

In a very different cultural context, the Amazon basin in South America, bioscientists have since the late 1970s sought out previously isolated and close-knit tribes such as the Karitiana Indians, the Surui- and the Yanomami peoples, to collect blood samples for research. This form of "biopiracy" has, arguably, replaced colonizers'

[10] In 2009, a man who had donated a kidney to his wife (2001) demanded to have his kidney physically returned to him, or be paid compensation money. He claimed that his estranged wife who had been involved in several extra-marital affairs after receiving her husband' kidney had refused him access to their children, and sought to use the economic value of his donated organ as a bargaining plea in the couple's divorce settlement.

and missionaries' practice of helping themselves to indigenous peoples' carved figures and decorated weapons in exchange for pieces of cloth or other objects which the buyers deemed as having little value. In 1996, another bioprospecting team arrived, promising medicine in exchange for more blood samples, which were of great interest to the community of genetic researchers studying disease transmission over generations.

Contact with representatives of the Western World also resulted in access to the Internet. To the Karitiana Indians' consternation, they discovered that their blood and information about their DNA code was being sold around the world for $85 per sample. And still no medicines had arrived in their settlements. Just as some people are against having photos taken of them because it involves a loss of "soul", so the Karitiana regarded the distribution of their blood and DNA as a violation of their integrity. Representatives of the Coriell Cell Depositories, a non-profit company, insisted that the samples had been collected in accordance with informed consent principles. However, it is questionable whether the Amazonian blood donors had sufficient consent competence, or any insight into the aims of Western medicine or modern bio-capitalist economics. From the Karitiana's and other Indian tribes' perspectives, they have been the victims of biological piracy or theft, Elgin–ed.

Disentanglement processes, through the strategic use of informed consent, like ways of redefining stolen art as compensation money, shows that the apparently neutral word "collection" sometimes masks over the highly questionable ways that these different kinds of depositories are put together.

Contributing Valuables: Seen from Different Positions

Recruiting donors of biologicals and collecting art are basically concerned with the transfer of objects from private individuals to public institutions. The willingness of those who part with their "valuables" depends on how they view their "opposite number". Does one lose something when one gives a gift, or does one instead gain something, in a longer time perspective, as a promise of something in store for oneself or one's progeny?

Aristotle argued that the categories with the aid of which we think are informed by and are formed on the basis of likenesses and different kinds of opposites. Applied to the concept "gifts", giving is the contrary of losing, the converse of receiving and the contradictory of taking.

This, apparently abstract, model is relevant for understanding and nuancing people's apprehensions when and if they are asked to participate in a research project by donating samples. Do people whose tissue samples are collected during medical consultations experience that they have freely given, or rather, surrendered parts of their biogenetic material? Judging from the tribe-spokesmen's reactions in the Amazon area, the sampling of blood was seen as an illegitimate appropriation of something of great value to them by culturally ignorant biocommercial representatives. Part of their identity had in their view been stolen and was forever

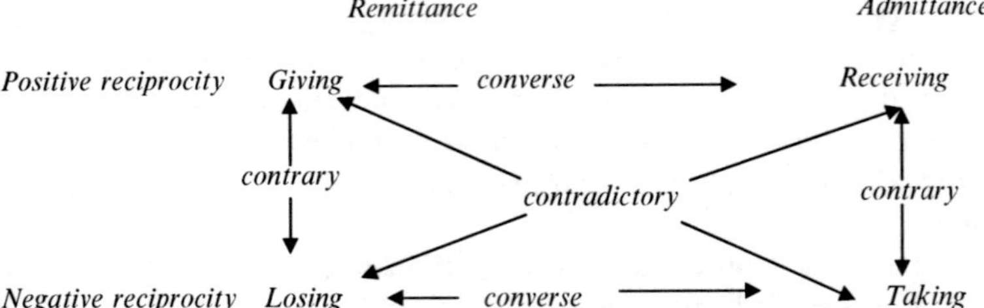

Reprinted with permission from table C.A. Gregory's chapter 33. 'Exchange and Reciprocity', ref. Companion Encyclopedia of Anthropology, edited by Tim Ingold. Published by Taylor & Francis UK (1996) on page 923, table 2.

un-returnable. Their blood had already been globally distributed as specimens to research biobanks.

Conclusion

Seeing collections of biologicals as sharing important properties with art galleries, goes to show that analogies are polysemic and depend on being interpreted in line with some, but not with other connotations, if they are to add to our understanding.

Analogies have a didactic potential which at the same time carry normative implications.[11] If we put aside the unethical ways in which some art collections have been established, it would seem that the custodians of biologicals as well as bioethicists and health politicians could profitably reflect on the way art donors and public art museum directors conduct, and have conducted, themselves.

The fate of the Elgin Marbles continues to raise heated discussion. When the Museum of Modern Art in New York returned Picasso's "Guernica", this was the result of the painter's instructions in his will. In cases when works of art *are* returned to their proper owners, as happened with the Gustav Klimt paintings in Vienna's Belvedere gallery in 2006, this has been because they were stolen goods, not gifts. In the case of the Amazonian Indians, the premises for informed consent, namely that the donor understands what s/he is doing, were not present. Considering their lack of informed consent competence, their sense of loss and the damage done to their dignity, the extraction of their blood must in hindsight be recognized as a form of theft.

Another issue has to do with the destruction vs. protection of the valuables contained in biological depositories and national art collections. In such collections it would be regarded as unethical to destroy objects, even if e.g. the artist who has donated one or several of his or her paintings later wishes to destroy them because he/she regards them as badly executed and/or unrepresentative of his/her present

[11] For this, see also chapter "Mapping the Language of Research Biobanking: An Analogical Approach".

point of view. In such, admittedly rather theoretical situations, the curators' evaluation would certainly overrule the artist's autonomy. Even if the artist has only *deposited* a painting in the Gallery, his/her work has already been judged as being of "national" value and hence his or her rights as depositor are perhaps not unlimited. The curators would in all likelihood claim that returning deposited works of art – when the artist's intention is to destroy them – would go against the public interest.

Although it is reasonable to protect contributors' biologicals for reasons of privacy, they also deserve to be protected for future research, and hence merit being stored as a national resource. The importance of salvaging these materials from destruction can be seen in the many documented examples of the way previously understood "waste" has later been transformed into "gold". Materials in biological depositories are subtly encoded, and it is up to future researchers to unlock the as-yet-unknowable information embedded in them. If we disregard those collections, which according to legislation *must* be stored, it is in my opinion very unfortunate that the collections of systematically gathered and catalogued biologicals are not always saved for posterity and potential research.

Present legislation, which enables biological contributors to recall and destroy their material, can be seen as representing a somewhat overzealous respect for autonomy. It presupposes donors who believe that their biologicals have some kind of inherent power – resembling the way sympathetic magic is thought to function; whereas it is primarily the power of potential mis-users, which ought to be at issue. Such safeguarding, however, is, according to current Norwegian legislation, the responsibility of the national bioethics committees. The increasing emphasis on accountability and public support necessitates winning the general public's trust on a long-term basis. This is contingent on strengthening the independence of regional, and national bioethics committees, and on raising public awareness about the sine qua non of altruistic gifting for research for the research community.

It remains to be seen whether future contributors to biobanks are able and willing to see similarities between, on the one hand, donating blood for life-saving transfusions in the operating theatre, and on the other, the potential of donating blood for research purposes aimed at preventing and curing various diseases. Whether Titmuss' model – in which people's altruism and their negative stance towards economic compensation were key ingredients – will prevail over the model exemplified by Mr. Moore, in which expectations of dividends from research results were a central theme, cannot be predicted.

Using art galleries as a heuristic tool for probing into the nature and aims of biobanks enables us to distinguish and see similarities between gifts, deposits, extractions, confiscations and thefts which together form the sometimes dubious mixture of many countries' national art galleries. This list of noble and ignoble ways of procuring art constitutes a timely warning for the pioneers within biobanking. As a corollary, it highlights the important position of bioethics committees and their dual role as gatekeepers and bioethically reflective gate openers.

References

Barth F (1993) Balinese Worlds. The University of Chicago Press, Chicago

Bourdieu P (1977) Outline of a Theory of Practice. Cambridge University Press, Cambridge, MA

Bourdieu P (1991) Language and Symbolic Power. Polity Press, Cambridge

Callon M (1998) Introduction: the embeddedness of economic markets in economics. In: Callon M (Ed.) The Laws of the Markets. Blackwell, Oxford, pp 1–57

Hambro Alnæs A (2001) Minding Matter. Organ Donation and Medical Modernity's Difficult Decisions (Doctoral thesis). Oslo University Press, Oslo

Hofmann B et al. (2006) Teaching old dog new tricks: the role of analogies in bioethical analysis and argumentation concerning new technologies. Theoretical Medicine and Bioethics 26:397–413

Holm S (2007) Head to head. Should genetic information be disclosed to insurers? Yes. British Medical Journal 334:1196

Jakobson R (1956) Two aspects of language and two types of aphasic disturbances. In: Jakobson R, Halle M (Eds.) Fundamentals of Language. Mouton, The Hague, pp 55–87

Kopytoff I (1986) The cultural biography of things. Commoditization as process. In: Appadurai A (Ed.) The Social Life of Things: Commodities in Cultural Perspective. Cambridge University Press, Cambridge, pp 64–91

Lakoff G, Johnson M (1980). Metaphors We Live By. University of Chicago Press, Chicago

Landecker H (1999) Between beneficence and chattel: the human biological in law and science. Science in Context 12:203–225

López JJ (2006) Mapping metaphors and analogies. American Journal of Bioethics 6:49–57

Mauss M (1990) The Gift: The Form and Reason for Exchange in Archaic Societies. Translated by WD Halls. Routledge, London

Nelkin D, Andrews L (1998) Homo Economicus. Commercialization of Body Tissue in the Age of Biotechnology. Hastings Center Report 28:30–40

Norway (2003) Act on Biobanks, LOV 2003–02–21 nr 12: Lov om biobanker (biobankloven)

Ortner S (1973) On key symbols. American Anthropologist, New Series 75: 1338–1346

Sankar P (2004) Communication and Miscommunication In Informed Consent to Researcyh. In Medical Anthropology Quartlerly. Washington. 18, 4:429–447

Titmuss RM (1970) The Gift Relationship: From Human Blood to Social Policy. Allen & Unwin, London

The Health Dugnad: Biobank Participation as the Solidary Pursuit of the Common Good[1]

Lars Øystein Ursin and Berge Solberg

Abstract Rosamond Rhodes and John Harris have recently argued that we all have a general moral duty to participate in medical research. However, neither Rhodes' nor Harris' arguments in support of this obligation stand up to scrutiny, and severe and convincing criticism has been levelled against their case. Still, to refute their arguments is not to refute the conclusion. There seems to be some truth to the view that when people are asked to take part in medical research, their choice is not completely morally neutral. In this chapter, we argue that the proper question to ask is when, rather than if, a certain moral duty to volunteer for medical research can be appealed to. To answer this question, we need a denser description of relevant research projects and their context rather than just describing medical research in general. Drawing on our study of participants in the Norwegian HUNT biobank, we use the normative implications of the Norwegian concept "dugnad" as an analogy to discuss the requirement of providing neutral information to potential biobank participants in order to promote their free and informed decision as to whether or not to take part. We suggest that normative recruitment is not just a question of principles and ethics. It is also a question of research design and the creation of the common good in the community where the research takes place.

---

[1] The original version of this chapter was published in the Nordic Journal of Applied Ethics 2 (2008), with the title "When is normative recruitment legitimate?" It is here reprinted by kind permission of the editors and Tapir Academic Press.

L.Ø. Ursin (✉)
Philosophy Department and Bioethics Research Group, Norwegian University of Technology and Science, Trondheim, Norway
e-mail: lars.ursin@hf.ntnu.no

J.H. Solbakk, et al. (eds.), *The Ethics of Research Biobanking*,
DOI: 10.1007/978-0-387-93872-1_15,

A General Duty to Participate in Medical Research

In an attempt to interpret anew the autonomy and obligations of participants in biobank research, Rosamond Rhodes in the article *Rethinking Research Ethics* (Rhodes 2005: 7) makes a frontal attack on contemporary research ethics. In bioethical literature, informed consent is argued for on the basis of an ambiguous concept of autonomy, Rhodes says. On the one hand, autonomy is taken as an *ideal* for the individual. The ideal, then, is that the foundation for an individual's choice is freedom and reflection. On the other hand, autonomy is taken to be a *norm* for how an individual's choices are to be understood.

While such a norm demands an assumption that an individual makes autonomous choices, according to Rhodes, the opposite position is prevalent in bioethics literature. The norm of autonomy is replaced by the ideal, which drastically restricts the kinds of people who can truly be said to be autonomous. In this manner, the kinds of people who are genuinely autonomous, able to give an informed consent, and to take part in research, are separated from the kinds of people who do not possess these qualities, and who are consequently excluded from research.

Rhodes accentuates autonomy as a social norm, and argues against the exclusion of groups of people as participants in research on the basis of an ideal of autonomy. Indeed, everybody should take part in research, Rhodes argues, because the vulnerable aspect in this context is the future patient rather than the present research participant. And to assume that it is against the will and interest of people to take part in a morally laudable and other regarding project such as improving medicine through research is for Rhodes deeply disrespectful (Rhodes 2005: 14). She states the implication of her views on autonomy regarding research participation thus: "So, in light of our appreciation of human vulnerability to injury and disease and our appreciation of the value of clinical research, reasonable people should endorse policies that make research participation a social duty" (Rhodes 2005: 15).

On the basis of these considerations, Rhodes puts forward a *novel proposal*: Her idea is that society, after thorough deliberation, should institute obligatory participation in medical research at regular intervals for all citizens. The choice is then not *if* you want to participate in a study or not, but *which* study to participate in. All studies would have to be approved by public medical authorities. This would draw attention to the approval process, and would require full disclosure of the study design, in order for institutions to be judged trustworthy by prospective participants. Projects should also be deemed of high quality and importance, and with few or no inexpedient burdens placed on participants. The granting of informed consent in this context would be part of the active exercise of one's autonomy, inside of a field restricted by law.

Rhodes intends to establish medical research as one of society's central tasks. And from this perspective, the demand that all research be of direct benefit to participants undermines its social and long-term purposes. In the regulation and evaluation of specific research projects, it is important to focus on the quality of the research, and to maintain legitimate trust on the part of participants. By making autonomy and participation the norm, the default position for Rhodes is that everybody can and will contribute to the common good resulting from medical research.

Sarah Chan and John Harris share Rhodes' view of the default position. They do not, however, think that this justifies conscription of participants to medical research (Chan and Harris 2008: 11). Harris likewise discusses in the article *Scientific research is a moral duty* the question of a putative duty to participate in research as a moral, and not a juridical or a political question (Harris 2005). In this article he emphasises two principles, both of which, he thinks, commit us to a moral obligation to participate in medical research. The first principle is our moral duty not to harm others. Harris argues that such harm is the consequence of declining to contribute to this kind of research. The second principle is the principle of justice, which results in the problem of the free rider.

Harris does not argue for any legal duty to take part in research, but holds that these principles make it ethically problematic to refuse participation. To participate is required, both to contribute to the common good, as well as to be able to respect oneself as a moral actor. On the basis of this, it is possible to presume that a safeguarded participation also would be in the interest of those deemed to be without full competence to consent. Harris concludes: "There is then a moral obligation to participate in research in certain contexts. This will obviously include minimally invasive and minimally risky procedures such as participation in biobanks, provided safeguards against wrongful use are in place" (Harris 2005: 247).

Perfect and Imperfect Duties

Although the views put forward by Rhodes and Harris touch upon something important, their arguments are far from unproblematic, as shown by the debate and criticism sparked by their articles (Beauchamp 2005; London 2005; Sharp and Yarborough 2005; Wachbroit and Wasserman 2005; McGuire and McCulloch 2005; Sharpsay and Pimple 2007; Brassington 2007). John Harris argues, for instance, that to choose not to participate in medical research conflicts with the principle of fairness. Non-participants are illegitimate free riders if they later benefit from the research in receiving improved health care. In making this argument, however, Harris overlooks the fact that even non-participants pay for the health care they receive through taxation or insurance premiums, and that they also have no choice but to benefit from research-based health care. Furthermore, it can be argued that one of the benefits of modern society is precisely a kind of institutionalised free riding in the form of division of labour. This makes it unnecessary (and unfeasible!) for everybody to take part in any kind of research from which we might possibly benefit.[2]

[2] In order to show that non-participants are free riders, Harris needs to show that in declining to take part, non-participants actually hamper the research in a decisive way. Chan and Harris compare non-participation in research with non-participation in immunization (Chan and Harris 2008: 4). Their analogy is flawed, however, since the non-participation of just some individuals disrupts herd

A similar kind of objection has been made to Harris' use of the principle of a duty to help others by taking part in medical research (Brassington 2007; Sharpsay and Pimple 2007). Sharpsay and Pimple invoke the Kantian notion of *imperfect* moral duties as the most precise way to describe the relevant obligation here, saying that "participation in medical research per se is not morally obligatory, but neither is it supererogatory; it is one way in which people may choose to discharge their imperfect obligation to help others" (Sharpsay and Pimple 2007). A *perfect* moral obligation always to help others would make our lives unmanageable, as we are finite beings with limited means. And because participating in medical research is but one of many ways to help others in need, it can at most be argued to be an imperfect obligation to take part.[3]

For Rhodes, consenting to take part in medical research is to contribute to the common good. The debate on informed consent for or a compulsory participation in medical research, must therefore take place in the context of a common understanding of the common good. But in a pluralistic and liberal society, such a consensus is not necessarily reached, or even aimed at (London 2005; Sharp and Yarborough 2005). Different answers will be obtained for questions such as: What are the merits of good health? What constitutes good health? Do biobank and other medical research promote public health in the right way?

Rhodes' system of mandatory research participation entails a limited obligation to take part in research projects. Even such a limited obligation is, however, hard to uphold. As argued by Robert Wachbroit and David Wasserman, "research participation should be seen as a valuable civic activity, like school tutoring, volunteer fire-fighting, and neighbourhood patrolling. Like those other activities, it is a way for individuals to serve a community from which they derive many benefits. It should be encouraged and praised like those other activities, but there is no reason to single it out as the subject of a universal duty" (Wachbroit and Wasserman 2005: 48–49).

This line of argument removes medical research from the prominent position that compels Rhodes and Harris to see it as subject to a duty to take part. The prominent position of medical research establishes both for Rhodes and Harris a duty to take part based on intergenerational fairness: We have an obligation not only to maintain the present level of medical care, but have an obligation to improve it through research for the sake of future generations, just as preceding generations by their research participation have made the present level of medical care possible.

Such an imperative to undertake research should stem from the moral obligation we have to help alleviate the suffering of today and tomorrow. But for the

---

immunity but not research opportunity. We will suggest a better way to view how the principle of fairness relates to biobank participation later in this chapter.

[3] Chan and Harris also seem to tend towards viewing the obligation in question as imperfect: "How much money 'should' you give to charity or to good causes, how hard should you work to discharge your obligation to your employer? The absence of a definable answer to this question does not make giving to charity or doing a fair day's work any less of a moral good; neither does the problem of how much research is enough invalidate the obligation to pursue it" (Chan and Harris 2008: 10). Both here, and in their blunt statement "there is an obligation to support all sorts of public goods" (Chan and Harris 2008: 5), it is hard to make sense of their view, if the notion of obligation implied is the perfect one.

research imperative to be a moral obligation, something we *must* do, failing to do medical research must not only harm people; it must also be indispensable in avoiding (future) harm. In his book *What Price Better Health,* however, Daniel Callahan questions both these assumptions. In countering the argument that more medical research is indispensable in avoiding (future) harm, Callahan reminds us that helping others by participating in medical research is but one aspect of our vision of a good society. Providing social security, proper education, family welfare and so forth – along with improving health care – is a necessary, but not sufficient condition for fulfilling this vision, it is also important not to mistake social and cultural problems for medical problems.

Callahan does not accept the assumption that we have a duty to develop more effective medical treatments for future generations. He quotes Hans Jonas in support of his view: "The destination of research is essentially melioristic (The belief that improvement of society depends on human effort.). It does not serve the preservation of the existing good from which I profit myself and to which I am obligated. Unless the present state is intolerable, the melioristic goal is in a sense gratuitous, and this not only from the vantage point of the present" (Callahan 2003). Callahan, like Sharpsay and Pimple, thus classifies medical research as an imperfect moral duty.

The Dugnad Concept

The questions of both intergenerational and intragenerational justice *are*, however, pertinent in the promotion of medical progress, unless one dismisses any duty to contribute to such progress, as Callahan does. And in opposition to Callahan, Rhodes aims to make medical research a common good that is part of a larger social contract.

Another way of thinking about this is that such an understanding can be created for every research project. It is the research project – through its design, context and intention – that has to construct and establish the common good, in order to justify normative recruitment. We will now explore this idea by taking a closer look at a specific medical research project and a particular way of describing participation in the project. This exploration aims to make possible a more nuanced view of the way in which participation in medical research should be taken to be a perfect or an imperfect duty – or no duty at all. The implications of the answer to this question regarding the recruitment of participants will subsequently be pursued.

The Norwegian health study and biobank research project HUNT[4] is referred to by policy makers as the largest health *dugnad* in Norway – or even in the world. HUNT is one of the largest existing projects in genetic epidemiology in the world. But what does the Norwegian word "dugnad" mean, and how does it relate to participation in health surveys and biobank research?

[4] HUNT is an acronym for "the Health Study of Nord-Trøndelag" in Norwegian.

The dugnad concept stems historically from pre-industrial Norwegian farm regions. In these regions, the farms were rather small, the produce was consumed by the farm people themselves, and the market for goods and labour was limited. To undertake tasks like roofing and haying, which were uncomplicated but required a great deal of labour over a short time, farmers had to rely on a circle of neighbours to take turns helping out. This kind of work was not paid, but the farmer who benefited from the work was expected to treat the people who came to help by serving good food and beverages on the day of the dugnad, and maybe even to host a party for his workers.

A standard definition is that "dugnad is when the neighbours of a farmer gather at his farm to help him, without getting paid, to accomplish a large task" (Østberg 1926). The traditional dugnad concept excluded communal and legal duties, and singled out the kind of informal duty to take turns in helping one another. The dugnad institution relied on a mutual understanding of reciprocity between economically equal farmers, and the "relation of reciprocity comprised of generations" (Norddølum 1976, 1980).

New technology, increased trade and social differentiation made the structural conditions for the traditional dugnad institution fade away in Norway in the first half of the twentieth century (Klepp 2001: 84). The dugnad concept, however, has survived and is still widely in use in Norway (Norddølum 1976: 72–73; Klepp 1982: 92). The activities nowadays called "dugnad" are different from the original dugnad work, but share certain aspects of the "good old" dugnad or maybe just the "dugnad spirit".

From an international perspective, the Finnish concept of "talkoot", and the American concepts of a "bee" or "barn raising", both have a similar meaning to the Norwegian concept of the "dugnad". The authors of this chapter learned this from the entry for "dugnad" in the Norwegian version of the *Wikipedia* – an international project which might be termed "the largest dugnad ever", not in the sense of a system of reciprocity, but in the sense of making people contribute to the common good motivated by personal pride and solidarity without any economic gain. The Wikipedia project illustrates that to invoke the "dugnad spirit" can be used to motivate and describe phenomena worldwide.

The *dugnad spirit* denotes that the values of liberty, equality and fraternity are actively promoted by a group and its members in freely committing themselves to work together as equals for the benefit of all. Present day dugnad is first and foremost associated with volunteering to do unpaid work for the common good. To be able to term something a dugnad, and to take part in a dugnad, is to make the activity morally praiseworthy. The dugnad spirit is then seen as a manifestation of an unselfish attitude that runs counter to a disintegrating society based on purely contractual relationships, and emphasises a spontaneous solidarity that is seen as both a moral ideal and the glue of society.

To benefit from or to take part in a dugnad should be motivated by a shared and acquired social conscience rather than by calculations of profit or from fear of sanctions. Helge Norddølum gives an example of an exploitation of the dugnad institution when a wealthy farmer in the Norwegian county of Valdres arranged a dugnad to build a mountain hotel (Norddølum 1976: 72–73). The dugnad principle of reciprocity was violated, as the hotel owners would not subsequently help participants build their own hotels. The dugnad spirit of solidarity was also illegitimately invoked, as these hotels were built by unpaid workers in order to profit the owners. The obligations associated with an economy of mutual dependence were taken advantage of by entrepreneurs operating in a market economy system. Nonetheless, as the example of deCODE shows, invoking a kind of dugnad spirit does not exclude economic profit from the dugnad result, if it is seen as beneficial for the community in which one regards oneself to belong.

In this chapter we discuss the dugnad model in relation to recruitment to biobank research. The salient feature of the model is the equality of the participants, an element of non-economical personal interest or gain in taking part, a system of reciprocity, the invocation of civic duties and communal solidarity, and the pursuit of a common good including communal prosperity. In addition, the tasks of a dugnad should not be complicated or risky in a way which places undue burdens on the participants. The ends and tasks of the dugnad should not be controversial. Only if it is reasonable to expect everybody to be able to attend, and have no moral qualms about attending, is it possible to blame people for not showing up.

The dugnad model invites a description of both the motivation and the justification for biobank recruitment which more nuanced and integrated than just pointing to aspects like ethical or legal obligations, altruism and gift donation, economical profit or personal interest. To invite to a dugnad places an obligation on the host to make sure that the dugnad criteria are fulfilled. The project should form part of a system of reciprocity which promotes communal solidarity and the common good. No requirement of special skills or potential for harm should prevent anybody from taking part. In this way a dugnad project should act as an incarnation of citizenship and the ethics of belonging to a community.

Biobank Participation

Does the analogy of dugnad serve as a means to achieving a more adequate description of what participation in medical research in general and biobank research in particular entails? We take the HUNT study as a starting point for a general discussion of the relevance and implications of introducing the dugnad analogy to this field.

Fully 110,000 people in the Norwegian county of Nord-Trøndelag have been or will be invited to take part in HUNT3, the third round of HUNT studies from 2006 to 2008. The HUNT cohort consists of a major part of the population of the county of Nord-Trøndelag. All citizens aged 13 and upward have been invited to participate in HUNT by completing a questionnaire on health-related issues, to undergo optional medical tests, and (from HUNT2 onwards) to allow a blood sample to be taken and included in the HUNT biobank.

In the previous HUNT1 study in the 1980s and the HUNT2 study in the 1990s, the participation rates were 88.1% and 71.3% of the adult population, respectively (Holmen et al. 2004). The participation rate in HUNT3 is expected to be about 60%. Even if the participation rate is declining, these figures show that the majority of the people of Nord-Trøndelag not only support the research project, but actually decide to take part. Steinar Krokstad, vice-chairman of the HUNT research centre, explains the willingness to participate in this way: "In Nord-Trøndelag, there is traditionally a strong belief in the power of cooperation and collective action. Cooperation has been strong, and when HUNT has invited people to participate in a health dugnad, they have shown up" (Krokstad 2004).

Krokstad goes on to state that "modern society is characterised by the disintegration of the community", and that the HUNT dugnad will contribute to counteract this development in a threefold way: Firstly, HUNT by itself promotes the dugnad spirit in its participants. Secondly, HUNT might be able to detect adverse health consequences of societal disintegration. And thirdly, HUNT promotes collective action for improved public health:

> The people of Nord-Trøndelag can be the first to benefit from new ways to better public health, through knowledge that can be communicated to the whole world in international journals. (...) Norway has developed from a poor country with a lot of poor health and living conditions to be a country with the best public health in the world. The Universal Health Insurance and the social security net that protects us from poverty are based on the old principles of equality, liberty and fraternity. And these institutions still contribute to good public health (Krokstad 2004).

The drop in the participation rate between HUNT1, HUNT2 and HUNT3 indicates that the dugnad spirit has declined in Nord-Trøndelag. In this chapter we will not speculate on reasons for this, but rather note that in HUNT (as in many other projects world wide), there is a need for normative recruitment in order to secure a high attendance rate. This means that a crucial question is whether normative recruitment is always wrong and incompatible with the ideals of modern research ethics, or if normative recruitment in a case like HUNT is legitimate.

In a focus group study with HUNT researchers, we asked whether biobank participants should have priority in receiving public health care over those who do not participate.[5] No one thought so, but one researcher expressed the general sentiment towards those who do not participate rather succinctly by remarking that

[5] The focus group participants comprised people who had given their consent to participate in the HUNT biobank (5 groups), former participants who had withdrawn their consent to take part in the biobank (3 groups), and researchers who were involved in or had an interest in HUNT (5 groups). The groups were recruited with the help of HUNT biobank. The focus group sessions took place in the fall of 2004 and the spring of 2005. The five discussion themes of the focus groups were: (1) The use (and abuse) of the biobank material. (2) Their own decision for giving consent/not giving consent, and the appropriateness of different kinds of consent. (3) Duty vs. autonomy in biobank research participation. (4) Ethical and practical consequences of doing genetic research vs. other kinds of medical research in HUNT. (5) Commercialization of the biobank research. The focus group participants discussed (rather freely) questions concerning the use of general consent to biobank participation, the adequacy of a putative duty to take part, ethical consequences of commercial use of HUNT biobank material, and their general hopes and fears concerning the biobank research of HUNT. The focus group study was designed by two ethicists (Berge Solberg and Lars

"they should maybe search their consciences". Another researcher elaborated on this remark when asked whether biobank participation should be a legal duty:

> I think that everybody has a moral duty to participate. And I think that Norwegians in general see it this way, and that the participation rate in HUNT shows that the people in Nord-Trøndelag see it this way. To participate should not be a legal duty, since it interferes with the private sphere. But I think there are few people who would oppose participation in HUNT, if the collective goods it entails are clearly stated, and that we all agree that such a study should be a part of our collective efforts to improve our health service.

The concept of dugnad has the potential both to clarify and obscure the balancing of privacy rights, civic duties and legal duties going on here. We will show how by identifying the determining factors present in the HUNT and the MIDIA[6] research project.

Is HUNT a Dugnad?

The word "dugnad" does not explicitly appear in the official information material for HUNT. But the *dugnad spirit* is evoked in the way that HUNT motivates people to participate. Thus this analogy seems to be clearly warranted. In an information folder for HUNT3 we read:

> Something very important for public health is happening in our county right now! You can contribute to vital research and increased knowledge about diseases which are of concern to us all. (...) We have every reason to be proud of HUNT. HUNT is the largest health survey of the world. (...) Please participate! Let's give each other an hour for better public health!

The request for giving "each other an hour for better public health" refers to the time it takes to complete the HUNT questionnaire and give a blood sample.[7] The participants contribute, from this perspective, mainly by giving their *time*. The risk of participation is conceived of as negligible, and the participants are not asked to make huge sacrifices: they will leave the research centre in the same shape as before – except without a few centilitres of blood.

From this perspective, the participants are primarily asked to do a bit of *unpaid work*: to show up and take time to answer questions and allow for health data and a blood sample to be obtained. It is *work* in the sense that participation is not for personal health purposes: no individual feedback is provided on the basis of biobank research findings. When the participants have done their share, the job is done. In this way, participants are considered to be contributing as *citizens* rather than as *patients*. Moreover, the work is *unpaid* in the sense that except for the free brief health check, there is no compensation given to participants.

Øystein Ursin) and a social scientist (John Arne Skolbekken), who also was the group moderator. For a presentation of further aspects of the focus group study, see Skolbekken et al. 2005.

[6] "MIDIA" is an abbreviation for "Environmental causes of type 1 diabetes" in Norwegian.

[7] See Collins of UK Biobank in Petersen 2006: 491.

HUNT could be said to be a dugnad in the modern sense of being a *gathering of people to do unpaid work for some kind of common good*. Of course, both the "gathering" and the "common good" might be said to be quite abstract in this case: Like Wikipedia contributors, the participants do not actually gather at one place. The common good is also vaguely conceivable rather than directly perceivable for the participants. Moreover, the participants are, as we will see, a bit uncomfortable regarding their contribution as "work". Given the fact that the free personal health check offered by HUNT motivates some people to take part makes it contestable to call their participation "work", and even debateable if the participation is wholly "unpaid". Moreover, the participants in both a traditional and a modern dugnad enjoy benefits like good food and beverages, but this kind of benefit is not of the same personal nature as an individual health check.

HUNT could also be said to be a dugnad in the traditional sense of offering an *intergenerational system of reciprocation between equal parties*: No HUNT participant is more important than another, everybody contributes in more or less the same way, and everybody can expect the same kind of possible benefit from the research from an intergenerational perspective. This emphasises how both the HUNT study and the traditional dugnad can be viewed as a kind of insurance institution. In this view, however, a major disparity would be that while stepping outside the traditional dugnad institution might have implied grave and direct social and economic consequences for a farmer in the nineteenth century, a person declining to take part in the HUNT study today should, as a matter of principle, expect no personal consequences from his decision in the future provision of health care. It is an important part of the HUNT recruitment policy, however, to appeal to the direct personal gain in getting a free health check. In this way, participation is not purely altruistic – there is "something in it for me", which makes it meet a basic criterion of the dugnad design.

The Opinion of Biobank Participants

In the focus group study with HUNT participants, we asked whether biobank participation should be considered a legal duty (Skolbekken et al. 2005). Like the researchers, none of the focus group participants thought this wholly appropriate. Biobank research is conceived of as interfering with the private, or autonomous, sphere of the citizen. To protect such a sphere is viewed as fundamental to the Norwegian constitutional State, separating it from totalitarian regimes. The ability to excuse oneself from participation in HUNT based on religious views and views of bodily integrity is seen as important. Making the right to health care somehow dependent on one's participation in medical research was definitely not endorsed by the focus group participants, because of the observed right not to participate, as well as the fact that everybody takes part in financing the universal Norwegian health service by paying taxes.

The general line of thought, however, echoing the opinions of HUNT researchers, is that even though a *legal* duty would be wrong, people should feel a certain *moral* duty to take part in HUNT. Everybody should participate in HUNT, one man says, because "the ideal is of course that everybody should contribute to the community, but then again you have the right to decide when it comes to your personal stuff". Generally the interests of the State and its citizens are perceived as identical when it comes to the aims of biobank research: It is in everybody's interest to promote health by improving our ability to prevent and treat diseases.

Biobank research is perceived as a low-risk way of participating in beneficial medical research. The participants have quite vague ideas of the potential embodied in the research; perhaps their children or future generations will benefit from HUNT (Skolbekken et al. 2005: 340). The motivation for their participation is altruistic and patriotic: They are proud to take part in a study for the possible benefit of the whole world, and take pride in the fact that such an altruistic project has been initiated by, and is being accomplished with the massive participation of, people from their own county (Antonsen 2005: 104).

The Importance of Solidarity

The main elements in HUNT that constitute a dugnad can easily be identified. Even though it is different from a traditional dugnad in some respects, it seems fair to say that HUNT is a dugnad, or at least is a project in the dugnad spirit. Does it or could it, however, have elements clearly incompatible with being a dugnad?

The participants in our study were not asked to relate the concept of dugnad to biobank participation, but their answers concerning the importance of taking part points to elements of the concept of the dugnad. Participation should not be a legal duty, nor should the question of participation be entirely neutral in moral terms. Participation should be morally laudable as a positive voluntary commitment to contribute to the common good.

On the other hand, the participants see commercialisation of biobank research as a possible threat to this aspect of the endeavour. To make medicine for the rich rather than the needy, and thereby to profit from the voluntary contributions of the inhabitants of Nord Trøndelag, would be at odds with the nature of the biobank project as they perceived it. This shows that solidarity is an essential motive for participation in biobank research, and that commercialisation might frustrate this motivation and fundamentally alter the nature of the enterprise.

This can be illustrated by comparing the HUNT project to the story of the dugnad in Valdres to build mountain hotels. With their goal of private profit, the Valdres hotel entrepreneurs violated the dugnad principles of reciprocity and solidarity, and therefore their framing of the project as a dugnad was illegitimate. In the eyes of participants, taking advantage of the potential commercial aspect of biobanking would transform the project in an essential way: The project would be about non-reciprocated private profit rather than about the mutual or common good, thereby exploiting participants if involvement is presented as a dugnad.

Interestingly, the principles that HUNT participants regard as both essential to the legitimacy of the study and as threatened by commercialisation, are the same as the principles the HUNT project has to adhere to in order to qualify as a dugnad: HUNT must be in pursuit of the common good in solidarity, from which all participants and their descendants equally benefit. It is, however, important to note that commercialisation per se is fully compatible with these principles, as long as commercial research is incorporated into the system of research ethics committees (Kettis-Lindblad et al. 2006), and if it just accelerates certain fields of research in addition to, rather than instead of, publicly funded research for the common good.

Normative Recruitment and the Helsinki Declaration

According to the Helsinki Declaration, the interests of the individual should always precede those of the society (§5). "The subject should be informed of the right to abstain from participation in the study or to withdraw consent to participate at any time without reappraisal" (§10). In §11 it is declared: "When obtaining informed consent for the research project the physician should be particularly cautious if the subject is in a dependent relationship with the physician /or may consent under duress". Taken together, these paragraphs seem to say that all recruitment to medical research must be normatively neutral: One in general should never argue that a person ought to forsake his or her own interests to participate in the interest of future health care (§5), and in particular should never argue that he or she has a particular obligation to participate given the relationship of dependence between the person and the provision of health care (§10–11).[8] As we have seen, participants and researchers in HUNT firmly reject the idea of refusing non-participants the same rights to future health care as the participants. And is the moral pressure of the dugnad model exactly what these paragraphs are meant to exclude?

The principles of the Helsinki Declaration are both meant to secure the autonomy of potential participants, and to protect them from harm. As touched upon above, the nature of biobank research makes the risk for physical harm negligible. The most important concern is thus to guarantee that no one is deceived or coerced to take part. The crucial question, then, is whether and when normative recruitment implies the deception or coercion of individuals, which would thereby make it illegitimate. Is it possible to defend an ideal of free and informed decisions by all potential biobank participants as to whether or not to take part, if participation in the research project in question is presented as morally laudable or obligatory? Is it legitimate to appeal to the dugnad spirit in recruiting people to HUNT?

The Helsinki Declaration, Harris and the HUNT participants all agree that a fundamental principle of medical research is that participation is voluntary, and that no

[8] See http://www.wma.net/e/policy/b3.htm

one is invited to take part in research with an unfavourable risk–benefit ratio. Granting this, one starting point is to say that any medical research should identify the dangers and the interests of the participants and society in the project, in order to be able to state these dangers and interests clearly in the invitation to take part. It would now be unethical for researchers to invite individuals to take part in a study in which they did not think the invited really should take part. In other words: The researchers who invite people to take part in a project not only generally *have* an interest in a high participation rate; it is more precise to say that the researchers always *should have* an interest in a high participation rate. Researchers should believe that it is in everybody's interest that everyone who is invited will choose to take part.

The dugnad analogy is demanding in its aim for a collective consensus on the need and legitimacy of the research, and the moral duty to take part. The crucial point, however, is that this puts a normative pressure *on the invited participants and the project designers alike*. To present a medical research project as a dugnad should in general be done with extreme caution, as it is a strong rhetorical device that might blur reflections on personal risk, as well as the nature of the common good involved. To put a normative pressure on the participants in this way therefore puts a huge normative pressure on the research institution and the relevant governmental bodies. They have to ensure and be sure that a project meets the criteria of being a dugnad. Only if these criteria are met is the invitation to take part in a research dugnad valid and the use of normative recruitment legitimate.

Given a transparent and informative process of voluntary recruitment, the research institutions are dependent on the trust of potential participants. This makes an appeal to the dugnad spirit a double-edged sword: If the research projects are conceived by participants to rightly deserve the dugnad label, it might improve the participation rate, but if the project is seen as not deserving the dugnad label, it might mean that the participants lose their trust in the project altogether. The fear that this might happen partly explains the reluctance of research institutes in Norway to invoke the dugnad spirit explicitly in their official documents and invitations.[9]

Rather than being a *simple* way to recruit people for research, normative recruitment is a *demanding* way to recruit volunteers for a transparent project dependent on trust. Normative recruitment might nevertheless be a way to make clear the mutual duties of a research-based health service, and its potential patients and research participants. This might promote rather than hamper the ability of participants to make an *autonomous* decision as to whether or not they should take part, as prescribed by the Helsinki Declaration. Normatively neutral recruitment might downplay ethical aspects of the research, such as urgency and justice, because people are simply invited in a neutral way and may participate if they want. Nobody has said that they should take part, so the motivation to autonomously question the ethical aspects of the relevant research is significantly lower.

---

[9] Likewise, the Governmental Regional Research Ethics Committees not easily approve of normative words like dugnad used for research recruitment.

When Normative Recruitment is Not Justified

Appeals to dugnad and the dugnad spirit need to be justified in the general design of a biobank like HUNT as well as in the specific projects using biobank material, or in targeted biobank projects. What does a specific biobank research project look like, if normative recruitment is not justified? The Norwegian MIDIA research project on environmental causes of type 1 diabetes is illustrative here.

The starting point for the MIDIA project was that people with a special genotype will have a higher risk of getting type 1 diabetes. About 2% of the population is in this group. In MIDIA, pregnant women were invited to let their future newborn children take the genetic test for type 1 diabetes. About 2,000 "high risk" children were then expected to be identified. These children would be followed by researchers for about 15 years. Their mothers and fathers were asked to deliver faecal samples every month until the baby was 3 years old. In addition, blood samples and questionnaires were to be delivered four times the first year and then once a year until the age of 15 (Rønningen et al. 2007: 2405).

Although MIDIA was huge, prestigious, with substantial national governmental funding and of international interest, it was found to violate the Norwegian Biotechnology Act. After having identified about 1,000 babies at risk, MIDIA came to be seen as highly controversial by Norwegians. Parents who were warned of an increased risk for their children based on the predictive genetic test expressed fear and anger about having this information. From their perspective, the fantastic experience of having a baby was tainted by the focus on a possible future disease, without any ability to prevent the disease (Mor til to døtre 2007: 1824).

The Norwegian Biotechnology Advisory Board considered the project in relation to the Biotechnology Act. They concluded that the predictive genetic testing of children for diseases that cannot be prevented is forbidden by Norwegian law (Foss 2007).

However, whether MIDIA was in accordance with Norwegian law or not, is not the main point here. The important thing is just to give an example of a research project putting substantial burdens on the shoulders of the participant. In its invitation letter, MIDIA used a language of normative recruitment: "Congratulations on the birth of a newborn citizen! [...] It may seem early, but we would still like to invite you and your little newborn citizen to make your first benevolent contribution to society".[10] The invitation letter refers to citizenship, to the relationship between a citizen and society, to benevolent contributions and the common good. The baby is referred to not as an individual but as a citizen, with the expression of sentiments and ideas about what good citizenship and civic duties amount to. As we have already made clear in this chapter, our argument is not that this is principally wrong. Rather we argue that the legitimacy of this kind of normative recruitment presupposes certain kinds of research designs – such as fulfilling the criteria for being a dugnad.

[10] See http://www.fhi.no/dav/D651389BCD.pdf

Our question is then: If the MIDIA project was more or less presented as a dugnad – was it in accordance with a "dugnad design"? MIDIA revealed the results of a baby's predictive genetic test to its parents. There are no preventive measures available for type 1 diabetes. This caused psychological stress and worry for some parents. Some parents were given information that they later wished they had rather remained ignorant of, and the right not to know was neglected. The need to provide faecal samples, blood tests and answer questionnaires on a continuous basis added to participants' inconvenience. For a 15-year old MIDIA participant, there was a 93% probability that she would not get diabetes, and she would have to live with the risk awareness for the rest of her life without being part of any research project.

In sum, it is easy to conclude that a project like MIDIA did not have a dugnad design. The inconvenience was substantive rather than negligible. It is not in accordance with dugnad criteria to subject invited participants to severe inconvenience or risk. The *empirical* factors of the study design are in this way crucial to assess the *ethical* question of the legitimacy of normative recruitment. In the MIDIA case, implicit references to civic duties and explicit references to citizenship and contributions to society functioned as an illegitimate rhetorical device.

Accounts of Duties

The aim of an account of duties to participate in medical research is to provide a middle ground between asserting a general duty to take part in medical research and a general principle of normatively neutral recruitment of participants – which implies that the potential participant should not feel any obligation to take part. While a general duty is argued for on the basis of a relationship of mutual duties between the health care provider and recipient, normatively neutral recruitment is argued for on the basis of fundamental principles of medical research ethics.

Daniel Callahan, as we have seen, is dismissive of the argument that we have a *duty* to conduct and participate in medical research to benefit future generations, in the way preceding generations have made our health care system possible. He must then hold either that there never really was such a social contract between generations, or that we stand in a radically different relation to our descendants concerning medical research than did our forebears. Both of these substantial claims are rather controversial, and, as we have seen, have not been met with approval among HUNT participants.

More promising than generating controversy over a general duty to participate in medical research seems to be to develop Rhodes' and Harris' sense of a prima facie moral obligation to take part in medical research as accurately as possible. Harris argued that people who do not participate in research are free riders who opt for the benefits from medical research without making a contribution. Contrary to Harris' argument, the division of labour in modern society is a form of an organised system of legitimate free riders. This argument can be turned on its head, however. Considerations of justice might be deemed relevant for individuals *specifically* called on to participate in this division of labour, like in the HUNT case. Infinite duties is then transformed into socially finite and perfect ones if part of a well-organised and limited system of medical research as the one described in Rhodes' "novel proposal".

But rather than just asserting a general duty to participate in such a system of research, the dugnad analogy illustrates the need for a description of how such a duty presupposes specific conditions regarding the research design. The research design has to meet conditions concerning both the nature of the involvement of the individual in terms of beneficence and non-maleficence. But it also has to make clear its contribution to the creation of the common good. The dugnad model presupposes a sensitivity and openness for debate on whether and how the research design actually promotes the creation of the common good, as conceived in the community in question. *Pace*, the purely apolitical accounts of duties of research participation promoted by Harris and Rhodes, points to a justification of normative recruitment which is sensitive to politics.

Citizenship and the Ethics of Belonging

Our discussion of the HUNT case in view of the dugnad analogy has shown that talking about moral obligations to participate in medical research essentially involves detailed descriptions of the research in question, including aspects like its organisation, its aims, its beneficiaries, its potential, its urgency and aspects of belonging and membership. The discussion of whether potential participants have a perfect or an imperfect duty to participate in medical research on the basis of a limited description of the research involved is not very promising. It is difficult to make a plausible case by asserting an individual's general duty to participate. A limited description of the relevant research also does a poor job of describing the moral *motivation* to take part in specific research projects.

A nuanced and situated description of the normative basis for individual participation in collective projects is vital to the discussion of moral motivations and obligations in this field. The dugnad analogy introduced in the HUNT case shows this in an illustrative way. People take part in dugnad, not just as individuals, but as members of a community. Their motivation is neither purely altruistic nor purely egoistic. It is more about a sense of belonging on different levels: We belong to a society where health is a common good. We belong to a patient group or a local community that may make a difference regarding health for future generations.

In this way we are members of communities that involve a kind of *civic* duty to participate. As members, or citizens, the right thing to do is to participate. In this way it might be said to be a kind of patriotic act, in Charles Taylor's sense, because it

> ...transcends egoism in the sense that people are really attached to the common good, to general liberty. But it is quite unlike the apolitical attachment to universal principle that the stoics advocated or that is central to modern ethics of rule by law. The difference is that patriotism is based on identification with others in a particular common enterprise. [...]

> Patriotism is somewhere between friendship or family feeling, on one side, and altruistic dedication on the other (Taylor 1995: 188).

In this way, patriotism can be viewed as highly relevant for participation in medical research. Patriotism and dugnad thus go hand in hand. This could imply a "politisation" of science. But there is nothing wrong with that. Rather, the opposite is true: When medical research is "politicised" through concepts like citizenship, community, belonging and patriotism, the question is also raised regarding the direction and development this community and this research should be headed towards. Opposition to biobank research is typically a political one, like the critique of biobank research representing a "geneticisation" of medical research – shifting the focus away from social inequality and health to a focus on genetic explanations. Such opposition does not lead to less civic engagement, but rather more. This challenges research communities for certain research projects to be able to defend normative recruitment, and to make an appeal to the common good.

Conclusions

The dugnad analogy offers the opportunity to understand how a specific research project should be designed to support an asserted moral obligation to take part. Ignorable risk, ignorable inconvenience and a common good that addresses each person as a member of a community rather than just an individual, are core elements in the dugnad design. Normative recruitment should be seen as legitimate in these cases. That the criteria essential to the legitimacy of HUNT coincides with the criteria to qualify as a dugnad shows the potential suitability of such an approach.

Normative recruitment is a powerful rhetorical device. Medical research is not in general a dugnad, and normative recruitment is not in general legitimate. An important message of this chapter is that as early as possible in the design phase of a project, researchers should reflect on the relationship of their project to the community of potential participants and to the common good. This will imply a "politicisation" of medical research – but that would be for the better. Ethics separated from politics is anaemic. And anaemic ethics for biobanking benefits neither biobank research nor the participants.

Acknowledgments The authors would like to thank Nancy Bazilchuk, Bjørn Hofmann, Bjørn Myskja, Rune Nydal, John-Arne Skolbekken, Jan-Helge Solbakk, and two anonymous reviewers for their valuable comments to and linguistic improvements of this chapter. All translations from Norwegian were done by Lars Øystein Ursin and Berge Solberg.

References

Antonsen S (2005) Motivasjon for å delta i helseundersøkelser. Norsk Epidemiologi 15:99–109

Beauchamp T (2005) How not to rethink research ethics. The American Journal of Bioethics 5:31–33

Brassington I (2007) John Harris' argument for a duty to research. Bioethics 21:160–168

Callahan D (2003) What Price for Better Health? Hazards of the Research Imperative. University of California Press, Berkeley

Chan S, Harris J (2008) Free riders and pious sons – why science research remains obligatory. Bioethics Epub April 25, pp 1–11

Harris J (2005) Scientific research is a moral duty. Journal of Medical Ethics 31:242–248

Holmen J et al. (2004) Befolkningens holdninger til genetisk epidemiologi illustrert ved spørsmål om fornyet samtykke til 61.246 personer - Helseundersøkelsen i Nord-Trøndelag (HUNT), Norsk Epidemiologi 14:27–31

Kettis-Lindblad Å et al. (2006) Genetic research and donation of tissue samples to biobanks. What do potential sample donors in the Swedish general public think? European Journal of Public Health 16:433–440

Klepp A (1982) Reciprocity and integration into a market economy: an attempt at explaining varying formalization of the "Dugnad" in pre-industrial society. Ethnologia Scandinavica 12:85–93

London A (2005) Does research ethics rest on a mistake? The common good, reasonable risk and social justice. The American Journal of Bioethics 5:37–39

McGuire A, McCulloch L (2005). Respect as an organizing normative category for research ethics. The American Journal of Bioethics 5:W1–W2

Mor til to dotre (anonymous author) (2007) Gentesting - en belastning for pårørende. Tidsskr Nor Lægeforen 13:1824

Foss G (2007) Gentesting av barn i MIDIA-prosjektet. Genialt 3:12–15

Norddølum H (1976) Dugnaden i det førindustrielle samfunn. Dugnad 1:65–80

Norddølum H (1980) The "Dugnad" in the pre-industrial peasant community. An attempt at an explanation. Ethnologia Scandinavica 10:102–112

Petersen A (2006) The genetic conception of health: is it as radical as claimed? Health 10:481–500

Rhodes R (2005) Rethinking research ethics. The American Journal of Bioethics 5:7–28

Rønningen K et al. (2007) Miljøårsaker til type 1-diabetes. Tidsskr Nor Lægeforen 18:2405–2408

Sharp R, Yarborough M (2005) Additional thoughts on rethinking research ethics. The American Journal of Bioethics 5:40–42

Sharpsay S, Pimple K (2007) Participation in biomedical research is an imperfect moral duty: a response to John Harris. Journal of Medical Ethics 33:414–417

Skolbekken JA et al. (2005) Not worth the paper it's written on? Informed consent and biobank research in a Norwegian context. Critical Public Health 15:335–347

Taylor C (1995) Cross-purposes – the liberal-communitarian debate. In: Taylor C (Ed.) Philosophical Arguments. Harvard University press, Cambridge MA

Wachbroit R, Wasserman D (2005) Research participation: are we subject to a duty? The American Journal of Bioethics 5:48–49

Østberg K (1926) Dugnad. Norske Folkeskrifter 71:1–32

Embodied Gifting: Reflections on the Role of Information in Biobank Recruitment

Klaus Hoeyer

Abstract During the past 15 years, informed consent has become an intensely debated issue surrounding human genetic biobanks. While ethicists typically agree that informed consent is related to respect for autonomy, various ethicists view autonomy differently and hold different views of the underlying mechanisms and values at stake when promoting the use of informed consent. In this chapter I elucidate such differences and contrast them to an alternative vision of the informed consent procedure developed primarily by social scientists studying the political context for research biobanking. With point of departure in an anthropological fieldwork in conjunction to a Swedish research biobank, I evoke a particular version of the alternative image of the agency going into the process of biobank recruitment. I elaborate on this image by way of a literary analogy and argue that it can contribute new dimensions to the ethical debate of research biobank practices.

Introduction

For over a century, human specimens have been collected and stored in structured archives in combination with phenotypic information for the purpose of advancing medical research (Lawrence 1998). Such collections are today typically known as human biobanks. In the early 1990s, innovative techniques in genetic research made such collections useful for a series of new purposes besides creating a demand for new, large-scale biobanks. Accordingly, biobanks are no longer the result of efforts undertaken by individual researchers; today they are constructed as collective infrastructures for a broad range of future research. In tandem with these changes, the formerly mundane practices of biobanking have acquired increased ethical attention. We are confronted with epistemological as well as ontological questions concerning

K. Hoeyer
Department of Public Health, University of Copenhagen, Copenhagen, Denmark
e-mail: K.Hoeyer@pubhealth.ku.dk

J.H. Solbakk, et al. (eds.), *The Ethics of Research Biobanking*,
DOI: 10.1007/978-0-387-93872-1_16, © Springer Science+Business Media, LLC 2009

not only what biobanks or tissue samples are, but also about the role and identity of the people from whom the samples are taken. As suggested in the chapters 'Mapping the Language of Research Biobanking: An Analogical Approach' and 'The Use of Analogical Reasoning in Umbilical Cord Blood Biobanking' (both by Hofmann et al.) of this book we tend to make sense of new technologies by way of analogies.[1] With new technologies, such as large-scale, genetic research biobanking, new subject positions also emerge. Therefore, besides drawing analogies to understand the technologies at stake, it can be a task for the ethicist to employ analogies to elucidate new subject positions and thus facilitate contemplation of the issues at stake for the people delivering the human biological material. Are they *donors* (no longer engaged in research once a sample is taken), (ongoing) *research participants* (Tutton 2007) or participants in co-production (see chapter 'Trust, Distrust and Co-Production: The Relationship Between Research Biobanks and Donors' by Ducournau and Strand)? Who should we compare the sources of material with to make them comprehensible? How can we give them shape and identity in a way that facilitates contemplation of their interests and agency? In this chapter I will draw analogies to literary characters that embody some of the features assumed in the ethical debate to characterize the people from whom samples originate. By doing so, I purposely simplify a diverse set of actors into three ideal types that I hope may serve the heuristic purpose of elucidating different implicit assumptions in the ethical debate about the people from whom samples are taken.

The role of informed consent has become the key concern in relation to the recruitment process to research biobanks during the past 15 years, partly in reflection of the ambiguities revolving around the rights and duties of the people providing tissue samples to research. In fact, informed consent has largely captured the agenda of the ethical debate surrounding new biobank projects. Most ethicists agree that the informed consent requirement is related to respect for the autonomy of the person from whom a sample is collected (Weir 1998; Diest and Savulescu 2002; Hansson et al. 2006; McQueen 1998). However, ethicists differ as to *why* is informed consent important and *how* is it seen to relate to autonomy as well as *what* autonomy seems to imply. In this chapter I suggest distinguishing between two such modes of thinking about informed consent in biobank research, or two traditions, as I will call them. My sketch should be read more like an identification of ideal types than as a description of precisely delineated groups of work. I suggest that, on the one hand, the two traditions reflect different values and different theoretical genealogies: one is more closely affiliated with the politically infused patients' rights movement and one with Kantian ethics. On the other hand, they tend to share some assumptions concerning the decision-making process that informed consent is seen to feed into: decisions are expected to rely on the information delivered during the consent process and to represent a right of the patient or lay donor/participant.

I contrast these two modes of thinking with a position typical for social science studies of research biobanks. I argue that their view on the decision-making process is radically different, not least because empirical work suggests that potential

---

[1] For this, see also Hofmann et al. (2006a, b).

donors pay little attention to the information delivered. With point of departure in an anthropological fieldwork experience in conjunction to a Swedish biobank, I evoke an alternative image of the agency going into the process of biobank recruitment. I suggest that people faced with the option of providing informed consent in this particular political context occasionally experience the choice as a double bind: they are stuck between two situations they do not like. In this predicament they create a third option by consenting without being informed. I wish to elaborate on the agency going into this radical solution by way of an analogy to the work of the Swedish Nobel Prize winner Selma Lagerlöf, and I argue that this alternative image of the donor can contribute new dimensions to the ethics of research biobanking.

Prevailing Assumptions: Two Traditions in Medical Ethics

There are probably as many views on informed consent as there are medical ethicists. The topic has been so widely debated that it is hazardous to attempt an outline of two dominant understandings of this requirement in terms of how it has been discussed in relation to research biobanking. Nevertheless, this is what I wish to do. I do not claim that my description is fulfilling in terms of a comprehensive representation of the debate, but I use the distinctions I introduce heuristically to highlight similarities and differences that are under-discussed when informed consent is described as a solution to the ethical quandaries of research biobanking. To simplify such differences further I suggest a name for each of the two 'traditions': *empowerment* and *enrolment*. With the former I label a tradition affiliated with the patients' rights movement (emphasising the patient's right to abstain from research and treatment) and with the latter a tradition more closely related to moral philosophy, not least feminist, communitarian, and Kantian ethics (emphasising the mutual moral obligations of research institutions and research participants during enrolment into research). I wish to suggest that both traditions view informed consent as a process through which a potential donor or research participant acquires knowledge and understanding of the proposed research project and based on this information decides whether or not to agree to have a sample taken. It is possible, however, to distinguish between different perceptions of the task that the procedure is supposed to carry out, the criteria according to which the choice is supposed to be made, the values inherent in and sustained by the tradition, and, finally, who are seen to be the primary agents in biobank research recruitment.

In what I name the *empowerment* tradition, the consent requirement is expected to enhance the power of the individual. The procedure is a tool through which risks associated with the study are communicated and the right to abstain from participation is ensured. The reasons for emphasising this type of right are plenty and reflect a series of well-known ethical scandals in medical research and practice from the atrocities of the Second World War to the Tuskegee Syphilis Study, etc. (Faden and Beauchamp 1986; Rothman 1991). Though biobanking rarely leads to any physical injuries on the people providing the samples, every attempt to employ presumed consent or other noninformed consent procedures has been met with intense criticism by people arguing that donors are thus deprived their right to exercise their autonomy and to assess the relevant risks associated with a given study (Kaye and Martin 2000; Maschke 2006; Merz et al. 2004). Before people engage themselves in research with potential implications for their safety, they must make themselves aware of the risks and be ensured a right to abstain from participation. Information must *precede* the act of participation. The criteria according to which this information should be assessed are rarely specified, but it is stressed that declined participation should not have any implications for their future treatment options and that nobody should feel a pressure to explain their reasons for opting out of a given project. The unconditional right to abstain is sustained by the Helsinki Declaration and numerous other, more local, policies and guidelines for procurement of biological samples. The insistence on the right to abstain from participation regardless of one's reasons indicates that personal preference is reason enough; to respect autonomy is to respect what people want (Eriksson and Helgesson 2005: 1072). Autonomy is therefore understood as 'self rule; the ability and tendency to think for oneself, to make decisions for oneself about the way one wishes to lead one's life based on that thinking' (Gillon 2003: 310). The emphasis on the right to *abstain* from participation, which reflects the genealogy of the informed consent requirement as a protection of the vulnerable individual, implicitly points to values inherent in and sustained by the consent requirement. A certain amount of scepticism towards the research institutions is not only desirable but it is facilitated by way of the requirement. Some even take for granted that if only people were adequately informed they would all be much more sceptical (Sigurdsson 2001). Hence, the potential research participants are assumed in this tradition to act as sceptical individualists, and they ought to be in a position to protect themselves against the research institutions. Accordingly, the primary agent in the recruitment process becomes the potential participant actively utilizing the option of declined participation, and as a result of this mechanism researchers must suggest only projects that will be deemed worthwhile and harmless by the participants.

If we were to give literary shape to this sceptical individualist, we might draw an analogy to the neoclassical novel *Catch 22* by Joseph Heller where all sympathy rests with the poor soldier Yossarian who sticks to his own ambition of survival when faced with an insane wartime bureaucracy. Yossarian refuses to buy into discourses and absurd rules putting his life at constant risk; he claims the right to his own life and body and makes up his own mind concerning what he owes to his country. As a legendary literary figure Yossarian can facilitate our contemplation of biobank ethics by giving shape and character to otherwise abstract assumptions running through ethical debates.

The empowerment tradition has been criticised by ethicists who have pointed to the fact that most of us wish to benefit from the fruits of medical research, and therefore we also have an *obligation* to donate to research biobanks. It has been suggested that it is quite simply ethically inadequate to focus on the right to abstain from donation (Chadwick and Berg 2001). I gather various such types of criticism under the name of the *enrolment* tradition, which is more closely related to moral philosophy, in particular Kantian ethics, rather than policy guidelines and the patients' rights movements. From a Kantian perspective it is not sufficient that information about risks precedes the act of research participation; a proper *choice* must be made and this choice is not morally responsible if it simply rests on personal preference and risk assessment. The definition of autonomy as 'decisions... about the way one *wishes* to lead one's life' is therefore, from this perspective, inadequate (O'Neill 2002).[2] Kant explicitly stated that following any type of preference or inclination would be to act as an animal with no moral capacity: 'I shall therefore not follow my inclinations, but bring them under a rule' (Kant 1775–1780/1997: 126). From this perspective, informed consent not only empowers the individual, but it also attributes a responsibility to the potential donor for assessing the common good in a given project. Accordingly, the enrolment process is configured as a situation in which a research institution and an individual establish mutual understanding of the future at stake. In line with this reasoning Onora O'Neill has argued that too much emphasis on informed consent and autonomy-as-self-rule detracts attention from the need to ensure *trustworthy* institutions and valuable research projects (O'Neill 2002). Rather than focusing on the right to abstain from participation, donation can be viewed as a positive right to engage with projects deemed worthy of support. The classical work on social policy by Richard Titmuss has fed into this type of considerations by highlighting how good research projects can facilitate altruism and community feeling (Titmuss 1997). The primary agent in the recruitment process is therefore no longer the donor, but the research institutions responsible for creating valuable and trustworthy research. And in contrast to the empowerment tradition, which typically views the donor as a sceptical individualist, the enrolment tradition wishes the potential donor to act as a prudent collectivist.

A legendary literary figure that could give body and flesh to the assumptions about proper conduct among donors in the enrolment tradition might be Dostoevsky's Prince Lev Nikolayevich Myshkin from *The Idiot*. Also faced with, at times, absurd bureaucracy and impossible demands, this man, in contrast to Yossarian, refuses to think only of himself. He risks his life and health; he only has the general well-being of others as his goal and ambition, and, tragically, he neglects the (unfair?) demands posed by the ones he love in favour of more abstract visions of a generalised noble conduct. As a contrast to the intuitive sympathy readers feel with the basically selfish Yossarian, the image of Prince Myshkin pays homage to the serenity of perfect human agency and the world to be if only everybody pursued these ideals.

In Fig. 1 these differences are summarised and contrasted to ideas immanent in social science studies of the recruitment process to research biobanking. The findings and arguments from these studies are elaborated below prior to some more detailed observations from a specific study in Sweden exemplifying the need to complement the images of donors as either sceptical individualists or prudent collectivists with less logocentric ideas about the agency going into the act of tissue storage for medical research.

[2] For this, see also Thorgeirsdóttir (2004) and Whong-Barr (2006).

	EMPOWERMENT	ENROLMENT	ENACTMENT
What does information do?	Enhances the power of the individual (by allowing him/ her to refuse participation)	Enhances the responsibility of the individual (and the institution)	Has a power effect (e.g. by limiting the probability of litigation)
View on the decision-making process	Information precedes act	Information precedes choice	Action and context shapes perception of information
Criteria for choice	Information assessed in light of preferences	Information assessed in light of moral duty	The act embodies values reflecting the (emerging) institution
Implicitly emphasise the right to…	Decline participation (participant's right)	Chose donation (donor's right)	Facilitate gifting (institution's right)
Values inherent in or sustained by the perspective	Scepticism	Trust	Community sharing
Sources of biological material viewed as…	Participants	Donors	Givers (expecting reciprocity)
… and assumed to act as…	Sceptical individualists	Prudent collectivists	Embodied activists
Primary agent in research recruitment	Participants	Institutions	Donor and institution mutually constitutive
Literary analogy to donor representation	Yossarian (Heller, Catch 22)	Prince Myshkin (Dostoevsky, The Idiot)	Elsalil (Lagerlöf, Mr. Arne's Money)
Theoretical genealogy	Helsinki Declaration, policies and guidelines (patient's rights movement), medical ethics	Kant, philosophy, medical ethics	Nietzsche, Foucault, social sciences, critique of medical ethics
Proponents include	Gillon 2003; Greely 2007; Maschke 2006; Merz 1997; Sigurdsson 2001	Chadwick and Berg 2001; Lipworth et al. 2006; Thorgeirsdóttir 2004; Whong-Barr 2006	Busby 2004; Busby 2006; Haimes and Whong-Barr 2006; Hoeyer 2003; Skolbekken et al. 2005

Fig. 1 Three perspectives on informed consent and the role of agency in biobank literature

The Contribution from the Social Sciences: Empirical Studies

In line with the ethical commentary, social science studies of the recruitment process to research biobanks have tended to focus on the issue of informed consent, though generally by highlighting the social implications of, rather than the individual intentions with, research participation. Some social scientists, who share the values of scepticism prevalent in the empowerment tradition, have suggested extending consent procedures to better accommodate the social context of biobanking, for example, by way of community consent (Weldon 2004). Others, more in line with the values of the enrolment tradition, but with differently structured arguments, have repudiated the public focus on informed consent for deflecting attention from what is deemed more weighty issues (Brekke and Sirnes 2006) relating to, in particular, trust and benefit sharing (Busby and Martin 2006; Petersen 2005; Haddow et al. 2007).

The most obvious discrepancy between social science studies and the empowerment and enrolment traditions, however, can be identified by looking closer at the qualitative, empirical studies carried out by social scientists studying the recruitment process. Such studies imply a very different approach to the power and agency going into informed consent as a social practice. The studies typically report a remarkable absence of attention among potential donors to the information provided through informed consent procedures. People asked to donate a tissue sample for medical research rarely read or remember the information that is supposed to either empower them or facilitate responsible decision-making (Busby 2004; Busby 2006; Haimes and Whong-Barr 2004; Hoeyer 2003; Skolbekken et al. 2005). If the content of information sheets is not actively used, there is reason to question the agency going into the donation (i.e. what motivates research participation if not information about the study) as well as the power effect of establishing an informed consent procedure.

These empirical, qualitative studies carried out in different European welfare states share some features going beyond the finding that potential donors rarely read or remember the information provided (which is hardly surprising to an empirical ethicist; Kaufmann 1983; Sugarman et al. 1999). Again, I wish to simplify the shared features and describe them by means of a name that can be contrasted with empowerment and enrolment, namely *enactment*. From the enactment perspective informed consent is not primarily a powerful tool in the hands of the prospective donor; it has a *power effect* that reconfigures the institutions. For example, consent procedures limit the probability of litigation and as such limit the options and rights available to donors. Also, from this perspective agency is viewed very differently. Rather than assuming that acts are based on reasoning and logical evaluation of information alone – what can be called a logocentric notion of human agency – acts are seen to emerge within (and contribute to the shaping of) a social context (Hoeyer and Lynöe 2006). In the words of Nietzsche: 'there is no "being" behind the doing, effecting, becoming; "the doer" is merely a fiction added to the deed – the deed is everything' (Nietzsche 1887/1967: 481). The acting embodies values, but such values are not necessarily *criteria* for choice in the more logocentric notion of decision-making held by empowerment and enrolment proponents. Rather, values reflect social context – such as the National Health Service (NHS) (Busby 2006) or other institutional and religious arrangements facilitating tissue donation (Hoeyer 2003; Simpson 2004); values emanate from acts when for example donations create or confirm a sense of belonging to a particular community, of doing a good deed. Donor and institution can therefore be seen as mutually constitutive; they emerge in their particular configuration through the acting that is performed. Values identified in these studies typically relate to community belonging and hopes for scientific progress, but donors are not described as detached 'altruistic' sources of biological material. On the contrary, they are *givers* with all the reciprocity involved in gift giving. In the classical book *The Gift* Marcel Mauss argued that any gift involves a threefold duty: to give, to receive and to reciprocate (Mauss 2000). Titmuss (1997) adopted Mauss' argument for an altogether different purpose when he suggested that proper institutions might promote altruistic gifting (Tutton 2004). Altruism was not a concern of Mauss (Sigaud 2002); you give because you are obliged to do so, and the gift does a job for you by way of obliging others. There are no free gifts (Frow 1997). When people donate, they do so in reflection of the services they already have received (from NHS, etc.) and with the expectation that the gift incurs on the receiving institution an obligation to carry out proper research. They typically prefer *not* to receive money, partly because payment in their view would set researchers free to use the tissue to anything they like (Hoeyer 2005). Therefore, people act not as sceptical individualists or prudent collectivists who carefully assess the research protocol; they embody and propagate values through their act of gifting; they are embodied activists.

I will develop the notion of the embodied activist further as I now go into more detail with some findings from a study carried out in northern Sweden. My point will be to explain why either sceptical individualism or prudent collectivism is not always a desirable option for the potential donor, and thus to engage the reader in the dilemmas of the embodied activist who donates without studying the consent sheets provided. Thereby, I also wish to endow the notion of activist with connotations other than those of the patients' rights movement. To reach this end, I elaborate on a third literary analogy.

The Importance of Political Context: A Swedish Example[3]

Since 1985 inhabitants in Västerbotten County in northern Sweden have been invited at the age of 40, 50 and 60 to undergo a medical check-up at public healthcare centres as part of a preventive healthcare programme. During the examination, which takes approximately 2 h, a risk profile is produced based on medical test results and an extensive questionnaire concerning diet, alcohol and smoking habits, physical activity, etc. Participants are often anxious to know their results and profile, and the examining nurse provides lifestyle counselling based on these. The questionnaire is stored in Medical Biobank at the local university hospital, and every participant is invited to donate an extra blood sample along with the questionnaire for the purpose of medical research. Approximately 68,000 inhabitants have donated a total of 78,000 samples, making Medical Biobank into a research biobank of international reputation (figures from June 2003).

[3] For a more elaborate version of the argument presented in the following paragraph please consult Hoeyer and Lynöe (2006).

In 1999, a start-up biotech company, UmanGenomics, was founded by the University of Umeå and the county authorities in order to optimise the utilisation of the stored samples. During 12 months of fieldwork, conducted intermittently from June 2000 until February 2004, I studied the establishment of UmanGenomics, the company's policy, and the way in which the ethical problems were discussed and dealt with by the authorities and the company. Besides interviewing policy makers, biobank researchers and nurses responsible for the medical check-ups, I observed the medical examinations of 57 participants at five different healthcare centres and I, subsequently, interviewed them separately. It quickly became obvious that people did not utilize the information they were offered. They tended to trust the nurses and the system the nurses represented. Even when they expressed doubts about the ethics of the research agenda, donors rarely studied the information sheet before deciding whether or not to donate. Therefore, the donation must be viewed as something other than an information-based, intentional (logocentric) act.

According to principle ethics, 'an autonomous person who signs a consent form without reading or understanding the form is qualified to act autonomously, but fails to do so' (Beauchamp and Childress 2001: 58). The empowerment tradition would locate the fault in the missing impact of personal preferences on the decision; the enrolment tradition would see the act as having no *moral* worth. Such views, however, regard the donation as a mistake rather than acknowledging donors' willingness to advance a program of research without knowing its exact purposes. When asked why they agreed to provide a sample for research, most donors talked in very general terms about the benefits for society and about a shared responsibility for advancing medicine. 'Science is really needed, that's all I know,' said one woman. Nobody could specify expectations concerning the type of research to be executed on their sample; indeed, the following response was typical: 'Well, I guess it's medicine – I don't know. They may research whatever they want with my blood [laughs]. As long as it's positive'. Similarly, when asking whether there was anything donors would not like their samples to be used for, I received only very vague answers. The clear majority had never thought about it.

The earlier responses do not, necessarily, imply that people had no reservations concerning commercial genetic research, or research in general. Such reservations were, however, very general. They related more to societal problems and the long-term future uses of science: matters that would never feature in a consent form. This may help explain why people pay so little attention to the information sheet: it quite simply does not address the issues they deem most important. In a survey sent to 1,200 Medical Biobank donors, we asked respondents to rank the issues in tissue-based research they believe were the most important. We had a response rate of 81%, and, of these, only 4% marked 'Whether the donor is informed about the purpose of research' (Hoeyer et al. 2004). The highest ranking concerns were as follows: whether all population groups would have equal access to research results; whether research would be generally applicable; that corporate interests should not determine the research outlook; issues of eugenic uses of genetic knowledge. These are indeed issues that can only be addressed at a societal level, not by individual donors. They would never feature in a consent form's information sheet.

If this inability of consent forms to address the concerns of tissue donors explains the lack of interest in the information sheets, it still fails to explain the motivation to donate. People want to help, but fear the misuse of science. They are suspicious of corporate interests, but with their donation they seem to accept UmanGenomics. How, then, is the choice to donate made? Must it be either intentional or non-intentional? By acknowledging as earlier the importance of the historico-political context, analytical attention moves away from the individual agent's *decision* to donate and on to the *situation* in which transfer of blood and healthcare data takes place. A focus on transactions as opposed to individual donations highlights an exchange *relationship* rather than presumably rational deliberation by one independent person. Thus viewed, the motivation for tissue transfer exposes an inter-subjective negotiation. In what follows, I suggest that the historical legacy of the Swedish welfare state is central to an understanding of this inter-subjectivity, and that the introduction of informed consent signals a change in that exchange relationship.

The Swedish welfare state has a long tradition of very elaborate healthcare services. Some participants in our study associated health so strongly with the state that they doubted whether it was legal for private companies to do health-related research. The primary health service relies on healthcare centres rather than on self-employed GPs. Often, various health services, including pharmacies, are located at the centres; the nurses and doctors employed are occasionally referred to as the parents of the People's Home, *Folkhemmet*, which is an old metaphor for the welfare state. In the southern part of Sweden, a privatisation process of healthcare centres is taking place, and the very establishment of UmanGenomics shows an increasing commitment to expanding the involvement of private companies and market forces in public health services. This commitment is expressed in a range of organisational aspects of health provision and amounts to a transformation of the formerly state-centred health system, for the better and for the worse. The historical experience of active political engagement in the welfare of the people, however, still permeates the health services (Vallgårda 2003). The ambition of the welfare state to take responsibility even for very intimate aspects of people's lives has inculcated the expectation of a particular set of rights and duties (Kerr 2004). The state, as a result, may be held responsible for any failure to meet people's health requirements; citizens are obliged to follow the advice given by the state. This type of healthcare system provides an historical experience worth contemplating when seeking to understand the phenomenon of donors not caring to read consent forms during medical check-ups.

What I wish to point out here is the specificity of the choice presumably offered through the increased emphasis on informed consent, and that it is only a 'choice' in a certain respect because it eliminates other choices. The historical vision of the welfare state implied a mutual responsibility. When introducing informed consent as a key regulatory practice, the governmental responsibility is reinterpreted by the authorities, who view it as their duty to offer informed consent (prior to participation) as a proxy for assuming responsibility for the ethical consequences and outcome of medical research. Simultaneously, the scope of responsibility of authorities is reduced. The meaning of 'high ethical standards' begins to equal 'voluntary participation'. The morality of the mutual exchange relationship, in this sense, is replaced by a contractual relationship, where the research participant makes a personal agreement with particular researchers. Ostensibly, the choice carries no special demands or limitations; citizens are free to engage in the projects they find appropriate. This reinterpretation of state authority is inconsistent with the historical experience of public healthcare centres. People's behaviour in the custody of a publicly employed nurse – their willingness to donate blood and healthcare information – shows that most people do not see themselves as engaged in a contractual relationship subject to sceptical, individualist assessments. Instead, they still seem to perform a duty embedded in trust and mutual obligations. The choice offered when introducing informed consent as a central regulatory practice eliminates another choice: to entrust the authorities with the responsibility of the ethical problems associated with research.

To observers affiliated with the enrolment tradition, this trust in authorities may seem to imply a lacking sense of personal responsibility, perhaps even impermissible irresponsibility. Considering the moral anxieties espoused by donors concerning matters such as ensuring equal access to research results, balancing corporate interests with public health goals and preventing eugenic uses of scientific results, however, it is perhaps understandable if they want the authorities to take responsibility for the construction of good institutional conditions for research. But where is the political commitment to ensure all population groups equal access and other concerns attributed great importance by donors? Are donors made responsible for research outcomes through the consent process?

In short, when confronted with informed consent, donors inclined to trust the authorities are caught in what Bateson calls a *double bind* (Bateson 1972). If they accept, they engage in a contractual relationship that lightens the authorities' responsibility; if they decline donation, they cannot perform their duty as citizens, which similarly erodes the mutual obligations of the old welfare state. For citizens who simultaneously wish to encourage good research and fear the implications of insufficient balancing of corporate interests, etc., this is a lose/lose situation. The specificity, or peculiarity, of the choice offered through informed consent is not limited to the options presented for choice. Rather, the choice itself effectively replaces a mutually obliging, ongoing exchange relationship with the fait accompli of a contract.

Lagerlöf and Sacrifice as Power

The embodied act of donation without consideration of the offered information is a sort of solution to the double bind. Donors try to oblige the institutions beyond the contractual level through the logic of gifting. They *act*, but they are activists in a sense very different from the sceptical individualists. Some social scientists would be inclined to see the gifting as an effect of internalised power: that people are made to donate without realizing that it is contrary to their own interests. Power is seen to operate by making people choose what the institutions want – in line with how some social scientists read and use the work of Michel Foucault. This position, however, reduces the agency of donors to an imprint of institutions and it neglects the very values that these people claim to promote: scientific progress, equality and community. The people who donate rarely see their 'own interests' as tied to an atomic version of individuals detached from community progress. Therefore, we must try to evoke an alternative image of the donor that might communicate the *will* going into the donation. I recommend complementing the Kantian *will-of-conscious-intentions* with a more Nietzschen *will-as-doing*. But, if we should capture the essence of the Swedish donor described earlier, it is important to develop the masculine and preference-based will described by Nietzsche within a more feminist framework emphasising the will to care and community (Walker 1998). To reach this end, I wish to draw a literary analogy.

To give further shape to the embodied act of gifting described earlier, I therefore suggest thinking of this act in the company of the Swedish national literary icon, Selma Lagerlöf (1858–1940): the first woman writer in the world to receive the Nobel Prize.[4] A theme running throughout not only her books, but also her personal life, can be seen as an intense will to influence the world, that is, to *do* or *shape* the world. It is a will focused on the common good rather than personal, idiosyncratic preferences. To many, Lagerlöf was an early feminist inspiration as she made a living for herself, restored the family wealth that her father had wasted, resisted marriage to live with a woman, and insisted on her own criteria for good literature regardless of changing fashions. Many of her female characters are caught in situations leaving them few options to pursue their interests and longings, but as a forerunner of feminist literature Lagerlöf always resisted portraying them as victims of the society in which they live. In one of the most acclaimed Lagerlöf biographies, Henrik Wivel argues that the supernatural short story *Herr Arnes penningar* (Mr. Arne's money, 1904) is the key to understand this feature of her authorship (Lagerlöf 1904; Wivel 2005) and it is, as I will show, also vital in developing a new understanding of activism as embodied ambition rather than logocentric calculation.

The novel is set during a harsh winter in Bohuslän during the war-plagued sixteenth century when Bohuslän was part of Denmark. Three destitute Scottish soldiers approach a greedy and harsh man, Mr. Arne, and in disguise ask for his money. Mr. Arne fights to his death for the hidden gold and as a result almost his entire household is slaughtered in a merciless killing. A young woman, Elselil, manages to hide and as she survives she becomes torn by the memories of the callous killing of her foster sister. The soldiers take themselves to a town by the sea where they as noblemen spend the stolen money while waiting for the frozen sea to loosen its grip on the ships bound for Scotland. After Mr. Arne's death, Elsalil has nowhere to go and with help from a young fisherman she also takes herself to the town by the

[4] Lagerlöf has written a treasure of classical novels (*Gösta Berling's Saga, The Holy City*, and *The Emperor of Portugallia*) as well as a widely read children's book (*The wonderful adventures of Nils*), many of which have been adapted for motion pictures. When her characters develop throughout a story, as Nils Holgersson does in the children's book, they learn to care for others, to act rather than wait, and to share without compromising their integrity (Wivel 2005).

sea. Here, she falls in love with one of the soldiers, Sir Archie (a name that reads like an English version of Mr. Arne and points to the two mens' shared greed). At this point the ghost of the foster sister begins to appear in front of Elselil and Sir Archie. Elselil gradually understands that the man she loves has killed her sister and that her sister cannot rest in peace unless his wrongdoings are exposed. Sir Archie proposes to marry her as she has first longed for, but she cannot enjoy the pleasures of a marriage built on the blood of her sister. Elselil is caught in a *double bind*. She must do something to ensure justice for her sister but she cannot endure to see her lover tortured and executed. She embodies an intense will, and she acts to bring harmony in her longing for a world without the tension. She decides that the wrongdoings of the man must be exposed, but that she will ensure his escape to Scotland. She acts. As local soldiers following her hint at a tavern approach the couple to capture Sir Archie, she tries to make the passively waiting Sir Archie leave. He refuses, first because he has decided that marrying Elselil is his only protection against the ghost, and then – as he realizes that she has deceived him – because he wants to humiliate her. She agrees to follow him just to make him move and ends in shield-protecting Sir Archie against the soldiers coming to arrest him. In the untenable situation she does the impossible thing: she gives her body as a shield and she personally penetrates her heart with one of the weapons of the surrounding soldiers. In the scuffle, Sir Archie escapes and brings the body of Elselil along with him. With possessive love he wants to build her an eternal stone palace as a tombstone. This idiosyncratic version of a masculine prison for the slain female spirit never materializes, however. The winter will not loosen its grip of the ships. Sir Archie is captured and slain on his boat, and Elselil is carried back across the frozen sea as the heroine who brought justice and peace to her foster sister while protecting her love. He dies – not because of her, but because of his own stupidity, greed and the coldness of winter that approximates the coldness of his heart. Elselil incarnates the bodily gift that warms the world – the bodily agency aimed at care and justice. She remains free from the masculine dominance of Mr. Arne's greedy household and Sir Archie's stone prison. Her sacrifice is an act of power that shapes the world in her image – and as the procession moves towards land, a spring storm liberates the sea from the prison of ice.

Conclusions

In this chapter I have pointed out how ethicists who agree on the importance of informed consent in research biobanking might do so for different reasons. I have elucidated and simplified two ways of thinking and then contrasted them with an approach taken in certain empirical qualitative studies carried out by social scientists. The main discrepancy emanates from the fact that while the former two traditions expect information to be an integral part of the decision-making, the latter must try to explain the fact that few people actually pay close attention to information sheets. Such differences are summarised in Fig. 1. I have explained the embodied gifting that I observed in my own fieldwork in Sweden as a response to a double bind reflecting a particular historico-political context, and the nature of this response was exemplified further by way of literary analogy to the work of Selma Lagerlöf.

What is the point of making us think about a character like Elselil? First of all, I wish to acknowledge the agency going into the act of giving 'parts of you' in the shape of phenotypic and genotypic data. Such acts are inadequately represented when seen as resting on either a deficit of 'real' information and understanding or as mere imprints of power structures on donors' prey to false consciousness. With their donation people enact values and such values, I think, should be respected. In line with the arguments of the enrolment proponents, there is an immense task in ensuring the trustworthiness of research institutions by way of explicitly addressing the issues of concern to the donating public. While informed consent is a valuable tool for providing the individual with means to assess the risk of, for example, invasive surgery or clinical trials, the majority of the concerns at stake in research biobanking operate at a very general level where consent procedures are of little help. Like Elselil, donors are actually left with few options. But many of them continue to care about the direction medical research is taking. Currently, medical research infrastructures are undergoing restructuring with new funding regimes, changed rules for patenting, public/private partnerships, etc. All of these changes will affect the most pressing issues of the donating public. By thinking more carefully about the embodied agency and the need to respect this as *will* and not as some detached, unexplainable 'altruism' (implying that once the sample is taken anything can be done to it), I hope that we might come up with better solutions for regulatory frameworks than those focusing only on the needs of the sceptical individualist.

I do not wish to *replace* empowerment and enrolment with enactment thinking. Rather, by adding a new analogy I wish to *expand* the repertoire with which we contemplate the subject positions emerging with contemporary research biobanking. Biobanks will enrol a number of people in different social and cultural settings. Probably, some of these people will act as sceptical individualists, some as prudent collectivists, some as embodied activists, while for some people we need to produce additional images. It is in fact an important task in the process of ethical reflection to encompass the diversity in subject positions. Also, I believe it is important not to let one representation of the power involved in the informed consent process to take precedence over all others, but rather to move between the different modes of inquiry and the insights they deliver. Informed consent, in principle, provides potential research participants with both options and responsibility. Normatively speaking, we have to consider the responsibility of the individual and the research institutions, as pointed out by the enrolment tradition. Also, clearly, empowerment theorists are right: informed consent does provide the potential research participant with leverage in the recruitment process and this leverage can be of crucial importance when seeking to ensure vulnerable groups against undue pressures. The social scientists are also right, however, when they point to the power effect of consent procedures in terms of responsibilization of the research subject and organizational concerns about litigation. Being informed is both a powerful tool in the hands of the research subject and a tool of power for the research institutions. In terms of Hans Herbert Kögler we need to encompass both hermeneutics and post-structuralism – because being informed is a double-edged sword (Kögler 1996). The empowerment and enrolment traditions help us understand how in the process of a recruitment process mutual understanding of the research purpose may evolve and the good reasons for wanting this to take place. The enactment tradition makes us appreciate some of the unintended implications of the practices established to reach these ends. By expanding, in place of monopolizing, the modes of inquiry we stand a better chance of addressing the needs of the people contributing to, and the ones dependent on results from, future research biobanks.

References

Bateson G (1972) Steps to Ecology of Mind. University of Chicago Press, Chicago

Beauchamp TL, Childress J (2001) Principles of Biomedical Ethics. Oxford University Press, Oxford

Brekke OA, Sirnes T (2006) Population Biobanks: The Ethical Gravity of Informed Consent. BioSocieties 1:385–398

Busby H (2004) Blood Donation for Genetic Research: What Can We Learn from Donors? In: Tutton R, Corrigan O (Eds.) The Gift: The Donation and Exploitation of Human Tissue in Research. Routledge, London, pp 39–56

Busby H (2006) Consent, Trust and Ethics: Reflections on the Findings of an Interview Based Study with People Donating Blood for Genetic Research for Research Within NHS. Clinical Ethics 1:211–215

Busby H, Martin P (2006) Biobanks, National Identity and Imagined Communities: The Case of UK Biobank. Science as Culture 15:237–251

Chadwick R, Berg K (2001) Solidarity and Equity: New Ethical Frameworks for Genetic Databases. Science in Society 2:318–321

Diest PV, Savulescu J (2002) For and Against: No Consent Should Be Needed for Using Leftover Body Material for Scientific Purposes. British Medical Journal 325:648–651

Eriksson S, Helgesson G (2005) Potential Harms, Anonymization, and the Right to Withdraw Consent to Biobank Research. European Journal of Human Genetics 13:1071–1076

Faden R, Beauchamp TL (1986) A History and Theory of Informed Consent. Oxford University Press, Oxford

Frow J (1997) Gift and Commodity. In: Frow J (Ed.) Time and Commodity Culture. Essays in Cultural Theory and Postmodernity. Clarendon Press, Oxford, pp 102–217

Gillon R (2003) Ethics Needs Principles – Four Can Encompass the Rest – and Respect for Autonomy Should Be "First Among Equals." Journal of Medical Ethics 29:307–312

Greely HT (2007) The Uneasy Ethical and Legal Underpinnings of Large-Scale Genomic Biobanks. Annual Reviews of Genomics and Human Genetics 8:343–364

Haddow et al. (2007) Tackling community concerns about commercialisation and genitic research: A modest interdiciplinary proposal. Socila Science & Medicine 64:272–282

Haimes E, Whong-Barr M (2004) Levels and Styles of Participation in Genetic Databases: A Case Study of the North Cumbria Community Genetics Project. In: Tutton R, Corrigan O (Eds.) Genetic Databases: Socio-Ethical Issues in the Collection and Use of DNA. Routledge, London, pp 57–77

Hansson MG et al. (2006) Should Donors Be Allowed to Give Broad Consent to Future Biobank Research? The Lancet Oncology 7:266–269

Hoeyer K (2003) "Science is Really Needed – That's All I Know." Informed Consent and the Non-Verbal Practices of Collecting Blood for Genetic Research in Sweden. New Genetics and Society 22:229–244

Hoeyer K (2005) The Role of Ethics in Commercial Genetic Research: Notes on the Notion of Commodification. Medical Anthropology 24:45–70

Hoeyer K et al. (2004) Informed Consent and Biobanks: A Population-Based Study of Attitudes Towards Tissue Donation for Genetic Research. Scandinavian Journal of Public Health 32: 224–229

Hoeyer K, Lynöe N (2006) Motivating Donors to Genetic Research? Anthropological Reasons to Rethink the Role of Informed Consent. Medicine, Health Care and Philosophy 9:13–23

Hofmann B et al. (2006a) Teaching Old Dogs New Tricks: The Role of Analogies in Bioethical Analysis and Argumentation Concerning New Technologies. Theoretical Medicine and Bioethics 27:397–413

Hofmann B et al. (2006b) Analogical Reasoning in Handling Emerging Technologies: The Case of Umbilical Cord Blood Biobanking. The American Journal of Bioethics 6:49–57

Kant I (1775–1780/1997) Lectures on Ethics: Immanuel Kant. Cambridge University Press, Cambridge

Kaufmann C (1983) Informed Consent and Patient Decision making: Two Decades of Research. Social Science & Medicine 17:1657–1664

Kaye J, Martin P (2000) Safeguards for Research Using Large Scale DNA Collections. British Medical Journal 321:1146–1149

Kerr A (2004) Genetics and Citizenship: Exploring Transformations. In: Stehr N (Ed.) Biotechnology Between Commerce and Civil Society. Transaction Publishers, New Bruswick, pp 159–174

Kögler HH (1996) The Power of Dialogue: Critical Hermeneutics After Gadamer and Foucault. The MIT Press, Cambridge, MA

Lagerlöf S (1904) Hr. Arnes penge. Gyldendal, Köbenhavn

Lawrence S (1998) Beyond the Grave – The Use and Meaning of Human Body Parts: A Historical Introduction. In: Weir RF (Ed.) Stored Tissue Samples. Ethical, Legal, and Public Policy Implications. University of Iowa Press, Iowa, pp 111–142

Lipworth W et al. (2006) Consent in Crisis: The Need to Reconceptualize Consent to Tissue Banking Research. Internal Medicine Journal 36:124–128

Maschke KJ (2006) Alternative Consent Approaches for Biobank Research. The Lancet Oncology 7:193–194

Mauss M (2000) The Gift. The Form and Reason for Exchange in Archaic Societies. Routledge, London

McQueen MJ (1998) Ethical and Legal Issues in the Procurement, Storage and Use of DNA. Clinical Chemistry and Laboratory Medicine 36:545–549

Merz JF (1997) Psychosocial Risks of Storing and Using Human Tissues in Research. Risk: Health, Safety and Environment 8:235–248

Merz JF et al. (2004) "Iceland Inc."?: On the Ethics of Commercial Population Genomics. Social Science & Medicine 58:1201–1209

Nietzsche (1887/1967). On the Genealogy of Morals. The Modern Library, New York

O'Neill O (2002) Autonomy and Trust in Bioethics. Cambridge University Press, Cambridge

Petersen A (2005) Securing Our Genetic Health: Engendering Trust in UK Biobank. Sociology of Health and Illness 27:271–292

Rothman DJ (1991) Strangers at the Bedside. A History of How Law and Bioethics Transformed Medical Decision Making. Basic Books, New York

Sigaud L (2002) The Vicissitudes of The Gift. Social Anthropology 10:335–358

Sigurdsson S (2001) Yin-Yang Genetics, or the HSD deCODE Controversy. New Genetics and Society 20:103–118

Simpson B (2004) Impossible Gifts: Bodies, Buddhism and Bioethics in Contemporary Sri Lanka. Royal Anthropological Institute 10:839–859

Skolbekken J-A et al. (2005) Not Worth the Paper It's Written on? Informed Consent and Biobank Research in a Norwegian Context. Critical Public Health 15:335–347

Sugarman J et al. (1999) Empirical Research on Informed Consent. An Annotated Bibliography. Hastings Center Report, Special Supplement 1–42

Thorgeirsdóttir S (2004) The Controversy on Consent in the Icelandic Database Case and Narrow Bioethics. In: Árnason G et al. (Eds.) Blood and Data. Ethical, Legal and Social Aspects of Human Genetic Databases. University of Iceland Press, Reykjavik, pp 67–77

Titmuss R (1997) The Gift Relationship. From Human Blood to Social Policy.The New Press, New York

Tutton R (2004) Person, Property and Gift: Exploring Languages of Tissue Donation to Biomedical Research. In: Tutton R, Corrigan O (Eds.) Genetic Databases: Socio-Ethical Issues in the Collection and Use of DNA. Routledge, London, pp 19–38

Tutton R (2007) Constructing participation in genetic databases. Citizenship, governance and ambivalence. Science, Technology and Human Values 32:172–195

Vallgårda S (2003) Folkesundhed som Politik. Danmark og Sverige fra 1930 til i Dag. Aarhus Universitetsforlag, Aarhus

Walker MU (1998) Moral Understandings. A Feminist Study in Ethics. Routledge, London

Weir RF (1998) Stored Tissue Samples. Ethical, Legal, and Public Policy Implications. University of Iowa Press, Iowa City

Weldon S (2004) 'Public Consent' or 'Scientific Citizenship'? What Counts as Public Participation in Population-Based DNA Collections? In: Tutton R, Corrigan O (Eds.), Genetic Databases: Socio-Ethical Issues in the Collection and Use of DNA. Routledge, London, pp 161–180

Whong-Barr M (2006) Informed Consent and the Shaping of British and US Population-Based Genetic Research. In: Guston D, Sarawitz D (Eds.). Shaping Science and Technology Policy: The New Generation of Research Madison. University of Wisconsin Press, Madison, pp 291–311

Wivel H (2005) Snedronningen. En litterær biografi om Selma Lagerlöf. Gyldendal, København

Conscription to Biobank Research?

Søren Holm, Bjørn Hofmann, and Jan Helge Solbakk

Abstract There are certain social activities that are conceived as so necessary and important that we legally oblige people to participate in them. Most societies have at some point in time had general conscription of male adult citizens into the armed forces; some societies require people to serve on juries and some have compulsory immunizations. Finally, in most societies, taxation exists as a kind of socially accepted, although at the same time often disliked, kind of compulsion. In this chapter the analogies of conscription and taxation are pursued to see to what extent participation in biobank research could be perceived as so important that some kind of conscription for research would be justifiable.

Introduction

There are certain social activities that are conceived as so necessary and important that we legally oblige people to participate in them in person. Most societies have at some point in time had general conscription of male adult citizens into the armed forces; some societies require people to serve in juries, while some have compulsory immunizations. In many of these cases it is possible to avoid the conscription by conscientious objection, but conscientious objection does importantly require that a person can give an acceptable reason based on some deep personal belief or value explaining why it would be not only inconvenient, but deeply morally problematic for that person to participate in the activity. Conscientious objection is not merely opting-out; it is opting-out for a socially acceptable and personally important reason.

Conscription differs from other kinds of compulsion by the state, for instance, taxation, by requiring a specific bodily performance by the conscripted person.

S. Holm (✉)
Cardiff Centre for Ethics, Law and Society, Cardiff University, Cardiff, UK
e-mail: holms@cardiff.ac.uk

J.H. Solbakk, et al. (eds.), *The Ethics of Research Biobanking*,
DOI: 10.1007/978-0-387-93872-1_17, © Springer Science+Business Media, LLC 2009

He or she has to be in a certain place and do certain things, whereas no specific performance is required by the person who is taxed. The taxes can be paid with money obtained in numerous different ways.[1] If a state, therefore, decided to use already stored health information about people for epidemiological research purposes without consent, it would be closer to taxation than conscription. But biobank research usually requires the person to give a blood or tissue sample and to allow future access to specific health information, and this does not fit well within a taxation model. Another difference to taxation is the radical future-oriented nature of biobank research. Taxation pays for many activities that society pursues at this very moment in time, and most of those who are taxed also benefit from these activities. However, building up a biobank is almost completely directed toward the future.

The element of specific performance in conscription usually involves interference with at least three areas of concern that are normally protected by rights in liberal societies, that is, autonomy, privacy, and bodily integrity. Furthermore, it has traditionally only been applied in relation to military defence or other activities perceived as essential to the functioning of society. It is thus not strange that conscription is only rarely used.

Can we draw an analogy between the kinds of conscription that society currently accepts and claim that participation in biobank research is so important that conscription for research is justifiable?[2] This is the question that this chapter seeks to answer.

In the first part of the chapter we will briefly consider the justification(s) for conscription outside of the biobank area. The second part will then look at the question whether there is a general moral obligation to participate in biomedical research. It will argue for the interim conclusion that there are certain situations where there is a moral obligation to participate in research. The third and final part will then consider whether and in what contexts this moral obligation can justify legal conscription.

But before moving on to the ethical analysis, it is important to note that the value of a biobank for epidemiological research is not based only on the biological samples it contains. These samples become valuable because they are linked to health and other information, and over time the major contribution from research participants becomes the health information that is obtained by the biobank from continuous linkage to clinical databases (e.g., cause of death, hospital event, or prescription databases). What the participants give, or perhaps sacrifice, is thus not only a biological sample to be analysed but an important part of their informational privacy in an area where at least some information is likely to be sensitive.

---

[1] If I pay taxes with money I have stolen, the person I took the money from cannot usually mount a legal claim that the state is "handling stolen goods" and is liable to pay the money back.

[2] In this chapter the discussion focuses on biobanks established to enable epidemiological research. We do not discuss biobanks that are established for therapeutic purposes such as blood banks, cell banks, or bone banks.

What Is the General Justification for Conscription?

In a liberal society conscription requires a strong justification because it constitutes a direct and often significant infringement of personal liberty. Conscription for military service is the paradigmatic case of conscription. In times of peace conscription requires a person to give up control of his own life for a specified time, receive military training, and perform military service that may entail some risk. And in times of war the time aspect becomes open ended and the risk a real risk of violent and painful death.

How is such a basic infringement of the liberty, bodily integrity, privacy, and self-determination of a citizen justified? There are essentially four elements to the justification:

- Necessity
- Reciprocity
- Equity
- Good consequences

Necessity as a justification comes to the fore in times of war or impending war. The defense of the nation requires great sacrifices and one of them is conscription for military service. Similarly one of the arguments for conscription to jury duty is that it is necessary for the administration of justice and one of the arguments for compulsory vaccination that it is the only way to ensure herd immunity.

Reciprocity is also a possible justification for conscription. There are two forms of reciprocity involved: (1) historical and (2) concurrent. Historically I can live in the well-ordered and peaceful society that I live in because others have defended it in the past and this creates an obligation on me to reciprocate and do what is necessary to defend it now. Concurrently, the fact that others now defend society against external threats creates an obligation on me to reciprocate.

Equity considerations explain why conscription is general and not just conscription of, for instance, the poor and the unemployed, and also why it is generally not possible to pay somebody else to take my place in the army. Historically many societies allowed conscripted persons to pay someone else to take their place, but in modern societies with a much stronger emphasis on equality this is not seen as an ethically acceptable practice.

Finally, conscription should only take place where it produces better consequences than a less coercive method for obtaining the same social good. If reciprocity and equity concerns can be met without resorting to conscription, this is the better option.

In the present context it is important to note that many countries in Western Europe are now moving away from military conscription to a professionalized military with contracted soldiers. The reasons for this are complex, but one which is relevant here is that these states now believe that their defensive aims can be achieved without conscription. Conscription is no longer seen as necessary and it is believed that the good consequences can be obtained by less coercive means.

Furthermore, given that the risk of death or injury is currently low, equity and reciprocity considerations play a minor role.[3]

In other forms of conscription the balance between the different elements of the justification is different. In conscription for jury service reciprocity and equity are more important than necessity. We could recruit jurors in other ways, but given the importance of formal justice in the justice system equity considerations have increased weight. And in compulsory vaccination the strongest justifications rely on necessity and good consequences. Compulsory vaccination is often only considered, if voluntary vaccination rates are too low to prevent epidemics.

An important lesson to be learned is thus that whether there is a sufficient justification for conscription for a specific social activity is context dependent. The strength of the justification varies over time and between different contexts. This is also evidenced by the fact that, while some countries have conscription for jury duty or compulsory vaccination, many others do not. Justifying conscription for biobank research will thus have to show that a consideration of these four elements provides sufficient justification in the concrete context.

The Moral Obligation to Participate in Biomedical Research

Is there a general moral obligation to participate in biomedical research? There are a number of different ways in which such an obligation can be established (Brassington 2007; Chan and Harris 2008; Evans 2004, 2007; Harris 2005; Shapshay and Pimple 2007). It follows straightforwardly from consequentialist theory that, if participation in biomedical research generates good consequences, then there is an obligation to participate. Very similar arguments can be based on considerations of beneficence where benefit to future patients generates the obligation. The empirical premises in these arguments in relation to the predictable benefits of biomedical research, that is, that biomedical research produces good consequences and health benefits are undoubtedly true in relation to medical research in general. Whether or not we believe that benefits in medical treatment or benefits in understanding of risk factors and public health are the main contributors to increased longevity and decreased morbidity in modern societies, both these developments rely on good research evidence. It is, however, much less obvious that the same claims can be made for each individual research project. Much research is probably of very little value and some may actually have long-term negative consequences. The arguments based on consequences can thus generate a general obligation, but translating that into a specific obligation is fraught with difficulty (see also later).

John Harris claims that an obligation to participate can be based in nonmaleficence, our obligation not to harm others (Harris 2005). If his arguments are valid, they are important, because many people hold that it is worse to cause harm to others than not to help them, or to put it differently that obligations of nonmaleficence

---

[3] The same argument makes conscription as a compulsory contribution of money to a government during a time of war less relevant as an analogy in our context.

are stricter than obligations of beneficence. Some also hold with Kant that our duty not to harm is a perfect duty, but our duty to benefit is only an imperfect duty. But Harris' argument is controversial and possibly even self-defeating. An important premise in Harris' argument is that there is no moral difference between actions and omissions that lead to the same result, that is, the so-called *acts-omission doctrine* is flawed. But this is very controversial, as, for instance, evidenced in debates about moral differences between active and passive euthanasia. It is furthermore the case that, if Harris is right, then the moral difference between nonmaleficence and beneficence is undermined because it becomes difficult to differentiate between actions instantiating a fulfillment of each obligation. If there is no moral difference between acts and omissions, then there are many circumstances where an agent harms people simply by omitting to benefit them.

Another possible justification is in reciprocity. The treatments that I now receive and that help me are only available because others have previously participated in biomedical research. Therefore, I have an obligation to reciprocate and participate in research now. Not participating would make me a Rawlsian free rider benefiting unjustly from the contributions of others to an important social good. A counterargument is that no obligation arises from past contributions because the moral value of the contribution of others in the past and the cost to them is neither increased nor diminished by my receiving the benefit. And if medical progress is only an optional goal, then I can dissociate myself from obligations of reciprocity simply by repudiating that goal (Jonas 1972). That others then continue to pursue it does not create any obligation for me, unless my nonparticipation increases the costs for them.

Conscription for Biobank Research?

How good is the analogy between military conscription and conscription for biobank research and what are the problems in transposing a moral obligation to participate into a legally enforceable obligation? Searching for good analogies is notoriously difficult and evaluating their force tricky (Hofmann et al. 2006a, b), but let us never the less try.[4] Biobank research is analogous to the maintenance of an effective defense force in a number of ways. In the long term the results from biobank research will benefit and contribute to the maintenance of society. If disease is considered as an external threat that modern welfare states have an obligation to protect us against as has been argued using a neo-Hobbesian approach (Ashcroft 2005; Erin and Harris 1993; Holm 1999), then research participants could, together with the researchers and health care staff, be conceptualized as the frontline soldiers in the battle against disease. It could also be argued that the methodological desirability of having comprehensive, representative, and unbiased recruitment to biobank research provides an analogy to the equity and good consequences of elements of the

[4] For this, see also chapters "Mapping the Language of Research Biobanking: An Analogical Approach" and "The Use of Analogical Reasoning in Umbilical Cord Blood Biobanking" by Hofmann et al.

justification for military conscription. A second analogy is that a general moral obligation to contribute to the activity underlies both military defense of society and biobank research.

But there are also clear disanalogies. Whereas there are very good methodological reasons to pursue comprehensive, representative, and unbiased recruitment, biobank research currently proceeds on a voluntary basis with some degree of deviation from optimal recruitment but still seems to produce valid scientific findings. It is thus difficult to make a strong necessity claim for conscription unless biobank scientists are either less than honest about the value of their current research or can convincingly show that they could do much better if conscription was introduced.

In this context it is interesting to note that there are other, arguably even more important, health care activities that require access to bodily materials but where we have no conscription. The link between the activity and health benefits is much more immediate in relation to organ and tissue donation and the maintenance of blood banks and an adequate blood supply, but no liberal society has currently implemented conscription in relation to these activities. They thus count as significant disanalogies. A possible explanation is that a personal right to bodily integrity is valued highly in the sphere of medicine, perhaps more highly than outside this sphere.

There are also significant problems involved in actually implementing a conscription system. It is obvious, as mentioned earlier, that there is no direct inference from a general obligation to participate in biomedical research to a specific obligation to participate in a particular biomedical research project, just as there is no direct inference from a general obligation of beneficence to an obligation to give money to any particular good cause. This creates the first problem for the idea that we could have legally enforced conscription for biobank research. Someone would have to decide in each specific instance whether conscription to this biobank pursuing this specific line of research is justified; this would in many instances be a decision made under fairly radical uncertainty. To make it, the decision maker would not only have to have information about the proposed biobank but also information about whether other current research projects are trying to answer the same scientific question.

The second problem for the conscription idea is that it is not particularly plausible that conscription is strictly necessary to achieve the aims that society wants to pursue through biobank research. Conscription may plausibly increase the speed with which a biobank can be established and its representativeness, but we know that reasonably representative biobanks can be established without conscription. Given the relatively limited physical involvement participants have in biobank research, we might as well try to achieve representativeness through allowing increased incentives for participation. If the perceived problem is that some say "no" to participation, one standard solution in market economies is simply to pay them more or incentivise them in some other way.

The third problem arises within the reciprocity and the "free rider" argument. Biomedical research requires many inputs to succeed. There are, therefore, many ways in which a person can contribute to research. A person may recognize an obligation based in reciprocity toward those who previously contributed to research and think that she has discharged this obligation fully by giving regular, large monetary donations to a cancer research fund. Is there an argument to show that she has not discharged her obligation fully and that she also needs to become a research participant? If there is such an argument, it would have to be symmetrical also showing that those who were "only" research participants had not discharged their obligations fully unless they were also monetary donors. This point can be expanded to a more general point about the fulfillment of obligations based in reciprocity. An obligation to participate in research is just one of the many obligations we have in reciprocity, and it is not obvious that these should be individuated in a way that makes this a discrete obligation that has to be discharged by actual, personal participation. This could be argued to be just one element of a much more general obligation to contribute to health care (possibly dischargeable by being a conscientious blood donor) or an even more general obligation to contribute to a good society (possibly dischargeable by many different kinds of voluntary contribution or by paying your taxes in full and without complaint).

We might try to solve this problem by saying that a person has a specific obligation to participate in research because she has benefited specifically from research, but this only hides the fact that no one actually benefits from research as such except in a very nebulous and general way. Everyone who benefits, for instance, by having their illness effectively treated, benefits from specific research done in the past and, if there is an obligation, it must be an obligation to support such specific research in the future. But such a more specific obligation may be void, if the person is not or is no longer a suitable research subject for this specific line of research, in case, for example, their illness may be completely cured. And it may also be difficult to discern what research is actually connected to the research that created the benefit.

The fourth problem is that enforcement of conscription may well have negative overall effects in relation to recruitment to biomedical research. News stories about unwilling research subjects being forced to participate in research are probably not likely to increase the number of people volunteering for research projects.

How can these various arguments be summed up? Unless a person is a very strict consequentialist or a very strict libertarian, there is no knock-down argument either for or against conscription in the context of biobanks. Conscription is not totally ruled out but there is clearly a considerable epistemic and evidentiary burden that must be discharged before it can be established that conscription for the establishment of a specific biobank is justified in theory and possible to implement in practice.

Conclusions

In this chapter we have argued for a number of conclusions. The first is that there is an analogy between military conscription and conscription to biomedical research in general and biobank research in particular, but that this analogy is not perfect. The second is that there is a general moral obligation to participate in biomedical research and that this obligation is robust and strong. Somewhat paradoxically, however, the third conclusion is that there will be few, if any, situations where there is sufficient justification for transposing the moral obligation into an enforceable legal obligation. The main reason for this is that, whereas my moral obligation is general, any legal enforcement would have to be very specific.

As a moral person I ought to consider seriously participation in biobank research if asked, and I should only refuse participation if I have very good moral reasons for refusing, but I should not be forced to participate. Hence, the analysis of the analogies and disanalogies shows that there is an imperfect duty to participate in biobank research, which may, perhaps, be called a "bio-duty."

References

Ashcroft RE (2005) Access to essential medicines: A Hobbesian social contract approach. Developing World Bioethics 5:121–141

Brassington I (2007) John Harris' argument for a duty to research. Bioethics 21:160–168

Chan S, Harris J (2008) Free riders and pious sons – Why science research remains obligatory. Bioethics doi: 10.1111/j.1467–8519.2008.00648.x

Erin CA, Harris J (1993) Aids: Ethics, justice, and social policy. Journal of Applied Philosophy 10:165–173

Evans HM (2004) Should patients be allowed to veto their participation in clinical research? Journal of Medical Ethics 30:198–203

Evans HM (2007) Do patients have duties? Journal of Medical Ethics 33:689–694

Harris J (2005) Scientific research is a moral duty. Journal of Medical Ethics 31:242–248

Hofmann B et al. (2006a) Teaching old dogs new tricks: The role of analogies in bioethical analysis and argumentation concerning new technologies. Theoretical Medicine and Bioethics 27:397–413

Hofmann B et al. (2006b) Analogical reasoning in handling emerging technologies: The case of umbilical cord blood biobanking. The American Journal of Bioethics 6:49–57

Holm S (1999) Is society responsible for my health? In: Bennett R, Erin CA (Eds.) HIV and AIDS testing, screening, and confidentiality. Clarendon Press, Oxford, pp 125–139

Jonas H (1972) Philosophical reflections on experimenting with human subjects. In: Freund PA (Ed.) Experimentation with human subjects. Allen and Unwin, London, pp 1–31

Shapshay S, Pimple KD (2007) Participation in biomedical research is an imperfect moral duty: A response to John Harris. Journal of Medical Ethics 33: 414–417

Ownership Rights in Research Biobanks: Do We Need a New Kind of 'Biological Property'?

Paula Lobato de Faria

Abstract This chapter first revisits the classical ongoing legal debate around ownership rights in human biological material, based on the two opposite perspectives – one that defends an absolute non-patrimonial view, denying the possibility of the existence of a property right in this field and the other that defends the existence of a property right over human bodily material and considers that denying participants in scientific research property right over their biological material may be a source of unfairness to them. Second, it analyses the consequences of the application of classical property rights to the biological material, such as the Portuguese Law does, advancing several arguments from in support of the conclusion that classical property rights do not adjust to the juridical characteristics of human biological material and its use in biobanks for research. The chapter ends up, in a third part, with a draft proposal of a new juridical construction for contemporary law, within property rights, that is, a new concept of 'biological property', which should be shaped by a balanced respect for both individual and scientific/society interests and a specific legal framework within property rights law that could reflect the norms of biolaw already applying in our societies to human biological material (e.g. principle of non-commercialisation and principle of informed consent). Because of its novelty and complexity the idea of a 'biological property' presented in this chapter is in need of further development. Only an international normative framework would be adequate to create and determine the juridical background of a new kind of property adjustable to human biological material and its significance in modern societies.

P.L. de Faria
National School of Public Health, Universidade Nova de Lisboa, Lisbon, Portugal
e-mail: pa.lobfaria@ensp.unl.pt

J.H. Solbakk, et al. (eds.), *The Ethics of Research Biobanking*,
DOI: 10.1007/978-0-387-93872-1_18,

Introduction

Biobanking for profit is still a relatively small industry in the EU (Hirtzlin et al. 2003a, b), and it is natural that legal problems arising from donors who see their biological material giving birth to huge amounts of profit to research entities are still perceived as a kind of science fiction. Opinions like the one given by the European Group on Ethics in Science and New Technologies on the Charter of Fundamental Rights (European Group on Ethics in Science and New Technologies 2000) that acknowledges the controversial nature of the issue of the commercialisation of human biological material, drawing attention to the necessity of a deeper discussion around it, have not yet led to any changes in the European legal framework.

European law is shaped by the fundamental principle of the prohibition against financial gain from the human body and its products, and also by the philanthropic view that the donation of human organs, tissues, and cells should be unpaid and seen either as a moral duty or as a public welfare service. The majority of voices representing the European legal doctrine[1] as well as national bioethics committees in Europe seem to follow this postulate with rigour, refusing to accept the idea of the existence of a property right between a person and her/his body.

Europe has to date not experienced any judicial cases with an impact similar to the *Moore* vs. *Regents of the University of California* case (1990), which very early alerted the US jurists to the potential unfairness of research profits that would not benefit the person who contributed the biological material for the research, or The *Washington University* vs. *W.J. Catalona et al.* (2005) case that only very recently (January 2008) was resolved, showing that the US judges have so far preferred to choose the solution that reflects the 'best general interest', thus rejecting the 'private property model' where one or few individuals could dominate the biological resources for research (Noiville and Bellivier 2007).

Nevertheless, it is relevant to continue to debate these issues in Europe to try to avoid the emergence of conflicts between the different actors involved in research biobanking and also to prepare the European legal and jurisdictional system to respond to possible court actions on the domain of ownership rights of donors of biological samples. Neither European nor North American doctrine and jurisprudence seem to agree on a formula that would define the proper 'ownership link' between a person and her/his biological material, be it donated (e.g. organs, blood, tissues and cells) or wasted (e.g. umbilical cord, placenta, urine, faeces); this makes obvious that we are dealing with a very controversial and difficult problem in the realm of biolaw (Annas 2004; Bovenberg 2006).

Even when a national law exceptionally decides to give a property right to the citizens over their biological material (such as is the case with the Portuguese law), conflicting situations do not seem to disappear showing that classical property rights are not compatible with the complexity of juridical, bioethical or social

[1] The majority of the authors cited in the list of references contest the application of property rights to human body materials.

idiosyncrasies associated with the collection and use of human biological material in research biobanking.

The aim of this chapter is threefold. First, to revisit the core controversy and the difficulties behind the apparent incapacity of the legal system to arrive at a solution that would provide a satisfactory answer to the problem of ownership rights between the person and her/his body products. Second, to analyse the legal and practical implications in research biobanking of the property rights formula introduced in the Portuguese law to handle this issue. Finally, to provide the legal doctrine of biolaw with a draft proposal of a new category, that of 'biological property', that would allow for a more appropriate legal framework for the juridical relations established in research biobanking between the subjects who have provided biological material and other stakeholders and institutions involved, e.g. researchers, universities, health institutions and industry.

The Classical 'Controversy': Questioning the 'Non-Commercialisation Principle'

As already mentioned, the European law is deeply imbued with the fundamental principle of the prohibition against financial gain from the human body and its products, as attested not only by European fundamental legal texts such as the European Convention on Biomedicine and Human Rights (art. 21), the Charter of Fundamental Rights of the European Union (§2, n. 2 of art. 3) and the Directive2004/23/EC, art.12 (European Commission 2004), but also by the insertion of the non-commercialisation of the human body rule in the national law of the majority of EU member States. However, in some European documents such as the aforementioned Report of the European Group on Ethics in Science and New Technologies on the Charter on Fundamental Rights of the European Union (CFREU) related to technological innovation, we may find that the principle of non-commercialisation is not as unquestioned and absolute as it may seem. In fact, the Group shows significant hesitation in accepting the 'too vague' prohibition of making financial gain from the human body because it considers that this prohibition runs in contradiction with for example the possibility of patenting of inventions derived from human elements, which is allowed by the European law, the payment for tissue banking services – it has always legally been permitted to the donors to receive compensation for the expenses and inconveniences related to the donation – (European Commission 2004), and, we may add, the general acceptance of selling several kinds of human materials (e.g. blood, milk and hair). It is interesting to note that in the same report the Group declares that it would have been preferable to specifically emphasize 'the protection of individuals against organ or tissue trafficking which would affect their dignity and rights', instead of the non-commercialisation of the human body rule, a suggestion that was not accepted in the final version of the Chart. In a former report (European Group on Ethics in Science and New Technologies 1998) the same Group already stressed the opinion that 'the issue of the commercialisation of human tissues', especially those 'which have been processed and prepared for therapeutic purposes', is 'controversial', mentioning that, the European attachment to the idea of altruism and non-profit in the donation of organs and tissues for research notwithstanding, in the USA an increasing number of patients whose cells provide genes that have been patented are already asking for some rewards in court – something which indicates that Europe could follow soon. These alerts do not seem to have caused much effect in the final versions neither of the CFREU nor of the already cited 2004 Directive on biobanking. The Portuguese National Council on Ethics commenting on the draft of the law on biobanks and genetic information (which stated that the 'stored material is the property of the people from whom it was obtained and of their direct family members') considered that 'it has not been usual in the sphere of biomedical law to accept the right to property in relation to human cells, tissues or organs' (Portuguese National Council on Ethics 2008). The Council based this opinion on the idea that the human genome is internationally considered common patrimony of humanity,[2] suggesting that references to the right of property should be avoided because they can constitute a potential 'source of conflicts' (Portuguese National Council on Ethics 2008). It is curious to note on this point that, in spite of these arguments, the Portuguese legislator did not change the norm. As I will analyse further on in this chapter, there is now in the Portuguese law a property right given to the person in what regards the health information and the biological material (Law no.12/2005, 26 January).[3] In any case, the Portuguese National Council on Ethics seems to have followed the position of the California Supreme Court in the *Moore* case or the US Supreme Court in the *Catalona* case in wanting to avoid the legal existence of an individual property right over one's biological material and to protect above all the European fundamental principles of the prohibition of commercialisation/financial gain from the human body and its attached or detached parts.

The Fear of 'Private Property' in Human Biological Material for Research

A common fear seems to prevail that, if judges had granted a property right to Moore or Catalona, this would constitute a precedent that could jeopardize medical research as a common good, giving too much power to donors or individual researchers. We need to revisit here the California Supreme Court sentence on the

[2] The argument that puts forward the human genome qualification as common patrimony of humanity is only acceptable in what refers to the 'phylogenic' part of the genome, that is the part that is common to the generality of the human species/collective and does not suit the 'ontogenic' part of it, that is the part that is unique to each human being/individual (Faria 1999: 200). For this, see also Universal Declaration on the Human Genome and Human Rights (UNESCO 1997).

[3] In the original: Lei no. 12/2005, de 26 de Janeiro, published in the official journal (Diário da República) no. 18, first series. This law defines the concepts of health and genetic information and the legal framework for biobanks.

appeal of *Moore* vs. *Regents of the University of California* (July 9, 1990) and The *Washington University* vs. *W.J. Catalona et al.* (2005) case where only very recently (January 2008) a final decision was made. Both cases show that US judges reject the 'private property model' in human biological material used for research. Each of them, however, showcases different sides of the controversy surrounding ownership rights in research biobanking. In the *Moore* case it was decided that Mr. Moore could not have a property interest in the 'Mo cell line', which was developed from his cells (without his informed consent) and patented by his physician Dr. Golde and another researcher who both had enormous financial gains from it. This case is most emblematic of the two main streams involved in the legal problem of ownership rights over ones' own body and body products, that is, a sense of unfairness and a legal ineptitude to solve it. The sense of unfairness is shared not only among the jurists who commented directly on the Moore case (Merz et al. 2002; Bovenberg 2006) but also by other authors who wrote generally on the subject of the right to remuneration or benefit sharing of donors of tissues and cells (Berg 2001; Holm 2004). Merz considers that '(...) fairness demands that profits be distributed among those who contributed to the research in an equitable manner' and 'strategies and policies that respect the contributions of the many involved parties need to be developed', while to K. Berg 'a state of unfairness would also exist if research on genes in a family led to marketable products and revenues for the pharmaceutical industry, unless the family was given something back'. If financial gain is obtained from genomic research based on the biological material of a whole population, no one opposes that the recipient of shared benefits should include the whole population. Hence, the authors argue that the same logic should be applied to research profits that have come, for example, from the genetic material of one individual or a small group of individuals.

With regard to the Moore case in particular, it has been argued that this decision creates a situation in which everyone is getting a portion of the profit except the person from whom the tissue originated, something which questions the court premise that a patient does not have sufficient property interests in the cells taken from her/his body and consequently asking 'how can property rights vest anyone then?' (Grandolfo 1992). Even if the interest of patients/individuals involved in biomedical research is normally altruistic or therapeutic, it does not sound fair or right to exclude them from the sharing of any profit that could not have been made by universities or the industry companies without the use of their biological samples.

In the *Catalona* case instead, we have a conflict that does not involve the profit sharing between a tissue donor and the researchers after the outcome of a successful scientific research, but the ownership of the tissue collections in itself. The final U.S. Supreme Court's decision on this case was released on the 22nd of January 2008. The Court let stand a unanimous 2007 ruling by the eighth Circuit Court of Appeal, which stated that prostate tissue and serum samples donated to Washington University can continue to be used by the institution for cancer research. The appeal court had affirmed the lower federal district court ruling that donors who gave tissue or serum samples to the University research cannot later compel the school to transfer ownership of the samples to another research institution. This decision was against that of W.J. Catalona (MD) who had argued in the lower courts that the University should transfer the tissues to him at his new place of employment. In this case the judge preferred to choose the solution that reflects the 'best general interest', rejecting the 'private property model' where one or few individuals could dominate the biological resources for research (Noiville and Bellivier 2007). It is interesting to refer on this point to a letter written by Catalona himself to the Editor of JAMA (Catalona 2005) where he argues that research participants have a federal level legal right to withdraw from research at any time and that they cannot waive this right, meaning that universities cannot assert sole right of ownership to samples that participants can withdraw at any time for any reason. Catalona was in this letter arguing against an article previously published in the same journal (Hakimian and Korn 2004). These authors replied in the same journal to Catalona, declaring:

> We stated our opposition to a regulatory or legal scheme that recognizes exclusive ownership interests in excised tissue specimens. We urge the expansive use of tissue resources, consistent with the reasoning articulated by the courts that the scientific value of these specimens is unique and irreplaceable, and that their potential contribution to the public library of knowledge should benefit all humankind.

This epistolary exchange in JAMA illustrates very clearly the controversy that has been going on for decades around the problem of ownership in human biological material and its use in research biobanking.

The Problem with Property Rights and the Human Body: 'Accepting or Not Accepting It': That Is the Question

The majority of the work that has been done in this field relates to the basic question whether or not the legal link between a person and her body can be of a proprietary kind. As stated by Karlsen et al.:

> The issue they want to address [i.e. the issue of ownership rights over the body and its parts] is by no means of a kind that lends itself easily to theoretical speculation. This has, perhaps, as much to do with the inherent intricacy of the issue itself as with the controversy it has managed to arouse.
>
> Karlsen et al. (2006: 215)

Few authors have gone further in trying to answer which kind of social and legal reality would emerge if the answer to the former question was affirmative. This is the case of authors such as Bovenberg who constructed several scenarios in which we could give a solution to the main controversies brought by the post-biotechnological commodification of biological material (Bovenberg 2006), and Bjorkman who went even further drawing a scheme of different kinds of rights to the different types of biological material (Bjorkman 2005, 2007).

These debates take place more rarely in the field of biolaw than in the fields of philosophy, bioethics or social sciences. Nevertheless, George J. Annas presented in the famous Genetic Privacy Act (1995) an early proposal for a federal US law, which defended the legal existence of a property right as the legal link between an individual and her own DNA. Later on (Annas 2004) he argued for the need to analyse differently the ownership rights and legal bounds between the person and her different types of biological material, mainly considering the nature of the uses or potential uses of these materials.

I agree with this position and also with those who state that 'the ruling framework of bioethical thinking is immanently committed to accepting what it most wishes to deny – that my body and body parts are my property' (Beyleveld and Brownsword 2000). In fact these authors argue very sensibly that article 22 of the Convention on Human Rights and Biomedicine (informed consent requirement) presupposes that 'there is property in our own bodies'.

On the other hand, I am very sceptical to the rhetoric that considers that the construction of the body as information is the strategy to overcome the legal qualification of human biological materials (Tallacchini 2005); it does not seem to suit a proper protection of an individual's right to self-determination over her body or body parts. The anonymity, isolation and purification of human body materials that would unify them all (independently of their individual sources) are rarely possible and valuable to scientific research. Some radical opinions even hold that 'if a living human being may not exercise dispositive control over his or her own body and its attached or detached parts, but someone else has the right to do so, we enter an area that closely resembles slavery' (Grandolfo 1992).

The realm of classical property rights might not be the one that best suits the rights that someone has in relation to their detached body parts but this can not be an argument to jurists to avoid trying to solve this problem.

The Portuguese Case: Considerations on the Legal Consequences of a Property Right Over One's Body

Against this current of doctrinal indecision, Portugal approved on the 26th of January 2005 the national legal framework for biobanks, that is, the Lei n. 12/2005 de 26 de Janeiro. §2 of article 18 states that the material stored in a biobank is 'the property of the people from whom it was collected and after their death or incapacity it is the property of their family members'. As mentioned in the introduction to this chapter, this piece of Portuguese legislation has been approved by the Parliament in spite of the prior opinion formulated by the National Council of Ethics against the use of property rights in relation to biological material. This implies that in Portugal, even if there were voices against the application of property rights to human biological material, these voices would no longer be valid in courts because either private or public biobank owners are obliged to comply with the legal principles. The law caused some reactions in the scientific community but there was no official opposition.

The mentioned disposition has not yet been used in any law case in Portugal, something that precludes us from knowing at this point how it will be used by judges and what consequences it will have in practice. However, it is certain that all the legal corollaries of property will apply with regard to the dispositional link between people and their biological material in the context of research biobanking. So, which corollaries are we then considering here?

First of all, we need to recall at this point some classical legal concepts such as 'disposition', 'right to property' and 'intellectual property' (Walker 1980; Prata 1989). 'Disposition' is a legal term that has two legal meanings:

- Synonymous of legal norm
- The form in which a right is exercised, which has as a consequence the lost, total or partial, absolute or relative of the particular right disposed.

The 'right to property' is the strongest right of ownership, best conceived of not as a single right but as a bundle of distinct rights, some or even many of which may be relinquished temporarily without loss of ownership. The kinds of rights which a right of property confers upon the objects of that right vary accordingly to the nature of the object, including the rights to possess, use, lend, alienate, use up, consume, abuse, let on hire, grant as security, gift, sell and bequeath the object. The right to property may exist in respect to both corporeal things (e.g. buildings, animals) and incorporeal things (e.g. copyrights, claims of damages, etc.). These categories are cross-divided into immovable objects (e.g. land) and movable things (e.g. animals, claims). The owner loses his property only if and when he uses up the object or transfers it without retaining any reversionary rights. Furthermore, the term 'intellectual property' refers to the kinds of property such as copyrights, patents and trademarks. The earliest use of the term 'intellectual property' appears to be an October 1845 Massachusetts Circuit Court ruling in the patent case Davoll et al. vs. Brown (1845) in which judge Charles L. Woodbury wrote that 'only in this way can we protect intellectual property, the labours of the mind, productions and interests as much a man's own as the wheat he cultivates, or the flocks he rears'.

Until the article of the Portuguese law that states the legal existence of a property right between the person and her biological material came into force the considerations over this subject were purely speculative and doctrinal. When a law acknowledges the ownership rights over our body as a property right, a new paradigm is emerging. We can no longer argue against it or for it. The property right is a legal reality and the considerations have to follow this starting point. So, even if we can see this law as an isolated case, it presents us with an interesting exercise that no longer is a discussion whether there is or not a property right but what will the implications and characteristics of this new property right be, considering all the legal, social and argumentative background of this issue.

Laura Underkuffler considers that the core interest asserted in the existence of a property right over human biological material is the vindication of personal decision making over one's own body and substances (Underkuffler 2003: 103–106). The same author argues that 'the competing interests in these cases, on the other hand seek to achieve states of affairs in which the body or its substances are publicly controlled or publicly used, in order to safeguard public health, or to enable others (through research or transplants) to live. As laudable as such public interests may be, they do not share the core values that the individual claims assert' (Underkuffler 2003: 105). This is very true as a first implication of the existence of a property right over one's biological material, that is, the assumption of a higher hierarchical position of individual rights within the context of research biobanking. It is not a coincidence that the political party that proposed the mentioned law in the Portuguese Parliament and wrote its draft was at the time pleading for the legalisation of abortion in the country.

Another interesting implication of the existence of a property right over one's body is that human biological material consequently has to be classified as a 'thing', because only 'things' (not 'persons') can be the object of a right to property; this may not comply with the androgynous status of DNA, which is at the same time a material (patrimony) and information (personal). It is true that the issues involved are brand new to a legal system that is still constructed according to ancient Roman categories like the one that divides the juridical world in 'things' and 'persons' and which cannot classify DNA as one or the other (Faria 1999: 193–203). There are a lot of divergences arising between those who defend the idea that biological material, including DNA, should be seen by the law as a 'thing' and those who absolutely reject this position and therefore hold that DNA is still more a 'person' than a 'thing'. To consider DNA as an object of a property right is then to cancel the dispute whether DNA is a thing or still part of the person.

Property rights' inherent powers apply, meaning that the person has the right to 'use, enjoy and dispose' her biological material (the 'jus utendi, jus fruendi and jus abutendi' of Roman law), the only limit to property rights being the principle of 'a right's abuse', which determines that 'it is not permitted to exercise one's rights when it manifestly exceeds the limits imposed by good faith and good practices, or by the social or economic aim of that right'. Or, if the right to property implies that the person has a right to enjoy the fruits (natural or man made) of her property, the owner of the biological material has a right to (at least) share the benefits resulting from research-industrial work over the same material.

Hence, when property rights apply to biological material, we are entitled to decide either to transfer the property of our biological material to biobanks or not. In case we do, we will have to declare it in a contractual form. Otherwise, the property of the material remains with the person from whom it was collected. This one has the right to withdraw that material from research at any time. Furthermore, it is possible that biobanks will have to share the benefits of the industrial outcome of the research done with somebody's biological material, but here again the person has to declare in a previous contractual form that she waives this right. If this waiving clause does not exist, we may consider that the outcome of a 'Moore case' in Portugal nowadays could have been different from the one in the USA.

Consequently, we may draw the conclusion that a classical property right to cover the link of one's own biological material has the merit of protecting the individual's self-determination but it underestimates the interest of science and the common good. In fact, if the individual claims on biological material are property rights, they will enjoy presumptive power over competing public interests. The conflict is real because the values of personal freedom and autonomy that such claims represent will almost never be shared by the public interests that oppose them (Underkuffler 2003). A classical property right applied to the ownership of human biological material does not allow the desirable achievement of a legal equilibrium between those two complementary interests. On the contrary, it may even endanger such interests by implicitly promoting a conflicting environment.

The Legal and Bioethical Construction of a New 'Biological Property'

In spite of the prevalent perception of unfairness with regard to the outcome of the Moore case, it has become evident that the classical qualifications of the law, such as property and personal rights, are not sufficient or adequate per se to adapt to new circumstances where someone's body products are the raw material to the industry and financial gains of others. If it seemed to be a shared agreement or perception that a person possesses exclusive 'dispositional' rights over her body and its products *before* they are removed or expelled from it, the same is not necessarily the case *after* this happens.

Several pathways have been pursued to try to overcome this dilemma, from proposing a model where DNA would be 'taxable property', as is the case with Bovenberg (2006: 192–204) who argues that 'a new tax on cell and tissue products derived from a donor, or set of donors, could provide a means of ensuring a fairer distribution of the fruits of regenerative medicine and the commercial use of tissue in general' to the rhetoric that considers that the construction of the body as information is the strategy to overcome the legal qualification of human biological materials (Tallacchini 2005).

I am aware of the complexity of the issue but I also think that it is time to create a legal and ethical framework that would avoid cases like Moore or Catalona. I will briefly argue that contemporary law needs a new kind of property right to adjust to the human body and its parts. This is the premise that leads me to propose a new kind of property right appropriate to human biological material and consequently also to introduce a new legal concept, that of 'biological property'. As already observed in relation to 'intellectual property', it is not novel that a new kind of property is invented legally. The kind of property definition I propose to introduce would be a juridical entity with an hybrid nature, balancing property and personality rights, thus allowing for the gap between a total identification of the human biological material as an untouchable subjective good (material property) and the total exclusion of these interests leading to an absolute free deliverance of human biological material to industry and research to be bridged.

Even when the law gives a property right to citizens over their biological material (such as the Portuguese law), this proprietary right, I argue, cannot have the same legal contours as classical property rights. The concept of property right, I suggest, would have the following characteristics:

- Object of the right: I call this *sui generis* property right 'biological property' because it is the ownership link between someone and its 'biological material' (there is a 'bio' link, i.e. there must be a DNA identification between the person and the object of property). Hence, the object of this right has to be a defined concept of 'human biological material' – all that contains human DNA.
- Distinction/criteria between 'in the commerce' (hair, milk, etc.) and 'out of the commerce' body products: Even if they use different arguments, Annas (2004) and Holm (2004) seem to approach a common theory in what concerns the need to regulate human products of biobanking differently according to the particular circumstances of each case. To the first quoted author the reason why the purchase and sale of human organs is prohibited is because it will probably put donors at risk of potentially coercive monetary inducements, and also because the 'altruistic/gift relationship' in organ transplantation is highly morally valued as a 'rare and praiseworthy event in medicine' (Annas 2004: 150). Nevertheless, not every donation of biological products obeys to the same premises as organ transplantation. If we adopt the transplantation analogy for all of them, we will most likely focus on the risks of the live donors and forbid commerce and sale. Annas considers that the dominant organ transplantation analogy is dysfunctional and misleading to be useful in the collection and banking of certain kinds of human biological materials such as placental blood or umbilical cord blood (Annas 2004: 150) because there are no such dangers as in organ donation. On the contrary, the blood analogy that allows some commerce and even to inform the donors if they want to opt for private banking is a much better framework in these cases.
- Respect of the individual, the familiar and the scientific/society good: Previous to the legal definition of a property right over our biological parts and material, there were already several other legal premises that can not be overridden. One example is the importance that my biological material can have to other people, such as family members, people who belong to the same genetic cluster or even humanity as a whole. Each biobanking activity should be well identified in terms of private or public nature, profit or non-profit finalities, therapeutic and/or research purposes, forms of identification of donors/subjects, protection of the confidentiality of identified donors/subjects, identification of financial sources, etc. This would allow the definition of the situations where the person can share benefits and the situations where the common good should prevail. Only after knowing all these elements, I believe, it will be possible to create an adequate legal framework in terms of ownership rights and possibilities of benefit sharing between all the actors involved in research biobanking.

Conclusions

To defend an almost absolute principle of non-profit with regard to human biological material seems to be the most comfortable legal and bioethical position in our society, since there are still no acceptable legal, bioethical or biopolitical solutions to permit donors of biological material to research biobanks' benefit sharing. Legislators, judges and bioethics committees prefer to adopt a precautionary position defending the non-use of property rights in this field because they are afraid that the commercialisation of the human body will be in the end of this pathway, with all the dangers that it implies, especially in some areas, e.g. selling of foetal tissue, embryos, etc.

The principle of non-financial gains from human body products has its roots in the so-called *transplantation model* (Annas 2004), i.e. the legal framework for the donation of organs and tissues for transplantation purposes and not for research purposes, in which the main concern is to protect potential donors from monetary coercive actions. Ownership regulations have to be considerate of the interests and values presented in research biobanking and its characteristics; it can transform biological material into a pitfall for scientists or into a tool to construct a fair and friendly research environment. Each kind of human biological material needs to have a legal regulation that adapts to the particular characteristics of its collection, conservation and purposes. Although the property rights framework is the only legal background in contemporary law that makes it possible to protect individual interests over one's biological material, classical property rights undermine a sound legal environment in research biobanking. Hence, a new kind of property right seems to be needed in contemporary legal systems, one which will be able to conciliate two apparently opposite legal interests (individual and public) being integrated by the legal framework of the international principles of biolaw. This new form of ownership right should have an international legal framework based on the idea that research biobanking is a universal need and interest. It is premature to predict that this chapter will in itself contribute to the universal creation of a new type of 'biological property' with its very specific characteristics. I have, nevertheless, tried to point out a pathway toward finding a solution to the ongoing quarrel over the issue of dispositional rights of human biological material.

Acknowledgment The author would like to thank Emmanuelle Rial-Sebbag (INSERM U558, Toulouse, France) for her precious contribution to the chapter.

References

Annas GJ (2004) American Bioethics – Crossing Human Rights and Health Law Boundaries. Oxford University Press, Washington

Berg K (2001) The ethics of benefit sharing. Clinical Genetics 59:240–243

Bovenberg JA (2006) Property Rights in Blood, Genes and Data – Naturally Yours? Martinus Nijhoff Publishers, Leiden, Boston

Cambon-Thomsen A (2002) Ethical aspects in different kinds of biobanks used for genomics in Europe. Presentation at the "Workshop on Biobanks – Practical, Ethical and Legal Aspects," Sept. 12–13, Uppsala, Sweden
Catalona WJ (2005) Letter to the editor. Journal of American Medical Association 293:1326
Davoll et al. Vs. Brown (1845) http://rychlicki.net/inne/3_West.L.J.151.pdf
European Commission (2001) Survey on opinions from National Ethics Committees or similar bodies, public debates and national legislation in relation to biobanks. Quality of Life Programme, Brussels
European Commission (2004) Directive 2004/23/EC, Directive on setting standards of quality and safety for the donation, procurement, testing, processing, preservation, storage and distribution of human tissues and cells. Brussels
European Group on Ethics in Science and New Technologies (1998) Opinion on the ethical aspects of human tissue banking. Report no. 11 on Human Tissue Banks
European Group on Ethics in Science and New Technologies (2000) Citizens rights and new technologies: A European challenge. Report on the Charter of Fundamental Rights related to technological innovationsas requested by the President of the European Commission, Brussels
European Group on Ethics in Science and New Technologies (1998) Opinion on the ethical aspects of human tissue banking. Report n. 11
Faria P (1999) Données Génétiques Informatisées – un défi au droit à la confidentialité des données personnelles de santé. Edn du Septentrion, Villeneuve d'Ascq
Faria P, Campos A (2001a) Banking of genetic material and data in Europe: Legal, Ethical and Economical Issues. Portuguese Ethical-Legal Contextual Report (for the EUROGENBANK consortium). National School of Public Health, New University of Lisbon, Lisbon
Faria P, Campos A (2001b) Banking of genetic material and data in Europe: Legal, ethical and economical issues. Spanish Ethical-Legal Contextual Report (for the EUROGENBANK consortium). National School of Public Health, New University of Lisbon, Lisbon
Faria P, Campos A (2005) A nova lei sobre informação de saúde, informação genética e biobancos – Guia das disposições mais importantes. Revista Portuguesa de Saúde Pública 23:97–101
Galloux JC (1989) De la nature juridique du matériel génétique ou de la réification du corps humain et du vivant. Recherche Juridique 3:519–550
Grandolfo GM (1992) The human property gap. Santa Clara Law Review 32:957–975
Hakimian R, Korn D (2004) Ownership and use of tissue specimens for research. Journal of American Medical Association 292:2500–2505
Harichaux M (1988) Le corps objet. In: Drai R, Harichaux M (Eds.) Bioéthique et Droit. PUF, Paris
Hirtzlin I et al. (2003a) An empirical survey on biobanking of human genetic material and data in six EU countries (on behalf of the EUROGENBANK consortium). In: Knoppers BM (Ed.) Population and Genetics – Legal and Socio-Ethical Perspectives. Martinus Nijhoff Publishers, Leiden, Boston
Hirtzlin I et al. (2003b) An empirical survey on biobanking of human genetic material and data in six EU countries. European Journal of Humans Genetics 11:475–488
Hofmann B et al. (2006) Analogical reasoning in handling emerging technologies: The case of umbilical cord blood biobanking. American Journal of Bioethics 6:49–57
Janger EJ (2005) Genetic information, privacy and insolvency. Journal of Law, Medicine and Ethics 33:105–108
Karlsen JR et al. (2006) To know the value of everything: A critical commentary to B. Björkman and S.O. Hansson's 'bodily rights and property rights'. Journal of Medical Ethics 32:215–219
Kennedy I, Grubb A (2005) Medical Law. Oxford University Press, Oxford
Knoppers BM (1999) Status, sale and patenting of human genetic material: An international survey. Nature Genetics 22:23–26
Labrusse-Riou C (1991) La maîtrise du vivant: matière à proces. Pouvoirs 56:106
Leclercq P, Catala P (1992) L'information est-elle un bien? In: Carbonnier J et al. (Eds.) Droit et informatique – l'hermine et la puce, Masson, Paris, pp 91–109

Lipinski TA, Britz JJ (2000) Rethinking the ownership of information in the 21st century: Ethical implications. Ethics and Information Technology 2:49–71

Mazen N (1988) Réflexions juridiques sur le matériel génétique de l'homme. In: Drai R, Harrichaux M. (Eds.) Bioéthique et Droit, PUF, Paris, pp 194–210

Merz JF et al. (2002) Protecting subject's interests in genetics research. American Journal of Human Genetics 70:965–971

Moine (1997) Les choses hors du commerce, une approche de la personne humaine juridique. LGDJ, Paris

Noiville C, Bellivier F (2007) Les collections d'échantillons biologiques entre propriété et contrat. Revue des Contrats 2:493–504

Portuguese National Council on Ethics (2008) www.cnecv.gov.pt

Prata A (1989) Dicionário Jurídico, 2nd edn. Almedina, Coimbra

Rynning E (2002) Legal issues on biobanking – Privacy versus freedom of research and property rights. Presentation at the "Workshop on Biobanks – Practical, ethical and legal aspects," Sept. 12–13, Uppsala, Sweden

Solbakk JH et al. (2004) Mapping the language of research-biobanks and health registries – From traditional biobanking to research biobanking – A project presentation. In: Árnason G et al. (Eds.) Blood and Data – Ethical, Legal and Social Aspects of Human Genetic Databases. University of Iceland Press, Center of Ethics, Reykjavík, pp 299–305

Tallacchini M (2005) Rethoric of anonymity and property rights. In: Human Biological Materials (HBMs). Law and the Genome Review, January to June, pp 153–175

Thouvenin D (1994) La personne et son corps: un sujet humain, pas un individu biologique. LPA 149:27

Underkuffler LS (2003) The idea of property, its meaning and power. Oxford University Press, Oxford

UNESCO (1997) Universal Declaration on the Human Genome and Human Rights

Walker DM (1980) The Oxford Companion to Law. Clarendon Press, Oxford

Westerlund L, Persson AH (2001) Civil reflections on the use of human biological material. In: Hansson MG (Ed.) The Use of Human Biobanks – Ethical, Social, Economical and Legal Aspects, Report I. Uppsala University, Uppsala, pp 61–82

The Washington University vs. W.J. Catalona et al. (2005) http://prostatecure.wustl.edu/

Legal Challenges and Strategies in the Regulation of Research Biobanking

Elisabeth Rynning

Abstract In this chapter, some of the legal challenges of research biobanking are discussed and illustrated by examples of possible analogies as well some comparative notes on the regulatory strategies adopted in the Nordic countries. Human biological material is compared with biological waste, raw materials, human beings, personal or nonidentifiable health data, and different kinds of public resources. It is concluded that the complex nature of human biobanks would seem to defy any attempt at a simplified regulatory analogy. Even so, it is clear that the application of more sophisticated analogical reasoning will still be valuable in the regulatory process and that policy makers must try to identify an appropriate combination of diverse approaches. While the international nature of biomedical research provides a strong incentive for more harmonised rules, the regulatory process is here further complicated by the plurality of religious, cultural, social and legal traditions, as well as issues of regulatory competence. Nevertheless, some degree of regional or even international consensus could certainly be reached with regard to less controversial areas and issues, and this potential must be further explored. The regulation of research biobanking should be perceived as an ongoing step-by-step process, rather than a problem that will soon be solved once and for all. In the short-term perspective especially, it must be expected that legal restrictions and administrative inconveniences may cause additional costs and delay or even prevent promising research. The long-term aim must be to serve the best interests of the public, by a careful balancing of the freedom of research against other fundamental rights and values.

E. Rynning
Department of Law, Uppsala University, Uppsala, Sweden
e-mail: Elisabeth.Rynning@jur.uu.se

J.H. Solbakk, et al. (eds.), *The Ethics of Research Biobanking*,
DOI: 10.1007/978-0-387-93872-1_19, © Springer Science+Business Media, LLC 2009

Introduction

For the past few decades, there has been a steadily increasing debate about the appropriate regulation and governance of research biobanking. Although the study of human biological material has always been an important part of medical research, new possibilities related to advances in the field of genetics have generated a growing demand for easily accessible biological samples and various types of associated data. The establishment of such research resources has been facilitated by the parallel development in information and communication technology, making it possible to store and process large quantities of data, with comparatively limited input of time, personnel, and money. It is widely believed that research on large-scale collections of human material and associated data will open up new prospects in improving the health of individuals and of humankind as a whole.[1]

At the same time, the creation and use of so called biobanks or genetic databases is still causing considerable regulatory problems. Researchers and biobank principals find it hard to establish what the relevant legal requirements pertaining to their activities really are. While expressing a wish for legal certainty, however, they are also anxious to avoid having their freedom of research and immaterial property rights threatened by unreasonable and impracticable demands and restrictions. Policy makers would, of course, be more than happy to provide appropriate and well-balanced regulations that offer adequate protection to all interests concerned, not least the freedom of research and the privacy rights of donors. Despite the fact that questions relating to further use of human tissue, as well as genetic testing and screening, have been on the agenda for decades,[2] many legislators are still finding it very difficult to determine what the rules should be.

One of the circumstances complicating the regulatory situation is that biomedical research is often based on extensive international collaboration and cross-border activities, whereas binding international rules are so far lacking in the specific area of biobank research. The fact that there is a vast variety of nonbinding declarations, recommendations, and ethical guidelines on biobanking, adopted by different organizations or advocated by groups of scientists, does not provide any immediate help to the researchers and biobank principals who are legally bound by the domestic laws of the countries in which they operate.[3] The variety of national regulatory strategies and material rules may thus obstruct cross-border biobank activities, but also make it more difficult for potential donors to foresee how their interests would be protected if samples were to be transferred to another country.

Even from the domestic perspective, however, the co-ordination and harmonisation of biobank regulation with legislation in related areas have proved to be highly problematic. In this chapter, some of the challenges involved in regulating human

---

[1] See for example OECD 2008 Draft Guidelines for Human Biobanks and Genetic Research Databases.

[2] For some early examples, see Milunsky and Annas (1976), Berg (1983), Knoppers and Laberge (1989), and WHO (1985) Community approaches to the control of hereditary diseases.

[3] Kaye (2006).

research biobanks will be discussed and illustrated by examples of possible legal analogies as well some comparative notes on the regulatory strategies adopted in the Nordic countries. The regulatory tools and issues of competency are also touched upon before the chapter is concluded by some words on the process of biobank regulation.

Human Research Biobanks: What Are They?

General Comments

As is often underlined, already the meaning of the term 'biobank' may in itself be debated. It seems to have come into use in the Nordic countries in the mid-1990s, starting with Denmark,[4] and is today widely used. Even so, there is no general consensus on the definition of a biobank, and a host of other terms are also used to label the same or similar concepts,[5] for example, tissue banks, collections of biological material, tissue collections, tissue repositories or bio-repositories, DNA banks, gene banks and genomic or genetic databases. Other alternative terms somewhat less common include bio-libraries and tissue libraries.

When the term biobank is used, it may not only refer to the actual biological materials, but also to the facilities where these are stored or to the institution responsible for the collection and storage of samples. An early Danish biobank definition was thus 'an institution where biological material and clinical information is collected and can be redistributed, either to serve the original donor or scientific, health administrative or health political purposes'.[6] Used as a verb, biobanking would normally refer to the organised collection and storage of human biological samples and associated data, in view of making them accessible for various biomedical or health-related purposes.

The German Nationaler Ethikrat defines biobanks as 'collections of samples of human bodily substances (e.g. cells, tissue, blood, or DNA as the physical medium of genetic information) that are or can be associated with personal data and information on their donors', and also underlines the twofold character of biobanks, as collections of *both* samples and data.[7]

Not all biobank definitions include the associated data, however. In the Icelandic legislation, a biobank is defined as 'a collection of biological samples which are permanently preserved', where biological sample means 'organic material from a human being, alive or deceased, which may provide biological information about him/her'.[8] The definition used in the Swedish Act on Biobanks in Health Care

[4] See for example Nielsen et al. (1996) and Hermerén (1997).

[5] Cf. Elger and Caplan (2006).

[6] Nielsen et al. (1996) and Riis (1997).

[7] Biobanks for research 2004: Opinion of the German National Ethics Council.

[8] Article 3 of the Act on Biobanks no. 110/2000.

comprises 'biological material from one or more human beings that is collected and preserved for an indefinite or limited period, and whose origin is traceable to an individual or individuals.'[9]

The definitions of population biobanks and genetic or genomic databases tend to have a stronger or even primary focus on the data aspects, for example in the HUGO Statement on Human Genomic Data bases, 2002: 'A genomic database is a collection of data arranged in a systematic way so as to be searchable. Genomic data can include inter alia, nucleic acid and protein sequence variants (including neutral polymorphisms, susceptibility alleles to various phenotypes, pathogenic mutations), and polymorphic haplotypes.' The OECD 2008 Draft Guidelines for Human Biobanks and Genetic Research Databases define human biobanks and genetic research databases as 'structured resources that can be used for the purpose of genetic research, which include: (a) human biological materials and/or information generated from the analysis of the same; and (b) extensive associated information.'

As a final example, the Council of Europe Recommendation Rec(2006)4 on research on biological materials of human origin includes also the long-term perspective in its definition of the particular concept of a population biobank:

'A population biobank is a collection of biological materials that has the following characteristics:

i the collection has a population basis;
ii it is established, or has been converted, to supply biological materials or data derived therefrom for multiple future research projects;
iii it contains biological materials and associated personal data, which may include or be linked to genealogical, medical and lifestyle data and which may be regularly updated;
iv it receives and supplies materials in an organised manner.'

It should be clear already from the examples provided that the language of biobanking is by no means a single universal language, but rather a group of languages or at least distinguishable dialects, where a word may have several different meanings and the same concept may be attributed different names. It may be self-evident that any regulatory project in this field would have to start by closely defining the proposed object of regulation, but also in international debate and co-operation it would be recommendable to beware of the risk for misunderstandings.

Human Material Removed and Stored for Various Purposes

This book is focussed on *research* biobanks, but human material and associated data may of course be collected and stored in biobanks for a number of other health-related purposes, such as patient safety and quality assurance in healthcare,

[9] Chap. 1, Sect. 2 of the Biobanks in Medical Care Act (2002: 297).

transplantation or transfusion, assisted procreation or the manufacturing of medicinal products. Many biobanks are in fact created and used for a mix of such purposes, which gives rise to the question whether or not the rules governing biobank research ever could or even should be universal. While a lot of the biobank debate of the 1990s would seem to have concerned issues related to the so-called *further use* of human tissue collected for diagnosis and treatment or leftover from surgical interventions, the samples stored in many of the large-scale biobanks or population-based genetic databases of the new millennium are normally intended for more long-term use in a variety of future projects that cannot be foreseen at the time when the samples and associated data are collected. Another category consists of biobanks originally established for use in a specific research project, but afterward preserved and used for other projects.

Different provisions could thus apply to human material removed, collected and stored for purposes related to for example:

- Diagnostics and treatment of individual donors/patients themselves
- Quality assurance and educational purposes in healthcare
- Assisted procreation or transplantation
- Health screening programmes
- Other health-related analysis, such as post-mortem examinations, protection against contagious diseases, insurance purposes
- Identification of individuals or familial relations in crime investigations, paternity cases, immigration matters, etc.
- The manufacturing of products

Research interests could concern the use of human materials originally collected and stored for:

- A particular research project or several specified projects
- Any of the purposes mentioned earlier, including *other* research projects
- Use in *future*, more or less unspecified, research projects

Depending on the original purpose for which the material was removed, collected or stored, different requirements for research-related use may be motivated. Whereas provisions on possible further use of different types of human materials would sometimes seem better placed in the legislation regulating the particular activity for which they were removed, it could also be argued that a comprehensive biobank legislation would have the advantage of providing overview and easier access to the relevant rules.

Type of Research Use Intended

From a regulatory perspective, the type of research for which the biobank materials are intended or used may also be highly relevant. As is clear from the above, many research biobanks are specifically focussed on the informational aspects of the biological material, in particular the genetic information. In this field, we could find both clinical research projects and longitudinal epidemiological studies. Some projects may involve large-scale genetic screening with diagnostic or treatment implications for individual research subjects, thus blurring the borderline between epidemiological research and clinical research and between public health measures and individual health care.[10]

Other forms of research with human biological material may be primarily aimed at studying different functions or applications of the tissues, cells or genes as such, in a natural or manipulated form. This could involve the development of cell lines, new treatments or products or experiments in reproduction. When biobanks are established or used for this kind of purposes, they would not normally be referred to as genetic databases or DNA banks, but more often as tissue banks or cell banks. Here, the human material is used rather as raw material than as a source of information, which could of course motivate somewhat different rules. It is not unusual for banks with for example human ova, embryos or embryonic cells to be subject to special legislation.

Research use of human biological materials and associated data may also vary in relation to the duration of the storage. Some biobank regulations may be applicable only to banks specifically set up for long-term research purposes, whereas others have a much wider time frame, from very short storage to virtually endless.[11] This gives rise to the question if samples collected and used for their intended purpose more or less directly should also be considered to constitute biobank materials, even if they are never or only for a very short time stored in what could be viewed as a biobank facility? Other issues concern the point of time at which the sample begins to fall under the biobank regulation and the time when it ceases to be part of the biobank, having for example been transferred to another biobank, or having been destroyed, rendered non-identifiable or turned into a product. In some contexts, the term virtual biobank is used to describe 'a controlled (access) database on samples stored at different locations'.[12]

Policy Makers Are Still Struggling

When deciding on the type of activities and materials or data that should be covered by biobank regulation, legislators need to identify and carefully asses the different interests concerned, in order to determine the extent to which the arising legal issues

[10] One example can be provided by the heated debate concerning the Norwegian large-scale cohort study MIDIA, aimed at identifying environmental triggers of type 1 diabetes. This study, which included screening of 100,000 newborns and 15 years follow-up of the 2,000 children with high-risk genotype, had started in 2001 after having been approved by both the competent Research Ethics Committee and the Data Inspectorate. Nevertheless, in 2007 the project was stopped by the Directorate of Health, since it did not comply with the prerequisites for predictive genetic testing of children, prescribed in the Act on Biotechnology. See the decision of the Norwegian Directorate of Health 10.12.2007, 07/2904–19.

[11] Compare for example the Icelandic and Swedish Biobank Acts.

[12] See the Web page of the PopGen Database, http://www.popgen.info/Glossary.cfm.

are already adequately covered by existing regulation. In most countries, it is likely that at least some aspects of research biobanking will be covered by legislation regarding for example scientific research, patients' rights, use of human gametes or embryos, bio-patents or the processing of personal data. Even so, it is equally likely that certain aspects will be left out, and with regard to issues that *are* covered, it must be determined if the applicable rules provide satisfactory results in the particular context of research biobanking, by offering a balanced protection in view of *all* interests concerned, taking into consideration also long-term societal goals. In assessing their domestic regulation, legislators must furthermore consider not only the need for internal consistency and harmony between different areas of their own legal system, making sure that similar rules are applied in similar situations, but also compliance with public international law and other external requirements.

One strategy could then be to complement existing, general legislation with a minimum of provisions specifically related to biobanking and another to introduce more comprehensive specific biobank legislation. The point of departure could be the type of human material and data, or the intended use or potential misuse. New provisions could be primarily focussed on material rules, for example allowing or prohibiting certain activities or results, under stipulated conditions, or they could be aimed at securing appropriate procedures and oversight. Which strategy or combination of strategies to choose may not be self-evident.

Legal Analogies and Comparative Law

External and Internal Legal Comparisons

One of the methods widely used both in legal science and in the development of new legislation is comparative studies, whereby inspiration is sought from the solutions discussed or implemented in other jurisdictions. Such comparative legal research is also a necessary part of any attempt to reach consensus for international regulation. In new areas of for example rapidly developing technologies, however, the available material for this kind of legal comparison can be limited, making it all the more important to study existing regulation of *similar* activities, relationships and interests. The consideration of domestic legal analogies is in fact indispensable in any development of new regulation, if internal harmony, coherence and consistency of the legal system are to be achieved. Also in the interpretation and application of existing laws, analogy is an important tool, since no legal regulation can specify in detail all the cases and situations that should be covered.

Legal analogies of course presuppose that some area of similar activities and interests can actually be identified, which is one of the core problems in the regulation of biobanks. There is no lack of possible analogies that may seem plausible at first sight, quite the contrary, but at a closer look none of the alternatives quite measure up to the diversity and complexity that characterises the field of biobanking. Efforts to make adjustments by combining two or more analogies tend to bring out conflicting principles. Nevertheless, a number of legal analogies have still been applied to biobank-related activities, comparing the human material to waste, raw material, human beings, personal or non-identifiable health data or a public resource. The analogies illustrate how human biological material has been perceived and regulated in different contexts and could thereby add to the understanding of the regulatory challenges and the strategies adopted in relation to biobanking.

Human Material as Biological Waste

A good place to start could be to look at the use of leftover or discarded human biological material, for example in the form of body parts or tissue removed in the course of a medical treatment, or blood samples having already been used for the intended diagnostic analysis. To the extent that the material is not needed for patient safety reasons, related to future care or treatment, the patient normally would not have any further use for the material and it could be regarded as biological waste, without value. In many countries, this kind of waste would be subject to certain rules regarding its disposal and destruction, for reasons related to the protection of environment, public health and safety. Even the patient himself or herself might not be allowed to bring his or her removed appendix home in a jar. An analogy based solely on the principles underlying such rules, however, would seem to imply that as long as discarded human biological material is handled in a competent way that does not constitute any danger to public health or environment; it should be free to use for research and perhaps also for other purposes. This policy has also been applied, at least to some extent. Although the waste analogy is not completely without relevance, however, it can quickly be discarded as insufficient since, indeed, in the words of George Annas, 'waste is not always what it seems'.[13]

Human Material as Raw Material

As opposed to being mere waste, certain types of human biological material have a long history of being attributed financial value, for example human hair. In the area of biomedicine, the human body is an important source of valuable raw material used to manufacture products or to be directly transplanted into other humans without previous processing. Not only materials such as blood or whole organs constitute precious resources, but even tissues that would otherwise have been regarded as biological waste could be sought after as raw material, for example placentas used for cosmetic products, medicines or stem cell research.[14] When human material is used as raw material or products, issues concerning patient and consumer safety become

[13] Annas (1999).

[14] See for example Annas (1988).

relevant, in addition to the previously mentioned more general requirements for the protection of environment and public health.

Another highly relevant legal aspect is that the use of human material in this context gives rise to questions related to the protection of property rights, in particular where products or inventions are developed from the material, as in the famous Moore case.[15] How should the human material be perceived at various stages of cultivation or manufacture?[16] Are for example cells and cell lines cultivated in vitro still just samples of human biological material, or do they constitute medicinal products? When does the biobank sample cease to exist, having been used up or turned into something else? Whether or not property rights could be part of an appropriate regulatory approach already to parts of the human body as such is a highly controversial issue.

Human Material as Part of the Human Being

Compared with other types of raw materiel, human biological material is undeniably different in the sense that it actually constitutes a part of a human being. This fact could motivate an analogy with the legal rules and principles applied in the protection of human beings, for example in relation to research, a field where an abundance of international and national regulations can be found.[17] Legal provisions in this field may be aimed at protecting the physical health and safety of the human being, which could of course be at risk at the removal of biological material, but may also focus on the bodily integrity and other privacy rights of the individual. Independent ethics review of research projects is often mandatory, and informed consent is a necessary precondition for virtually any research intervention.[18] Once the human biological material has been removed, however, the health-related risks would not be as prominent, but the protection of human dignity and privacy could still be of relevance to the prerequisites for lawful further use of any material removed.

Here, things start to get complicated. If human biological material as such does not merit quite the same protection as a living human being, which would perhaps seem a plausible conclusion, other familiar questions arise. What degree of legal protection should for example be offered to such human biological material as has a *potential* to develop into a full human being, i.e. embryos or human ova? And what

[15] Moore vs. Regents of the University of California, 793 P.2d 479 (Cal. 1990) Cal. Rep 146; see for example Laurie (2002) or for an early comment, Annas (1988).

[16] Cf. Rynning (2003a) at pp. 103–105.

[17] For extensive information on various rules and guidelines for research, see the Web resource CODEX of the Swedish Council for Research, http://www.codex.vr.se/codex_eng/codex/index.html.

[18] This is the first principle of the Nurnberg Code (1947), also laid down for example in the World Medical Association Declaration of Helsinki on Ethical Principles for Medical Research Involving Human Subjects (first adopted in 1964), in Article 4 of the UN International Covenant on Civil and Political Rights and in Article 16 of the Council of Europe Convention on Human Rights and Biomedicine (1997).

about research interventions carried out when they can no longer constitute a threat to the donor's life or health, for example the taking of material from dead bodies or tissue from aborted foetuses? It soon becomes obvious that the legal protection offered to the body of a deceased human being and to certain types of biological material, such as human embryos,[19] ova or foetal tissue is quite different from the rules pertaining to biological waste or raw materials for products. The fact that not all types of human biological materials are attributed the same status, depending on their origin, potential function and the context in which they are removed from the body brings out the question of what determines the protection that should be offered to different categories of biobank material. Additional issues may arise when original properties of human material are radically changed, as for example is the case with so-called iPS cells,[20] or when mixed living human–animal material is created, as in so-called cybrid embryos. Apparently, also interests related to society as a whole, as well as donor relatives, must be considered. Nevertheless, the mere fact that a biological sample originates from a particular individual could have privacy implications related to this person's wishes to decide about its use. Does this mean that an explicit and specific informed consent is always needed or could it sometimes be sufficient merely to provide an opportunity to opt out?

Another principle considered to be based on respect for human dignity is that, as opposed to raw materials and products, human *beings* cannot be owned, bought or sold.[21] To what extent is it reasonable or even necessary that this principle of non-commercialisation be applied also to parts of the human body, in order to protect not only the human dignity of individuals, but also other societal values?[22] Certain activities, which could be envisaged as commercialisation or commodification of parts of the human body, are often thought to put such interests at risk.[23]

Human Material as Health-Related Personal Data

The issue of privacy rights naturally brings us to the analogy with rules pertaining to health-related personal data. This analogy would indeed seem appropriate, when research biobanking is primarily viewed as a way to make biological and medical data accessible for research.[24] The processing of personal data, and in particular

[19] Cf. Article 18 of the Council of Europe Convention on Human Rights and Biomedicine, demanding particular protection for embryos in research.

[20] Induced pluripotent stem cells, which are derived from adult body cells but have the ability to develop into any cell type, including germ line cells.

[21] Cf. Articles 4 of the UN Universal Declaration on Fundamental Human Rights (1948) and the European Convention for the Protection of Human Rights and Fundamental Freedoms (1950).

[22] See for example Nelkin and Andrews (1998).

[23] See Article 21 of the Convention on Human Rights and Biomedicine. Opposite view presented for example in Beyleveld and Brownsword (2001), pp. 192–194 and Laurie (2002), p. 299 ff.

[24] Already in some of its recommendations from the early 1990s, the Council of Europe voiced the idea that human tissue should be considered a source of information and be protected in the same

sensitive data such as health-related information, constitutes another highly regulated area, also at the international level.[25] Protection of privacy in this field is focussed on confidentiality and autonomy, although the protection offered is generally weaker and use of personal data without consent is more accepted than is the case with bodily interventions. The right to withdraw consent need not necessarily mean that data already collected must be erased. Other important interests, related to for example public health and freedom of research, thus carry great weight in this context.[26] Requirements on ethics review may also be less strict with regard to studies involving only data.

As part of their privacy rights, individual research subjects normally have a right of access to health information collected about them, but there may also be a right *not* to know. This brings out questions about feedback concerning research results and how to handle incidental findings that could be of relevance to the individual donor.

When genetic information is studied, there may be different opinions on who should actually be considered as 'the person or persons concerned' by the research.[27] Does the study of a particular person's genetic composition 'concern' only the individual from whom the biological material has been obtained, or does it concern also his or her close genetic relatives, the extended family or even a larger population? Who should then give consent and who should be informed? There are also other aspects of genetic information that have contributed to the debate on whether or not genetic data should be treated differently from other types of information, for example its ability to predict future development and risks.

Human Material as Non-Identifiable Health Data

Could there be situations where it would be reasonable to make an analogy with non-identifiable health data, i.e. data that cannot be traced to any identifiable individual? In relation to such data, protection of the freedom of expression and freedom of information will often be the overarching interests, implying that the data could

---

way as other media carrying personal information; see Recommendation R (92) 1 of the Committee of Ministers to member States on the use of analysis of deoxyribonucleic acid (DNA) within the framework of the criminal justice system and Recommendation R (92) 3 of the Committee of Ministers to member States on genetic testing and screening for health care purposes. See also Hondius (1997).

[25] See for example the Council of Europe Convention for the Protection of Individuals with regard to Automatic Processing of Personal Data (1981) and the Recommendation (97) 5 of the Committee of Ministers to Member States on the protection of medical data; Directive 95/46/EC of the European Parliament and of the Council of 24 October 1995 on the protection of individuals with regard to the processing of personal data and on the free movement of such data.

[26] In accordance with Article 8.2 of the European Convention of Human Rights, infringements of the right to private life may be justified provided they are in accordance with the law and necessary in a democratic society for the protection of certain other important interests, such as public health.

[27] See for example Gertz (2004) and Gevers (2005).

be freely accessed, used and disclosed. If we disregard the fact that for research purposes, de-identification of samples may not be acceptable for reasons of quality control, follow-up, etc., the question remains whether it would be at all possible to de-identify human material containing DNA. The answer would seem to depend on the definition of identifiability applied. In the explanatory memorandum to the Council of Europe Biomedicine Convention, it is underlined that 'even when the sample is anonymous the analysis may yield information about identity.'[28]

The differences in legal protection between identifiable and non-identifiable data draw attention to a few additional issues. First, the process of de-identification is obviously a crucial point, since this is where the individual concerned definitely loses control over the use of the data.[29] Second, if the legal protection only covers information related to identifiable individuals, no particular protection is offered to the privacy interests of individuals belonging to an identifiable *group*, nor to any donor interest in not contributing to a certain activity. Lack of linguistic interoperability is of course an additional problem that has been repeatedly discussed with regard to the concept of identifiability, both at the domestic and the international level.[30]

Human Material as a Public Resource

Finally, one possible analogy could be made with the concept of public resources or public goods.[31] Since the human genome is often referred to as a common heritage of humanity, this could imply that information on that very genetic heritage should be considered public property. Public resources can be of many different types, such as historical or cultural heritages, natural resources, air and water, public roads and libraries, schools and hospitals. Legal rules pertaining to public resources may thus be quite diverse, but are normally focussed on protection of collective interests rather than individual ones.[32] Decisions on the use of public resources would typically be based on some form of community consent and equitable access. The resources are protected and preserved for future generations and not-for-profit principles are often applied. It is normally understood that public resources should be used for the public good, but this does not necessarily mean that all benefits gained from the use of a public resource are also shared within the community. With regard to human biobanks and genetic data, however, it has been argued that 'benefits *resulting from* the use of human genetic data, human proteomic data or

[28] Explanatory memorandum paragraph 135, with reference to relation to Article 22 of the Convention.

[29] Cf. Article 23 of the Council of Europe Recommendation Rec (2006) 4 on research on biological material of human origin.

[30] Elger and Caplan (2006).

[31] Cf. for example the HUGO Ethics Committee Statement on Genomic Data Bases 2002, Knoppers and Fecteau (2003), and Sheremeta and Knoppers (2007).

[32] Ossorio (2007).

biological samples collected for medical and scientific research should be *shared* with the society as a whole and the international community.'[33]

Examples provided on the possible forms of such benefit sharing include the provision of new diagnostics, facilities for new treatments or drugs stemming from the research; support for health services; and capacity-building facilities for research purposes.

Summing Up the Analogies: The Biobank a Legal Anomaly

Although the analogy with health-related personal data would seem appropriate in many instances, it is equally clear that it does not cover all aspects of research on human biological material. One possible objection to the analogy could be that the information embedded in a sample of human material is in a way both endless and unforeseeable, but with the possibility of digitalising DNA and thus transferring the biological information to a file or document; this particular difference between genetic data and biological material does seem to dwindle.[34] Nevertheless, it is undisputable that data cannot in themselves contaminate our environment or spread diseases threatening public health and safety, nor can they be used as raw material for transplants and products or have the potential to develop into a full human being. Samples of human biological material are not *just* carriers of data, but have traits of several other conceptual legal categories as well.[35] From the regulatory perspective, the complex nature of human biobanks would seem to make them into an anomaly, defying any attempt at a simplified analogy. Accordingly, policy makers regulating human biobanks must find a way to balance this combination of diverse legal analogies.

One of the primary goals in the regulation of research biobanking is often to facilitate *justifiable* use of human biological material, for research purposes. This would serve the public interest in gaining new knowledge and – eventually – improved public health, by protecting the freedom of research and guaranteeing research access to information and biological samples. Equitable access, solidarity and sharing of resources are thus important principles to consider in the regulatory process, which also go well with the concept of human biobanks as public resources. Nevertheless, these principles may sometimes conflict with the interests individual researchers could have in excluding others from access to their research materials for at least a certain period of time, especially with regard to biobanks with rare samples collected by the researchers themselves. It may thus be argued that driven too far, the principle of equitable access could reduce the incitements for certain types of research and create difficulties in attracting sponsors. Not only industry, but also many universities and individual researchers seem to find the protection of intellectual

---

[33] Article 19(a) of the UNESCO International Declaration on Human Genetic Data (2003).

[34] Rynning (2003a), at p. 116.

[35] See e.g. Sethe (2004).

property rights and financial rewards to be important incentives in their research activities.

The precondition that use of biobank materials must be *justifiable* draws attention to the presumption that research biobanking is arguably in the public interest and to the need for legitimacy. Since the creation and continued existence of large-scale research biobanks is very much dependent on the goodwill and political support of the donors, i.e. the public, it would certainly be unwise to 'jeopardize, for the sake of administrative convenience and short-term research gains, the interests, the wishes, and [...] rights of those who contributed to these resources',[36] thereby also putting at risk the long-term interests of biomedical science.

Public trust in biobank research involving genetic testing is believed to primarily depend on how the use of samples and data is undertaken and communicated, in particular with regard to the areas of 'informed consent, storage, data protection and the degree of anonymity of samples, the communication of study results and, where appropriate, of individual test results.'[37] These privacy issues are further complicated by the difficulties involved in the provision of adequate protection to all donors (including for example minors and unborn children) as well as to their families and other related groups or populations. Issues of confidentiality could also concern the potential access of third parties to samples and data, for different purposes. Even biobanks exclusively set up for research-related purposes could potentially be used as sources of information for example in a criminal investigation.[38]

Irrespective of the need for protection of confidentiality and other privacy rights, the public obviously has additional interests in the appropriate use of human biological material, related to more abstract and evasive moral values, such as our perception of human dignity in a wider sense. If our dignity as human beings and our respect for others could be put at risk by certain uses of human material, this should be taken into consideration and balanced against the public interest in scientific development. The problem is of course to decide which particular moral concerns should be recognised and what weight they should be attributed in the balancing against hopes for improved global public health and well-being. This is also a field where the plurality of moral and religious values will continue to constitute an obstacle to harmonised international or European rules in certain areas of research.

It is clear that human biobanks give rise to many controversial issues that are debated internationally as well as at the domestic level. Areas of controversy may concern for example the type and scope of donor and community consent, the rights of donors to know and not to know individual results, the role and form of ethics review, conditions for access to samples and data, property rights and commercial

---

[36] Greely (2007).

[37] McNally and Cambon-Thomsen et al. (2004a), at p. 23.

[38] Cf. Council of Europe Recommendation R (92) 1 of the Committee of Ministers to member States on the use of analysis of deoxyribonucleic acid (DNA) within the framework of the criminal justice system. For an account of the use of the Swedish PKU biobank in the investigation following the murder of Sweden's Minister of Foreign Affairs, Anna Lindh, see Wendel (2007), at p. 115.

interests in biobanking, and governance and monitoring.[39] In order to give just a few examples from the international debate, some comments will be provided concerning the relationship between the biological samples and genetic data, the justifiability of genetic exceptionalism and the ambiguity of the term anonymity.

Some Areas of Controversy

Biological Samples Are Not Data

The fact that human biological material has not traditionally been perceived primarily as a carrier of personal data, and is at all events not *only* a carrier of data, has caused uncertainty about the direct applicability of data protection legislation to biobank materials. To the extent that data protection rules are considered applicable, questions also arise concerning the interpretation of such provisions in the particular context of biobanks and the potential conflicts with other rules that could be of relevance.[40]

Not surprisingly, for example the implementation and interpretation of the so-called Data Protection Directive 95/46/EC,[41] with regard to biobank materials and associated data, varies considerably between different Member States of the European Union.[42] Although it would seem clear that personal genetic *data* fall under the Directive, few Member States have taken the position that this applies to the human biological material carrying the genetic information.[43] It is also the opinion of the Article 29 Data Protection Working Party that whereas human tissue samples are sources out of which personal data can be extracted, they are not personal data themselves.[44] The *extraction of information* from the samples therefore constitutes collection of personal data, to which the data protection rules apply, but the collection, storage and use of the tissue samples may be subject to separate sets of rules.

Despite the fact that biological samples are not data themselves, it can still be argued that human tissue should be protected to the same extent as other media carrying sensitive personal information.[45] If different rules are applied to biological samples, as compared with other carriers of personal data, this could lead

---

39 Cambon-Thomsen et al. (2007).

40 Rynning (2003a), at p. 117. See also the Article 29 Data Protection Working Party, Working Document on Genetic Data adopted in March 2004.

41 Directive 95/46/EC of the European Parliament and of the Council of 24 October, 1995 on the protection of individuals with regard to the processing of personal data and on the free movement of such data.

42 Rouillé-Mirza and Wright (2004) at p. 193 f.

43 As for the case of Denmark, however, see Hartlev (2005), pp. 236–237.

44 Opinion 4/2007 on the concept of personal data.

45 See for example the Council of Europe Recommendations mentioned in footnote 23; also the Article 29 Data Protection Working Party, Working Document on Genetic Data adopted in March 2004.

to unwanted and peculiar consequences, for example inconsistencies in the rules pertaining to DNA in digitalised versus natural form. Nevertheless, additional protection would often seem to be called for with regard to other aspects of the use of human material.

Genetic Exceptionalism

One controversial question is whether or not genetic data are actually different from other types of health data, to such an extent as to merit special legal protection of the data or their source, i.e. the human biological material. This kind of so-called genetic exceptionalism is argued in for example the UNESCO International Declaration on Human Genetic Data (2003), where it is claimed that human genetic data have a *special* status, due to the following characteristics:

i they can be predictive of genetic predispositions concerning individuals.
ii they may have a significant impact on the family, including offspring, extending over generations, and in some instances on the whole group to which the person concerned belongs.
iii they may contain information the significance of which is not necessarily known at the time of the collection of the biological samples.
iv they may have cultural significance for persons or groups.[46]

A similar view is held by the Article 29 Data Protection Working Party, referring also to the fact that identification of individuals by the genetic print presents an additional unique nature of genetic data, since 'genetic data are likely to reveal information on several people while making it possible to identify only one of them'.[47] Although particular legal protection is found to be both needed and justified, however, the Working Party points out that one of the first guarantees conditioning the use of genetic data should be to avoid attributing to these data a universal explanatory value.

The primary argument normally brought forward against genetic exceptionalism is that the characteristics of genetic information are by no means unique, but could all be found in different types of other health-related information.[48] In accordance with this line of argument, beliefs that genetic information is exceptional would rather seem to be based on the mistaken idea that our genetic composition determines our future. By responding to this misconception with exceptionalistic regulation on genetic information, policy makers could thus implicitly provide support for genetic determinism and 'the notion that genetics exerts special power over our lives.'[49]

[46] Article 4 (a) of the UNESCO International Declaration on Human Genetic Data (2003).

[47] 2004 Working Document on Genetic Data, p. 4.

[48] McNally and Cambon-Thomsen et al. (2004b), pp. 32–34.

[49] Murray (1997).

In the 25 recommendations on genetic testing presented by a group of European experts, it is argued that 'the sentiment that genetic data are different from other medical information [...] is inappropriate.'[50] Nevertheless, the group of experts acknowledges that 'some genetic information, as part of medical information, has specific dimensions which are not necessarily common to all medical information.'[51] They find that current efforts to establish laws and regulations that apply specifically to genetic testing and data handling are an understandable response to public concerns, but insist that they are 'only acceptable as a stepping stone to more considered and inclusive legal and regulatory frameworks that encompass all medical data and testing'.

It would seem that the justifiability of genetic exceptionalism is partly a question of how this concept is defined, since beliefs about genetic exceptionalism may be of regulatory significance without necessarily being true.[52]

The Ambiguity of Anonymity

Since the question of identifiability could have considerable impact on the privacy protection offered with regard to data as well as biological materials, it is all the more problematic that lack of terminological interoperability in this field continues to prevail and that the definitions adopted are often incomplete. A few examples from well-known European and international soft law instruments illustrate this quite clearly.

In the Council of Europe Recommendation from 2006, biological materials are thus divided into two main categories based on whether they are considered identifiable or not. Identifiable are those biological materials which, alone or in combination with associated data, allow the identification of the persons concerned *either directly or through the use of a code*.[53] If the user of the biological materials has access to the code, the materials are referred to as 'coded', whereas the term 'linked anonymised materials' is applied when the code is under control of a third party and not accessible to the user of the biological material. Non-identifiable materials, in the recommendation also referred to as 'unlinked anonymised materials', are those biological materials which, alone or in combination with associated data, do not allow, with reasonable efforts, the identification of the persons concerned. In the Explanatory Report, a reference is made to the definition of 'identifiable' in Directive 95/46/EC on the protection of personal data. In the terms of this Directive, however, a person can be *indirectly* identifiable for example with reference to 'one or more factors specific to his physical, physiological, mental, economic, cultural or

[50] McNally and Cambon-Thomsen et al. (2004a), pp. 8–9.

[51] McNally and Cambon-Thomsen et al. (2004b), p. 35.

[52] Ample evidence of this is provided e.g. in Kakuk (2008).

[53] Article 3.

social identity', without any use of a code.[54] The question of how this possibility of indirect identification should be interpreted has been subject to endless discussion already with regard to data.[55] Since the possibility of indirect identification would seem even more relevant with regard to human biological material containing DNA – as has been previously pointed out in another Council of Europe document[56] – it is unfortunate that the issue is not pursued in this Recommendation.

In accordance with the terminology used in this Council of Europe Recommendation, anonymous human biological materials and data can thus be either non-identifiable or identifiable, depending on whether they are unlinked or linked to identifying information by the use of a code. This does in a way highlight the ambiguous use of the term anonymous, which in some contexts may be understood to mean merely 'coded' and in others 'non-identifiable.'

The Council of Europe terminology would not seem to have been considered relevant to the definitions recommended by the European Medicines Agency (EMEA), in their note of guidance on definitions for genomic biomarkers, etc. In the EMEA definitions that were finalised during 2007 and came into operation in May 2008, anonymised data or samples do *not* allow for the donors to be identified. Here, genomic data and samples are divided into four categories: 'identified', 'coded', 'anonymised' or 'anonymous'.[57] Whereas this terminology makes an additional distinction between anonymised and anonymous samples, there is still no category for samples that are only indirectly identifiable by *other* means than coding, for example by use of information on the time and place where the sample was collected and various features of the donor. This type of samples or data would not really seem to fit the description of the category 'identified', which refers to samples or data labelled with personal identifiers such as name or identification numbers (e.g. social security or national insurance number) that make them 'directly traceable back to the subject'. Should indirectly identifiable samples then be considered anonymous? 'Anonymisation' in the EMEA guidance refers to the deletion of the link between the subject's identifiers and the unique code of such samples or data as have initially been *coded*, and it is intended to prevent subject re-identification. 'Anonymous' data and samples are defined as never having been labelled with personal identifiers when originally collected, and no coding key having been generated, but it is also made clear that there is no potential to trace anonymous genomic data and samples to individual subjects.

The slightly older UNESCO Declaration on Human Genetic Data (2003) distinguishes between three categories of data, with regard to their degree of

[54] Article 1 a.

[55] See e.g. Romeo Casabona (2004). See also the Article 29 Data Protection Working Party Opinion 4/2007 on the concept of personal data.

[56] Cf. paragraph 135 of the Explanatory Memorandum to the Biomedicine Convention, concerning Article 22.

[57] ICH Topic E15 Definitions for genomic biomarkers, pharmacogenomics, pharmacogenetics, genomic data and sample coding categories. Note for Guidance on Definitions for Genomic Biomarkers, Pharmacogenomics, Pharmacogenetics, Genomic Data and Sample Coding Categories.

identifiability: (a) data 'linked' to an identifiable person (i.e. data that contain information, such as name, birth date and address, by which the person from whom the data were derived can be identified), (b) data 'unlinked' to an identifiable person (i.e. data that are *not* linked to an identifiable person, through the replacement of, or separation from, all identifying information about that person *by use of a code*) and, finally, (c) data 'irretrievably unlinked' to an identifiable person (i.e. data that cannot be linked to an identifiable person, through destruction of the link to any identifying information about the person who provided the sample).

Comparing these three documents, adopted by highly influential European and international agents and often referred to, it is surprising to find such diversity in the terminology used. Samples or data that have been coded, and where the user of them does not have access to the key, could thus be defined as 'linked anonymous', 'coded' or 'unlinked', depending on which guiding document is consulted. In the same way, samples or data that are labelled or associated with personal identifiers, such the name or social security number of the person concerned, could be defined as 'directly identifiable', 'identified' or 'linked'.

In addition to this terminological confusion, all three documents lack clarity with regard to the categorisation of samples and data where the person concerned could be indirectly identified by the user, or by somebody else, with reasonable effort. Such indirect identification of donors would not seem too difficult to achieve even in large-scale biobanks.[58]

Varying Strategies in Domestic Regulation: Some Examples from the Nordic Countries

Is There a Nordic Approach to Biobank Legislation?

Before discussing the possibility of harmonised global or European rules for research biobanking, it could perhaps be of interest to briefly consider the differences and similarities found in regulations developed within a more limited geographic area, such as the Nordic countries. The five autonomous nation states of Denmark, Finland, Iceland, Norway and Sweden form one of the oldest and most comprehensive regional co-operation areas in the world, sharing a considerable part of their cultural and historical heritage.[59] The Nordic languages are also closely related, with the one exception of Finnish (but then Swedish has been the second official language of Finland for nearly a century). One particular field of modern day co-operation, dating back more than a 100 years, has been law reform and the drafting of new legislation. The Nordic countries also pride themselves of being unique in

[58] See for example Greely (2007).

[59] For more information on Nordic co-operation, see the Web page of the Nordic Council, http://www.norden.org/start/start.asp.

terms of well-developed population records and other databases, which has made them very well suited for epidemiological research.

Considering this tradition of co-operation and shared Nordic values, there seems to have been surprisingly little regulatory coordination in the areas of health care and biomedical research.[60] There is for example a great variety in the time of introduction, regulatory method applied and material contents of the Nordic laws on topics such as the status and rights of patients, assisted procreation,[61] genetic testing or ethics review of research. The list of differences in approach and timing could be made much longer and include also ratifications of the Council of Europe Biomedicine Convention (Denmark in 1999, Iceland in 2004 and Norway in 2006, while Finland and Sweden are still lacking). Bringing the focus back to the specific area of biobank regulation, however, Iceland here took the lead with the Act on Biobanks that entered into force in January 2002, having introduced the Act on a Health Sector Data Base some years before. Within a period of just a few years, the other Nordic countries all adopted some kind of biobank legislation.

Although it is not possible to provide any comprehensive comparative overview of the Nordic biobank laws and regulations in this context, a few comments could still be motivated in order to illustrate some of the variations seen both in approach and in material rules. It should be noted, however, that revisions of the biobank legislation are taking place in several Nordic countries. In a Norwegian Bill that was passed in June 2008, the Act on Biobanks is incorporated in a new Act on Health Research, which will enter into force at a date to be decided later.[62] In this new version, several provisions are changed in a more liberal direction, in order to facilitate research. Finland would seem to be considering a change of strategy and is now discussing the introduction of more comprehensive biobank legislation.[63] An investigator has recently been appointed to revise the Swedish biobank legislation, which has been subject to a lot of criticism due to a number of different shortcomings.[64] In this context, it is also of relevance that the regulatory landscape surrounding the biobank provisions has changed since the biobank laws were first passed. In Sweden, for example, an Act on Ethics Review of Research Involving Humans entered into force in 2004 and an Act on Genetic Integrity in 2006.[65] Most recently, there is a new Act on Patient Data[66] as well as legislation based

[60] An overview of Nordic law in the field of biotechnology, in the form of tables updated as of April 2005, can be found in the publication Legislation on Biotechnology in the Nordic Countries – An overview, TemaNord 2006: 506.

[61] On the Nordic differences in this field, see Burrell (2006).

[62] *LOV 2008–06–20 nr 44: Lov om medisinsk og helsefaglig forskning (helseforskningsloven)*; see Government Bill Ot.prp. nr. 74 (2006–2007).

[63] Information provided at the Web page of the Finnish Ministry for Social Affairs and Health, http://www.stm.fi/Resource.phx/hankk/biopankki/index.htx.

[64] Terms of reference issued by the Swedish government: Direktiv 2008:71, Översyn av lagen (2002: 297) om biobanker i hälso-och sjukvården m.m.

[65] *Lag (2006: 351) om genetisk integritet.*

[66] *Patientdatalag (2008: 355).*

on Directive 2004/23/EC on human tissues and cells.[67] In all the Nordic countries, the complex relation to other legislation has implications for the applicability and interpretation of the specific biobank regulation.

Type and Scope of Nordic Biobank Legislation

The Nordic countries thus all decided to regulate the use of human biobank materials at approximately the same time,[68] but the legal solutions preferred vary considerably. Iceland chose to introduce as a comprehensive Act on Biobanks, a strategy also adopted in Sweden and Norway, where such acts entered into force in 2003.[69] Denmark and Finland, on the other hand, have so far addressed the biobank issues by introducing complementing provisions in already existing, more general laws. The reasons for this difference in approach could at least partly be found by comparing the coverage in previous legislation of issues related to biobank research and the perceived special problems arising from biobank activities. At the start of the new millennium, both Denmark and Finland already had legislation on ethics review of research as well as on patients' rights.[70] In Denmark, these Acts were complemented with certain biobank-related provisions, but the general data protection legislation had previously been declared directly applicable to biobank samples.[71] In Finland, the revised Act on the Use of Human Organs and Tissue for Medical Purposes was supplemented with new provisions on the collection and further use of human tissue.

Iceland also had legislation on patients' rights and a ministerial regulation on scientific research in the health sector,[72] but still felt that a special Act on Biobanks was needed to regulate certain specific activities. The Icelandic Act on Biobanks

[67] *Lag (2008: 286) om kvalitets- och säkerhetsnormer vid hantering av mänskliga vävnader och celler*, implementing Directive 2004/23/EC of the European Parliament and of the Council of 31 March 2004 on setting standards of quality and safety for the donation, procurement, testing, processing, preservation, storage and distribution of human tissues and cells.

[68] For some of the early legal debate in the Nordic countries, see e.g.: Nielsen et al. (1996), Arnardottir et al. (1999) and Eriksson (2003). For those familiar with the Nordic languages, more comprehensive analysis of the Danish, Norwegian and Swedish biobank laws can be found in the following works: Hartlev (2005), Halvorsen (2006) and Rynning (2003b).

[69] Swedish Biobanks in Medical Care Act, *Lagen (2002:297) om biobanker inom hälso- och sjukvården m.m.*, in force 1 January 2003 and the Norwegian Act on Biobanks, *LOV 2003–02–21 nr 12: Lov om biobanker (biobankloven)*, in force from 1 July 2003.

[70] See the Finnish Act on the Status and Rights of Patients, *Lag om patientens ställning och rättigheter 17.8.1992/785* and the Danish Act on the Legal Status of Patients, *Lov om patienters retsstilling LOV nr 482 af 01/07/1998* (later incorporated in the 2008 Health Act, *Sundhedsloven*); the Danish Act on Ethics Review of Scientific Research, *Lov om et videnskabsetisk komitésystem og behandling af biomedicinske forskningsprojekter LBK nr 69 af 08/01/1999* and the Finnish Act on Medical Research, *Lag om medicinsk forskning 9.4.1999/488.*

[71] See report from the Danish Ministry of Interior and Health (2002) Redegørelse om biobanker, pp. 49–52.

[72] The Icelandic Act on the Rights of Patients No. 74/1997 and the Icelandic Regulation on Scientific Research in the Health Sector No. 552/1999.

is in fact the only one of the Nordic biobank laws that exclusively deals with *long-term* research biobanks, where samples are to be kept for more than 5 years. It can be assumed that the heated debate on the Icelandic Health Sector Data Base Act and the plans of deCODE Genetics to collect blood samples from all of the Icelandic population had some influence on this legal development.[73] Icelandic research biobanks of a more temporary character are still governed by the rules on scientific research and the processing of personal data.[74]

A different point of departure is chosen in the Biobank Acts of Norway and Sweden, where the choice of comprehensive legislation would seem to have been based primarily on a wish for uniform and accessible rules to govern the majority of research biobanks as well as banks for diagnostic or treatment purposes. Whereas the original Norwegian Act is applicable to all such biobanks, however, the Swedish Act covers only samples from identifiable donors, and only banks that have been set up within the professional activities of a health care provider, thus excluding for example biological materials that have been collected directly by a pharmaceutical company or a research institution, purely for research purposes.[75] Provided the samples in such biobanks can be traced to an identifiable donor, however, the research would still fall under the Act on Ethics Review and be subject to the same consent requirements, but a number of additional safeguards laid down in the Act on Biobanks would not apply. Needless to say, this distinction, which was criticized even before the Act was passed, has created considerable problems related to interpretation and implementation of the Swedish rules. Furthermore, no criteria for the crucial concept of donor identifiability have been presented.

The comprehensive Icelandic, Norwegian and Swedish Biobank Acts all contain special provisions regarding registration and monitoring of biobank activities and in particular the Icelandic legislation closely regulates the rights and administrative duties of the 'biobank licensee'. The Swedish and Norwegian Biobank Acts also prescribe certain restrictions with regard to international transfer of biobank material, even within the EU.

Issues of Consent and Withdrawal

At first sight, the legal prerequisites related to the collection or use of human biological material for research purposes may not seem very divergent. The main rules in all the Nordic countries thus include requirements for ethics review and informed donor consent, but with regard to use for new purposes of samples already banked, the rules show a wider variation. It should also be noted that since the Icelandic Act is aimed exclusively at long-term biobanking, the informed consent required for collection and storage under this Act may be quite broad, or even what is called 'open consent', indicating that the aims, purposes, methods, etc. of the future projects for

[73] Arnardottir et al. (1999).

[74] See Articles 2 and 15 of the Act on the Rights of Patients No. 74/1997.

[75] If the samples in such biobanks can be traced to an identifiable donor, they would still fall under the Act on Ethics Review and would thus be subject to the same consent requirements, etc.

which the samples may be used, are unknown at the time when they are collected.[76] When the samples are later used in research, no additional consent is needed.

The other Nordic countries would not seem to accept the kind of open consent used for Icelandic research involving samples from long-term biobanks, but this does not mean that a fully informed and specific consent will always be a necessary prerequisite for research on biobank materials. Although consent is formally the default requirement when banked materials are to be used in a project for which the donor's specific consent has not already been obtained, all the Nordic countries provide exemptions from this requirement. In Denmark and Sweden, as in Iceland, the competent ethics committee may thus allow the use of biobank materials for new research-related purposes without consent, under certain circumstances. In Norway, such exceptions must also be authorised by the Ministry of Health and Care Services and in Finland by the National Authority for Medicolegal Affairs. The prerequisites for the exemptions also vary somewhat between the countries.

In this context, it should be noted that all the Nordic countries, except Sweden, view the preservation of biological samples from patients, for purposes related to their own future health care needs, as a more or less integrated part of the medical record keeping, which is carried out independent of the patients' wishes.[77] Since such samples may prove to be of interest also to researchers, the rules pertaining to further use could in practice result in research use without any consent at all. In Denmark, a special opt-out register has been introduced, where patients can register their wish not to have their samples in clinical biobanks used for other purposes than those related to their own health care.

As a contrast, the Swedish Biobank Act provides a good illustration of the difficulties involved in applying the *same* consent requirements to biobanking for health care purposes as to banking for research purposes. In order to meet reasonable prerequisites for banking of biological samples in view of research use, the consent requirements are unusually strict from the health care perspective. Explicit, specified informed consent is thus formally required also for the preservation of samples for purposes related to the future health care of the donor (with the exception of routine samples that are only preserved for a shorter time, normally not more than 2 months). At the same time, the 'informed consent' required for the *storage* of clinical samples also in view of future research is in fact so unspecific that it will normally be quite insufficient as consent to the participation in a future research project. The administrative burdens placed on health care providers thus seem somewhat disproportionate. In practice, however, the Swedish health care providers have adopted a pragmatic approach to this demanding opt-in system and actually apply what is in principle an opt-out system, where the patient has to send in a so called no-thank-you-counterfoil if he or she does not wish the sample to be stored.[78] Even

[76] Helgasson (2004).

[77] See, for example, Section 11 of the proposed Norwegian Act on Health Research and the report from the Danish Ministry of Interior and Health (2002) Redegørelse om biobanker, pp. 195–196.

[78] An English version of this counterfoil can be accessed at the biobank webpage of the Swedish Association of Local Authorities and Regions, http://www.biobanksverige.se/getDocument.aspx?id=79.

so, Sweden would seem to be the only Nordic country where a patient has an unconditional right to have a biobank sample destroyed or de-identified even when it is being kept solely for the secure health care of the donor himself or herself.

With regard to samples intended for *research* purposes, however, all the Nordic countries except Denmark[79] allow donors to withdraw their consent at any time and have the samples destroyed or at least de-identified.[80] In Norway, the donor may even have the information extracted from the sample erased.

Access and Benefit Sharing

Whereas Denmark considers the data protection legislation to be directly applicable to biobank samples and the Norwegian biobank definition includes data that are associated with the biological material, the other countries apply more separate sets of rules to data and biological material. At least in Sweden, this has direct implications also for the accessibility of the materials. Although researchers are able to invoke their constitutional right of access to data in the files and documents of public agencies and institutions (albeit subject to applicable privacy restrictions),[81] there is no corresponding right of access to samples stored in biobanks set up by such agencies, even if the donors were to accept it.[82] In the case of digitalised DNA, however, the rules pertaining to data would of course apply.

Neither would any of the other Nordic countries seem to provide legal guarantees for equitable research access to biobank samples. The fundamental prerequisite that biobanks should be used in ways that are conducive to the public good, however, is established in the introductory sections of both the Icelandic and the Norwegian Biobank Act.

The concept of biobanks as shared resources is not completely absent in the Swedish legislation either. The fact that a biobank may constitute a valuable public resource is for example referred to in the *travaux préparatoire* in relation to the provision regulating the closing down of a biobank, that is no longer considered significant to the purpose for which is was established.[83] Authorisation to close down the bank and destroy the samples must thus be sought from the National Board of

[79] According to the position of the Danish National Committee for Biomedical Research Ethics, donors should not have the right to have their samples and data removed from use in an ongoing research project, whereas they should otherwise be allowed to have their banked samples and data removed; see the Committee yearbook 2003 (Den Centrale Videnskabsetiske Komités Årsberetning 2003), p. 37 f.

[80] Section 7 of the Icelandic Biobank Act; Section 7 of the Finnish Act on the Use of Human Organs and Tissue for Medical Purposes and Section 14 of the proposed Norwegian Act on Health Research.

[81] This general right of access to official documents is regulated in the Freedom of the Press Act, and it covers also documents in for example public universities and hospitals; see Rynning (2004), p. 383.

[82] Rynning (2003a), p. 116f.

[83] Government Bill 2001/02:44 Biobanks in Health Care, pp. 54–55.

Health and Welfare, who will consider if there is no public interest in preserving the specimens. The Act on Biobanks also offers a possibility for researchers who are refused access to the samples in a biobank, to have the refusal (re)considered by the National Board of Health and Welfare, but the opinion of the Board is not binding to the biobank principal.

All the Nordic countries in principle support the standard of non-commercialisation of human biological material as such and explicitly prohibit transactions of human material in view of financial gain. Although this principle is by no means without exceptions,[84] the laws do not address the possibility of benefit sharing. The area of lawful financial incitements for the donation or sharing of biobank samples is thus a matter of interpretation, but would seem to be rather limited outside the scope of actual cost coverage. When the biobank sample is turned into a product or some kind of research results, however, the principle of non-commercialisation no longer applies. Unfortunately, the relevant criteria defining such transition remain unclear.[85]

The Challenge of International Harmonisation

These brief comparative comments illustrate how different the legal approaches to biobank issues can be, and how the focus and strategies applied in the regulation of biobanks may vary quite widely, even within a comparatively homogenous region such as the Nordic countries. In the global and even the European perspective, harmonisation projects must be expected to involve more complex challenges, considering the existing diversity of cultural and religious values as well as the pluralism of legal traditions. In this context, questions also arise concerning the regulatory competency of different international or supra-national organisations, and the regulatory tools available.

Regulatory Tools and Competency

Hard Law and Soft Law

When different types of regulation and governance are discussed, one distinction often made is the one between formally binding hard law regulation and various types of so-called soft law, in the form of codes of conduct, declarations, resolutions, recommendations, guidelines, policy documents, etc.[86] Even though soft

[84] For example in Sweden, exceptions apply to blood, hair, breast milk, teeth, de-identified embryonic cell lines, as long as they are not already part of a biobank; see Chap. 8, Sect. 6 of the Act on Genetic Integrity.

[85] Rynning (2003a), pp. 103–105; Halvorsen (2006), pp. 219–220.

[86] This section is based on Rynning (2009).

law regulation is normally classified as rules that are not legally binding as such, the boundary between hard law and soft law is not distinct and the two types of regulation often interact in different ways. Soft law instruments may thus have certain – indirect – legal effects and are aimed at and may produce practical effects.[87] Sometimes soft law can be viewed as a transitional mode of regulation, a precursor to binding legal instruments, but it may also be used as an independent, alternative steering mode, conveying power to actors that have only limited influence in traditional regulatory processes.[88] Soft law regulation can be introduced also in areas where 'the legal competence [of the regulatory body] is weak or nonexistent',[89] and is thus a tool available to non-governmental organisations[90] as well as other actors with limited regulatory powers, for example the OECD. The status of soft law documents will thus be dependent on context as well as time. Guidelines issued for example by a ministry or by a competent public authority are likely to have a formally stronger standing than a code of conduct issued by a professional organisation, but the latter type of guideline may become indirectly binding, if considered to express the professional standard required by law, i.e. 'good practice', or if it is explicitly referred to in binding regulation. In the area of public international law, soft law documents may also be used as tools for the interpretation of binding instruments.

Although questions may well be raised concerning the democratic legitimacy of soft law,[91] such instruments also have certain advantages. Soft law is thus considered to leave more room for flexibility and rapid reactions and is believed to be particularly useful 'when dealing with complex and diverse problems that are characterized by uncertainty'.[92] This type of regulation is therefore often found in areas of human rights and environment. However, since soft law in itself does not provide any legal sanctions, such regulation alone is not sufficient where important interests must be guaranteed appropriate judicial protection.

International Regulation

The fact that it is primarily the duty of the nation states to provide an appropriate legal framework for protecting the human rights of individuals and other important interests related to the private or the public sector, by no means excludes the possibility of international regulation. In areas where harmonisation is called for, the introduction of international norms could on the contrary be highly desirable as tools

[87] Senden (2005).

[88] Mörth (2004b), at p. 198.

[89] Frykman and Mörth (2004), at p. 163.

[90] Consider for example the impact of the World Medical Association Declaration of Helsinki: Ethical Principles for Medical Research Involving Human Subjects (adopted by the WMA General Assembly in 1964, revised in 1975, 1983, 1989, 1996 and 2000).

[91] Frykman and Mörth (2004).

[92] Mörth (2004a), at p. 3.

for both voluntary and mandatory adjustments of domestic laws. The appropriate level of the regulation is a question of legal competency, but also of suitability and principles of subsidiarity. However, in particular where a certain topic or issue does not fall within the competency of any supranational organisation, the development of *binding* international rules obviously presupposes the existence of consensus.

At the global level, a number of normative documents in one way or another addressing biobank-related issues have been adopted by inter-governmental bodies, for example the WHO 1998 Proposed International Guidelines on Ethical Issues in Medical Genetics and Genetic Services, the UNESCO Declaration on the Human Genome and Human Rights (1997) and the UNESCO Declaration on Human Genetic Data (2003). A more recent document in this the field is the OECD 2008 Draft Guidelines for Human Biobanks and Genetic Research Databases. All these documents, however, belong to the area of soft law and accordingly do not have any formally binding status.

At the European level, there are two particularly important actors of relevance to the regulation of research biobanking: The Council of Europe and the European Union. While the Council of Europe is without doubt the most important and influential European body with regard to the protection of human rights, the European Union – with its primary focus on economical cooperation and development – is in many ways a more powerful institution. Even so, since the EU only has limited regulatory competency in areas of health care, research ethics and human rights, the powers of this supra-national organisation would not seem to fully encompass the legal issues arising from research biobanking.

The Council of Europe

The Council of Europe quite early took an interest in the use of human biological material, both as a source of information and as material used for transplantation and other applications. Recommendations from the Committee of Ministers in the late 1970s include topics such as harmonisation of rules on the taking and use of human substances for transplantation, as well as international exchange and transportation of human substances.[93] With the growing advances in genetics, several recommendations on the use of genetic analysis and genetic screening for different purposes were adopted in the early 1990s,[94] followed by a recommendation

[93] Recommendation R (79) 5 of the Committee of Ministers to member States concerning international exchange and transportation of human substances and Recommendation R (79) 5 of the Committee of Ministers to member States concerning international exchange and transportation of human substances.

[94] Recommendation R (90) 13 of the Committee of Ministers to member States on prenatal genetic screening, prenatal genetic diagnosis and associated genetic counselling, Recommendation R (92) 1 of the Committee of Ministers to member States on the use of analysis of deoxyribonucleic acid (DNA) within the framework of the criminal justice system and Recommendation R (92) 3 of the Committee of Ministers to member States on genetic testing and screening for health care purposes.

on human tissue banks in 1994.[95] Not until 2006, however, was a recommendation adopted that explicitly addresses issues related to the use of human biological material in research.[96] All these recommendations belong to the area of soft law, as opposed to the Council of Europe Conventions, which are formally binding to the ratifying parties. In this category there is not only the firmly established European Convention on Human Rights and Fundamental Freedoms (ECHR 1950), which provides both basic principles and a procedural framework for enforcing the rights, but there is also the Convention on Human Rights and Biomedicine (1997), with the Additional Protocol concerning biomedical research (2005). Although the Biomedicine Convention is not as widely ratified as the ECHR,[97] and lacks an adequate system for enforcement, it still has had considerable impact on the legal development both within Europe and at the global level.[98]

When it comes to research biobanking and the distinction between materials collected for a specific project and those stored for future research, however, it is not altogether easy to determine the outcome of the multi-layered set of rules provided by the combination of legally binding provisions in the ECHR, the Biomedicine Convention and the Protocol on Research, with the 2006 Recommendation.[99] The explanatory memorandum to the Protocol on Research states that this instrument is not applicable to the removal and storage of human material for *future* research, whereas Article 11 of the Recommendation on biological material prescribes that the Protocol *should* be applied. Since for example the consent requirements of the Protocol are more strict and detailed than those of the Recommendation, it becomes unclear what the applicable requirements really are. On the question of how to balance the donors' right to know with their right not to know of information collected about their health, uncertainty also prevails. References to the general provisions in Article 10 of the Biomedicine Convention, indirectly made in Article 25 of the Recommendation via article 26 of the Protocol on Research, do not provide any real guidance.

Despite its shortcomings, which include also the inadequacies mentioned earlier with regard to the definition of non-identifiable data, the Recommendation on research on biological material of human origin is still an important document, as the first official and comprehensive European instrument on this topic. Nevertheless, due to its vagueness and the lack of comprehensive examples, doubts have been

---

[95] Recommendation R (94) 1 of the Committee of Ministers to member States on human tissue banks. The definition of human tissue in this recommendation is very narrow, however, thus covering 'all constituent parts of the human body, including surgical residues but excluding organs, blood and blood products as well as reproductive tissue, such as sperm, eggs and embryos. Hair, nails, placentas and body waste products are also excluded.'

[96] Recommendation Rec (2006) 4 of the Committee of Ministers to member states on research on biological materials of human origin.

[97] At present (December 2008), the Biomedicine Convention has been ratified by 22 member states of the Council of Europe, although signed by more than 30.

[98] Roucounas (2005) and Gadd (2005).

[99] See for example Nys (2008).

raised as to the usefulness of this Recommendation to drafters of binding biobank regulations.[100]

The European Union

Despite the existence of several legal EU instruments that concern certain aspects of biobanking, there is none specifically addressing research biobanking. To the extent that biobank materials are understood to constitute carriers of personal data, certain activities must thus comply with the requirements of Directive 95/46/EC on the protection of personal data. When biobank research on human biological materials lead to new inventions, Directive 98/44/EC on the legal protection of biotechnical inventions could be relevant. If biological samples are used in pharmaceutical trials, there is Directive 2001/20/EC on the implementation of GCP in the conduct of clinical trials to consider, and experiments aimed at using biobank materials or products made of such materials for human application would fall under Directive 2004/23/EC on human tissues and cells, or Regulation (EC) No. 1394/2007 on advanced therapy medicinal products. There is also a very recent proposal for a Directive on standards of quality and safety of human organs intended for transplantation.[101]

All these legal instruments are in one way or another related to EU competency topics such as the free movement of goods and services, consumer safety or public health.[102] As already mentioned, the more specific areas of national research policies and biomedical ethics do not as such fall within the lawmaking competency of the EU, unless it could be successfully argued that harmonisation is necessary for the development of 'a single research area', as a part of the internal market.[103] Indeed, all the aforementioned documents do in fact to some extent touch upon also for example requirements of consent, privacy and ethics review, thereby illustrating the complexity of EU legal development. Even so, it is still unlikely that European consensus could be reached on the more controversial issues related to policies for research involving human subjects or human biological material. In this field, the EU must therefore primarily rely on alternative means of governance to

---

[100] Harmon (2006).

[101] Proposal for a Directive of the European Parliament and of the Council on standards of quality and safety of human organs intended for transplantation, presented by the Commission December 12, 2008.

[102] They are thus based on Articles 95 or 152 of the EC Treaty. Even such a human rights inspired document as the Data Protection Directive thus has the *primary* aim of facilitating the free flow of personal data; see Recital 8 of the Directive.

[103] Cf. Hervey and McHale (2004), pp. 238 and 281, with reference also to Articles 163 and 164 EC Treaty. In recently adopted Council Conclusions on The Launch of the 'Ljubljana Process', it is declared that 'Europe now needs to develop a common vision and effective governance of the European Research Area (ERA).' See outcome of proceedings Competitiveness Council of 29–30 May 2008.

influence the policies developed in research biobanking, for example so-called open coordination and funding policies, etc.[104]

The advisory body called the European Group on Ethics in Science and New Technologies (EGE), which expresses opinions and adopts recommendations on various issues, can also be regarded as a means of indirectly influencing the policies of Member States. The opinions submitted by the EGE include topics such as human tissue banking and umbilical cord blood banking, as well as research involving human embryonic stem cells,[105] and the group has also addressed issues related to the patenting of products or other inventions based on human biological material.[106] Although the function of the EGE is to provide expert advice to the European Commission, the opinions submitted are of course considered also by various domestic bodies in the Member States.

Addressing the Regulatory Challenges

Human biological material *combines* a number of powerful characteristics, which in different ways distinguish it from other research materials. It is thus an important source of personal data, but at the same time constitutes an extraordinary type of raw material that can be used in its original form or processed into different types of products, and may under certain circumstances even be developed into new human beings. Needless to say, balancing the various and highly important interests concerned by research activities involving such materials is no simple task. It would be unrealistic to expect that an appropriate balance could be achieved without the revision and re-consideration of several less successful attempts, in a process based on extensive and serious public debate.

Research biobanks come in many forms and shapes and are sometimes closely linked to biobanks set up for other purposes. The unlimited potential for multiple use, new use and cross-use of different types of biobank material is another reason for the difficulties involved in finding *the* appropriate set of rules. From a very narrow regulatory perspective, it would certainly be easier if the various types of biobanks could be clearly distinguished from each other and the different uses kept separate, but this is not the reality that has to be dealt with. Therefore, the diversity of biobanks and their potential uses make it very important to be clear about

---

[104] Hervey and McHale (2004), at p. 239 f, 412 ff.

[105] Opinion n°11 – 21/07/1998 – Ethical aspects of human tissue banking, Opinion n°12 – 23/11/1998 – Ethical aspects of research involving the use of human embryo in the context of the fifth framework programme, Opinion n°19 – 16/03/2004 – Ethical aspects of umbilical cord blood banking and Opinion n°22 – 13/07/2007 – The ethics review of hESC FP7 research projects.

[106] Opinion n°2 – 12/03/1993 – Products derived from human blood or human plasma, Opinion n°3 – 30/09/1993 – Opinion on ethical questions arising from the Commission proposal for a Council directive for legal protection of biotechnological inventions, Opinion n°8 – 25/09/1996 – Ethical aspects of patenting inventions involving elements of human origin and Opinion n°16 – 07/05/2002 – Ethical aspects of patenting inventions involving human stem cells.

the intended scope of different provisions suggested and adopted. Without a clear and consistent terminology, it will remain difficult even to discuss the alternative solutions available.[107]

The international nature of biomedical research undeniably provides a strong incentive for harmonisation of the rules pertaining to research biobanking, but here the complex balancing of diverse interests is further complicated by the plurality of religious, cultural, social and legal traditions. It is unlikely that all such differences could be overcome even in the long-term perspective, but some degree of regional or even international consensus could certainly be hoped for with regard to less controversial areas and issues. It is therefore an important task to identify those areas of consensus and explore the potential for further development of harmonised norms. One indispensable tool in the search for acceptable uniform standards is the comparative research performed in numerous European projects, primarily or partly aimed at finding ways to overcome different regulatory problems in research biobanking.[108] Although the field of biobank research seems to defy any attempt at simple analogies, it is clear that the application of more sophisticated analogical reasoning will still be valuable in the regulatory process.

Continuous interaction between developing soft law and hard law instruments, based on comparative and multi-disciplinary research as well as extensive public debate, is arguably the only way forward in this highly complex area of regulation. It should also be openly recognised that the regulation of research biobanking must be perceived as an ongoing step-by-step process, rather than a problem that will shortly be solved once and for all. In this process, a certain degree of diversity in domestic regulation will provide valuable opportunities to explore different models and approaches and may accordingly be an advantage in the long-term perspective.

The participation of biomedical scientists in the development of adequate standards for biobank research is of course essential. In the short-term perspective, however, life science researchers and biobank principals will have to accept that certain legal restrictions and administrative inconveniences may delay their work, cause additional costs and occasionally even prevent promising research. These are the implications of democracy, where the best interests of the public can only be served by a careful and independent balancing of the freedom of research against other fundamental rights and values.

---

[107] Elger and Caplan (2006).

[108] In addition to the project behind this book (Mapping the language of research biobanks), see for example projects such as ELSAGEN (Ethical, Legal and Social Aspects of Human Genetic Databases, http://www.elsagen.net), GeneBanC (Genetic bio and dataBanking: Confidentiality and protection of data, http://www.genebanc.eu), PRIVILEGED (Privacy in Law, Ethics and Genetic Data, http://www.privileged.group.shef.ac.uk), and BBMRI (Biobanking and Biomolecular Resources Research Infrastructure, http://www.bbmri.eu).

References

Annas, G.J. (1999) Waste and Longing – The Legal Status of Placental-Blood Banking, New England Journal of Medicine 340: 1521–1524.

Annas, G.J. (1988) Whose Waste Is It Anyway? The Case of John Moore, The Hastings Center Report, Vol. 18, No. 5, pp. 37–39.

Arnardottir, O.M. et al. (1999) The Icelandic Health Sector Data Base, European Journal of Health Law 6: 307–362.

Berg, K. (1983) Ethical Problems Arising from Research Progress in Medical Genetics. In: Berg, K. & Tranøy, K.E (eds.) Research Ethics, Alan R. Liss, New York, pp. 261–275.

Beyleveld, D. & Brownsword, R. (2001) Human Dignity in Bioethics and Biolaw, Oxford Univeristy Press, Oxford.

Burrell, R. (2006) Assisted Reproduction in the Nordic Countries – A Comparative Study of Policies and Regulation, TemaNord 2006:505, Nordic Committee on Bioethics, http://www.norden.org/pub/sk/showpub.asp?pubnr=2006:505.

Cambon-Thomsen, A. et al. (2007) Trends in Ethical and Legal Frameworks for the Use of Human Biobanks, Eur Respir J 30: 373–382.

Elger, B.S. & Caplan, A.L. (2006) Consent and Anonymization in Research Involving Biobanks: Differing Terms and Norms Present Serious Barriers to an International Framework. EMBO Rep 7(7): 661–666.

Eriksson, S. (2003) Mapping the Debate on Informed Consent. In: Hansson, M. & Levin, M. (eds.) Biobanks as Resources for Health, Research Program Ethics in Biomedicine, Uppsala University, Uppsala, pp. 165–196.

Frykman, H. & Mörth, U. (2004) Soft Law and Three Notions of Democracy: The Case of the EU. In: Mörth, U. (ed.) Soft Law in Governance and Regulation: An Interdisciplinary Analysis, E. Elgar Publisher, Northampton, MA, pp. 155–179.

Gadd, E. (2005) The Global Significance of the Convention on Human Rights and Biomedicine. In: Gevers, J.K.M., Hondius, E.H. & Hubben, J.H. (eds.) Health Law, Human Rights and the Biomedicine Convention, Martinus Nijhoff Publishers, Boston, MA.

Gertz, R. (2004) Is it 'Me' or 'We'? Genetic Relations and the Meaning of 'Personal Data' under the Data Protection Directive. Eur J Health Law 11: 231–244.

Gevers, S. (2005) Human Tissue Research, with Particular Reference to DNA Biobanking. In: Gevers, J.K.M., Hondius, E.H. & Hubben, J.H. (eds.) Health Law, Human Rights and the Biomedicine Convention, Martinus Nijhoff Publishers, Boston, MA, pp. 231–243.

Greely, H.T. (2007) The Uneasy Ethical and Legal Underpinnings of Large-Scale Genomic Biobanks. Annu Rev Genomics Hum Genet 8: 343–364.

Halvorsen, M. (2006) Norsk biobankrett – Rettslig regulering av humant biologisk materiale, Fagbokforlaget, Bergen.

Harmon, S.H.E. (2006) The Recommendation on Research on Biological Materials of Human Origin: Another Brick in the Wall, Eur J Health Law 13: 293–310.

Hartlev, M. (2005) Fortrolighed i sundhedsretten – et patientretligt perspektiv, Forlaget Thomson A/S, Copenhagen.

Helgasson, H.H. (2004) Informed Consent for Donating Biosamples in Medical Research – Legal Requirements in Iceland. In: Árnason et al. (eds.) Blood and Data. Ethical, Legal and Social Aspects of Human Genetic Databases, University of Iceland Press & Center for Ethics, Reykjavik, pp. 127–134.

Hermerén, G. (1997) Protecting Human Integrity. In: Sorsa, M. & Eyrfjörö, J. (eds.), Human biobanks – Ethical and Social Issues (Nord 1997:9), The Nordic Committee on Bioethics, Copenhagen, pp. 17–36.

Hervey, T.K. & McHale, J.V. (2004) Health Law and the European Union, Cambridge University Press, Cambridge.

Hondius, F.W. (1997) Protecting Medical and Genetic Data, European J Health Law 4: 361–388.

Kakuk, P. (2008) Gene Concepts and Genethics: Beyond Exceptionalism, Sci Eng Ethics 14(3): 357–375.

Kaye, J. (2006) Do We Need a Uniform Regulatory System for Biobanks Across Europe? Eur J Hum Genet 14: 245–248.

Knoppers, B.M. & Laberge, C. (1989), DNA Sampling and Informed Consent, CMAJ, 140: 1023–1028.

Knoppers B.M. & Fecteau, C. (2003), Human Genomic Databases: A Global Public Good? Eur J Health Law 10: 27–41.

Laurie, G. (2002) Genetic Privacy – A Challenge to Medico-Legal Norms, Cambridge University Press, Cambridge.

Legislation on Biotechnology in the Nordic Countries – An overview, TemaNord 2006:506, http://www.norden.org/pub/ovrigt/ovrigt/uk/TN2006506.pdf.

McNally, E. & Cambon-Thomsen, A. et al. (2004a) 25 Recommendations on the Ethical, Legal and Social Implications of Genetic Testing, European Commission, Brussels.

McNally, E. & Cambon-Thomsen, A. et al. (2004b) Ethical, Legal and Social Aspects of Genetic Testing: Research, Development and Clinical Applications, European Commission, Brussels.

Milunsky, A & Annas G.J. (eds.) (1976) Genetics and the Law, Plenum Press, New York.

Murray, T.H. (1997) Genetic Exceptionalism and 'Future Diaries': Is Genetic Information Different from Other Medical Information? In Rothstein, M.A. (ed.), Genetic Secrets: Protecting Privacy and Confidentiality in the Genetic Era. Yale University Press, New Haven.

Mörth, U. (2004a) Introduction; Conlusions. In: Mörth, U. (ed.) Soft Law in Governance and Regulation: An Interdisciplinary Analysis, E. Elgar Publisher, Northampton, MA.

Nelkin, D. & Andrews, L. (1998) Homo Economicus Commercialization of Body Tissue in the Age of Biotechnology, The Hastings Center Report, Vol. 28, No. 5, pp. 30–39.

Nielsen, L. et al. (1996) Health Science Information Banks: Biobanks: Danish Central Scientific-Ethical Committee, Danish Council of Ethics, Danish Medical Research Council, Copenhagen.

Nys, H. (2008) Research on Human Biological Materials and the Council of Europe: Some Unanswered Questions, Overlaps and Empty Boxes. European Journal of Health Law 15: 1–6.

Ossorio, P.N. (2007) The Human Genome as Common Heritage: Common Sense or Legal Nonsense? J Law, Med Ethics 35: 425–437.

Riis, P. (1997) Biobanks Yesterday, Today and Tomorrow:Purposes, Types and Controversies. Human Biobanks – Ethical and Social Issues. In: Sorsa, M. & Eyrfjörö, J. (eds.), Human Biobanks – Ethical and Social Issues (Nord 1997:9), The Nordic Committee on Bioethics, Copenhagen, pp. 37–41.

Romeo Casabona, C.M. (2004) Anonymization and Pseudonymization: Legal Framework at a European Level. In: Beyleveld, D., Townend, D., Rouillé-Mirza, S. & Wright, J. (eds.) The Data Protection Directive and Medical Research Across Europe, Ashgate Publishing, Aldershot, pp. 34–49.

Roucounas, E. (2005) The Biomedicine Convention in Relation to Other International Instruments. In: Gevers, J.K.M., Hondius, E.H. & Hubben, J.H. (eds.) Health Law, Human Rights and the Biomedicine Convention, Martinus Nijhoff Publishers, Boston, MA.

Rouillé-Mirza, S. & Wright, J. Comparative Study on the Implementation and Effect of Directive 95/46/EC on Data Protection in Europe: Medical Research. In: Beyleveld, D. et al. (eds.) The Data Protection Directive and Medical Research Across Europe, Ashgate Publishing, Aldershot, pp. 189–230.

Rynning, E. (2003a), Public Law Aspects on the Use of Biobank Samples – Privacy Versus the Interests of Research. In: Hansson, M. & Levin, M. (eds.) Biobanks as Resources for Health, Research Program Ethics in Biomedicine, Uppsala University, Uppsala, pp. 91–128.

Rynning, E. (2003b) Offentligrättslig reglering av biobankerna – en utmaning för lagstiftaren. In: Wolk, S. (ed.), Biobanksrätt, Studentlitteratur, Lund, pp. 49–175.

Rynning, E. (2004) Processing of Personal Data in Swedish Health Care and Biomedical Research. In: Beyleveld, D. et al. (eds.) Implementation of the Data Protection Directive in Relation to Medical Research in Europe Aldershot: Ashgate Publishing, Aldershot, pp. 381–402.

Rynning, E. (2009) Legal Tools and Strategies for the Regulation of Chimbrids. Taupitz, J. & Weschka, M. (eds.), CHIMBRIDS – Chimeras and Hybrids in Comparative European and International Research, Springer, Berlin, pp. 80–88.
Senden, L. (2005) Soft Law, Self-Regulation and Co-regulation in European law: Where Do They Meet? Electron J Comp Law, 9.1: 23.
Sethe, S. (2004) Cell Line Research with UK Biobank – Why the New British Biobank Is Not Just Another Population Genetic Database. In: Árnason, G. et al. (eds.). Blood and Data – Ethical, Legal and Social Aspects of Human Genetic Databases, University of Iceland Press & Center for Ethics, Reykjavik, pp. 313–319.
Sheremeta, L. & Knoppers, B.M. (2007) Beyond the Rhetoric: Population Genetics and Benefit-Sharing. In: Phillips, P.W.B. & Onwuekwe, C.B. (eds.). Accessing and Sharing the Benefits of the Genomics Revolution, Springer, Berlin, pp. 157–182.
Wendel, L. (2007) Third Parties' Interests in Population Genetic Databases: Some Comparative Notes Regarding the Law in Estonia, Iceland, Sweden and the UK. In: Häyry, M. et al. (eds.) The Ethics and Governance of Human Genetic Databases – European Perspectives, Cambridge University Press, Cambridge, pp. 108–119.

Legal Documents and Other Public Materials

The United Nations

Universal Declaration on Fundamental Human Rights (1948)
International Covenant on Civil and Political Rights (1966)
UNESCO Declaration on the Human Genome and Human Rights (1997)
UNESCO International Declaration on Human Genetic Data (2003)
WHO (1998) Proposed International Guidelines on Ethical Issues in Medical Genetics and Genetic Services (WHO/HGN/GL/ETH/98.1)
WHO (1985), Community approaches to the control of hereditary diseases. Report of a WHO Advisory Group, Geneva, 3–5 October 1985 (document HMG/WG/85.10)

The European Union

Consolidated EC Treaty, Official Journal of the European Union C 321 E/39, 29.12.2006
Regulation (EC) No 1394/2007 of the European Parliament and of the Council of 13 November 2007 on advanced therapy medicinal products and amending Directive 2001/83/EC and Regulation (EC) No 726/2004
Directive 95/46/EC of the European Parliament and of the Council of 24 October 1995 on the protection of individuals with regard to the processing of personal data and on the free movement of such data
Directive 98/44/EC of the European Parliament and of the Council of 6 July 1998 on the legal protection of biotechnological inventions
Directive 2001/20/EC of the European Parliament and of the Council of 4 April 2001 on the approximation of the laws, regulations and administrative provisions of the Member States relating to the implementation of good clinical practice in the conduct of clinical trials on medicinal products for human use

Directive 2004/23/EC of the European Parliament and of the Council of 31 March 2004 on setting standards of quality and safety for the donation, procurement, testing, processing, preservation, storage and distribution of human tissues and cells
Proposal for a Directive of the European Parliament and of the Council on standards of quality and safety of human organs intended for transplantation, presented by the Commission December 12, 2008 (COM(2008) 819 final)
Council Conclusions on The Launch of the "Ljubljana Process," outcome of proceedings Competitiveness Council of 29–30 May 2008, 10231/08 RECH 200 COMPET 216
European Medicines Agency (EMEA) ICH Topic E15 Definitions for genomic biomarkers, pharmacogenomics, pharmacogenetics, genomic data and sample coding categories. Note for Guidance on Definitions for Genomic Biomarkers, Pharmacogenomics, Pharmacogenetics, Genomic Data and Sample Coding Categories, EMEA/CHMP/ICH/437986/2006
Article 29 Data Protection Working Party, Working Document on Genetic Data adopted in March 2004, 12178/03/EN WP 91
Article 29 Data Protection Working Party Opinion 4/2007 on the concept of personal data, 01248/07/EN WP 136
European Group on Ethics in Science and New Technologies (EGE)
Opinion n°2 – 12/03/1993 – Products derived from human blood or human plasma
Opinion n°3 – 30/09/1993 – Opinion on ethical questions arising from the Commission proposal for a Council directive for legal protection of biotechnological inventions
Opinion n°8 – 25/09/1996 – Ethical aspects of patenting inventions involving elements of human origin
Opinion n°11 – 21/07/1998 – Ethical aspects of human tissue banking
Opinion n°12 – 23/11/1998 – Ethical aspects of research involving the use of human embryo in the context of the 5th framework programme
Opinion n°19 – 16/03/2004 – Ethical aspects of umbilical cord blood banking
Opinion n°22 – 13/07/2007 – The ethics review of hESC FP7 research projects
Opinion n°16 – 07/05/2002 – Ethical aspects of patenting inventions involving human stem cells.

The Council of Europe

European Convention for the Protection of Human Rights and Fundamental Freedoms (1950, ETS no. 005)
Convention for the Protection of Individuals with regard to Automatic Processing of Personal Data (1981, ETS no. 108)
Convention for the Protection of Human Rights and Dignity of the Human Being with Regard to the Application of Biology and Medicine: Convention on Human Rights and Biomedicine (1997, ETS no. 164)
Explanatory Report to the Convention on Human Rights and Biomedicine, 1996
Additional Protocol to the on Human Rights and Biomedicine, concerning Biomedical Research (2005, ETS no.195)
Recommendation R (79) 5 of the Committee of Ministers to member States concerning international exchange and transportation of human substances
Recommendation R (90) 13 of the Committee of Ministers to member States on prenatal genetic screening, prenatal genetic diagnosis and associated genetic counselling
Recommendation R (92) 1 of the Committee of Ministers to member States on the use of analysis of deoxyribonucleic acid (DNA) within the framework of the criminal justice system
Recommendation R (92) 3 of the Committee of Ministers to member States on genetic testing and screening for health care purposes
Recommendation R (94) 1 of the Committee of Ministers to member States on human tissue banks

Recommendation (97) 5 of the Committee of Ministers to Member States on the protection of medical data

Recommendation Rec (2006) 4 of the Committee of Ministers to member states on research on biological materials of human origin

The OECD

OECD 2008 Draft Guidelines for Human Biobanks and Genetic Research Databases

Denmark

Act on the Legal Status of Patients, *Lov om patienters retsstilling LOV nr 482 af 01/07/1998*

Act on Ethics Review of Scientific Research, *Lov om et videnskabsetisk komitésystem og behandling af biomedicinske forskningsprojekter LBK nr 69 af 08/01/1999*

Ministry of Interior and Health (2002) Redegørelse om biobanker. Forslag til retlig regulering af biobanker inden for sundhedsområdet. Betænkning nr. 1414

Den Centrale Videnskabsetiske Komité, Årsberetning 2003

Iceland

Act on the Rights of Patients No. 74/1997

Regulation on Scientific Research in the Health Sector No. 552/1999

Act on Biobanks no. 110/2000

Finland

Act on the Status and Rights of Patients, *Lag om patientens ställning och rättigheter 17.8.1992/785*

Act on Medical Research, *Lag om medicinsk forskning 9.4.1999/488*

Act on the Use of Human Organs and Tissue for Medical Purposes, *Lag om användning av mänskliga organ, vävnader och celler för medicinska ändamål 2.2.2001/101*

Germany

Biobanks for research, 2004, Opinion of the German National Ethics Council, http://www.ethikrat.org/_english/publications/Opinion_Biobanks-for-research.pdf

Norway

Act on Biobanks, *LOV 2003–02–21 nr 12: Lov om biobanker (biobankloven)*

Act on Health Research, *LOV 2008–06–20 nr 44: Lov om medisinsk og helsefaglig forskning (helseforskningsloven)*

Government Bill Ot.prp. nr. 74 (2006–2007) *Lov om medisinsk og helsefaglig forskning (helseforskningsloven)*

Decision of the Norwegian Directorate of Health 10.12.2007, 07/2904–19, available at the web page of the Directorate, http://www.shdir.no/vp/multimedia/archive/00019/SHdirs_svar_p__FHIs__19665a.pdf

Sweden

Act on Gentic Integrity, *Lag (2006:351) om genetisk integritet*

Swedish Biobanks in Medical Care Act, *Lag (2002:297) om biobanker inom hälso- och sjukvården m.m*

Act on Patient Data, *Patientdatalag (2008:355)*

Act on Standards of Quality and Safety for Human Tissues and Cells, *Lag (2008:286) om kvalitets- och säkerhetsnormer vid hantering av mänskliga vävnader och celler*

Government Bill 2001/02:44 Biobanks in Health Care

Direktiv 2008:71 Översyn av lagen (2002:297) om biobanker i hälso- och sjukvården m.m., http://www.sou.gov.se/kommittedirektiv/2008/dir2008_71.pdf

Other Materials

World Medical Association Declaration of Helsinki: Ethical Principles for Medical Research Involving Human Subjects (adopted by the WMA General Assembly in 1964, revised in 1975, 1983, 1989, 1996 and 2000)

HUGO Ethics Committee Statement on Genomic Data Bases (2002)

PopGen Database, http://www.popgen.info/Glossary.cfm

Web page of the Nordic Council, http://www.norden.org/start/start.asp

Web page of the Finnish Ministry for Social Affairs and Health, http://www.stm.fi/Resource.phx/hankk/biopankki/index.htx

Web resource CODEX of the Swedish Council for Research, http://www.codex.vr.se/codex_eng/codex/index.html

Annexation of Life: The Biopolitics of Industrial Biology

Jan Reinert Karlsen and Roger Strand

Abstract From the new imperialism of the second half of the nineteenth century to the biocolonialism at the beginning of the twentieth century, expansionist policies went from annexing land to annexing life. Arguing that the panacea of human ailments professed to be achievable by unlimited expansion into the molecular space of human life comes with a yet unrecognized prize, the dynamics of this historical shift is analyzed as a driving force out of political institutional control. By reconnecting the knowledge-based economy, industrial biology, and biocolonialism into a dynamic complex, the purpose of this chapter is to contextualize postgenomic research biobanking and to bring out its destructive propensity to topple existing moral and political institutions.

> Expansion is everything", said Cecil Rhodes, and felt into despair, for every night he saw overhead "these stars ... these vast worlds which we can never reach. I would annex the planets if I could.
>
> S.G. Millin (Arendt 1951: 124)

Introduction

The century separating the ruler drawn maps of the late nineteenth century and the high-throughput nucleotide maps of the human genome sequence variation published at the turn of the millennium (Sachidanandam et al. 2001; Venter et al. 2001) not only marks a shift in the *direction* and *scale* of expansionist policies, but also more importantly, a shift in their *nature*. At the beginning of this period the idea of "expansion" undergoes a *semantic* and *axiological* transformation that reverberates into later transformations to come. According to Hannah Arendt, what

J.R. Karlsen (✉)
Section for Medical Ethics, Faculty of Medicine, University of Oslo, Oslo, Norway
e-mail: j.r.karlsen@medisin.uio.no

J.H. Solbakk, et al. (eds.), *The Ethics of Research Biobanking*,
DOI: 10.1007/978-0-387-93872-1_20,

distinguishes the "new imperialism"[1] from other forms of expansionist policies is the new *meaning* and *value* given to its core idea:

> Expansion as a permanent and supreme aim of politics is the central political idea of imperialism. Since it implies neither temporary looting nor the more lasting assimilation of conquest, it is an entirely new concept in the long history of political thought and action. The reason for this surprising originality – surprising because entirely new concepts are very rare in politics – is simply that this concept is not really political at all, but has its origin in business speculation, where expansion meant the permanent broadening of industrial production and economic transactions characteristic of the nineteenth century. (Arendt 1951: 125)

Refracted through this concept, *annexation of land* becomes a rational as well as natural policy: By providing (a) unrestricted access to natural and cultural resources, it connects (b) new markets and capital to a homeland-based system of mass production, thereby inducing (c) reinvestment rather than accumulation of surplus capital, which, as a side effect, (d) increases the mutual dependency between national economies and international banking and investment capital (Hobson 2005). A volatile dependency, one should add, is that there is a seemingly simple arithmetic limitation to policies utilizing annexation of land as a vehicle for permanent economic and industrial growth: The Earth's surface is approximately 510 million km^2, of which 149 million km^2 is above the sea level.[2] From the conference held in Berlin in 1884[3] to its war torn ruins in 1945, expansionism had gained a momentum too powerful for nation states to handle – so powerful, that when it led humankind toward its most destructive climaxes, it was unable to stop.

The rigor mortis of imperialism did not conclude expansionism, however. Rather its direction changed toward new frontiers, new terrains too miniscule to be rendered in the scales of geographical maps – "the endless frontier", in the words of Vannevar Bush (1945). From the war torn ruins of Berlin to the eerie ruins of Hiroshima and Nagasaki, one frontier was being closed and a new one was opened. When humankind created the fake suns over Hiroshima and Nagasaki, expansion became irrevocably tied to basic science and its applications. While the "Scramble for Africa" had been rather loosely orchestrated, often driven by private colonial entrepreneurship, the new scramble for technoscientific objects, became matters of national priority, planning and security from the very onset. The prototype of industrial science, "The Manhattan Project," had demonstrated what could be achieved when reductionist science is planned, organised, and funded at an industrial level. The industrialization of physics was soon followed by the industrialization of medicine, as the conceptualisation of health becomes disjoined from the

---

[1] The new imperialism or new colonialism is a period of intensified annexation that lasted from the second half of nineteenth and first half of the twentieth century. The reason for extending the period to WW2 is that Nazi Germany's Lebensraum policies and colonization of Eastern Europe and Russia, on central points, coincides with the idea of expansionism of the new imperialism.

[2] At its height, the British Empire encompassed 20% of all the earth's landmasses, some 30 million km^2.

[3] The Berlin Conference ended the so called "Scramble for Africa" by dividing what remained of African land between the Great European Powers.

experience of the sick (Canguilhem 1991), and conjoined to the ends of biomedical expansionism. Health, it is thought, can only be achieved if there is a final solution for human ailments and disease (Illich 1975).

In the same year as the roadmap for decolonization was laid forth by the General Assembly (United Nations 1960) and territorial expansion was formally closed, acclaimed physicist Richard Feynman publicly announced that "there's plenty of room at the bottom!" (Feynman 1960). A new space had been established for expansion – "in principle" at least. If it is not volumes, but *surfaces* that are chemically and biochemically active, then reducing the scale of operation from *kilometer* to *nanometer* increases the area per mass beyond our powers of imagination, enabling immense creative opportunities for nanotechnology and biotechnology.

The colonial rhetoric of *pristine emptiness*, now at the molecular level, was informing "progressive" policies on nanotechnology as well as biotechnology from the very beginning. In Feynman's words: "In the year 2000, when they look back on this age, they will wonder why it was not until year 1960 that anybody began seriously to move in this direction" (1960). Standing approximately at the vantage point from where Feynman would have us look back, we argue that the historical direction taken is not merely a shift in scale, but a recasting of the very space in which technological dominion (Kay 1992) and annexation of living Nature are now being pursued. It is at the moment when this space is constructed within human life itself, and when policies are made to facilitate expansion into this space, that the problems specific for biocolonialism begin to emerge.

The Knowledge-Based Economy

While the carefully orchestrated publication of the human genome sequence variation was announced as one of the most momentous scientific achievements in human history (White House 2000), the public spectacle surrounding the result eclipsed an even greater feat. Over half a century after physics, biology had been up-scaled to an industrial level. The Human Genome Project provided an organisational, logistical, and technological *paradigm* for integrating, rearranging, and harmonising the interdisciplinary research, investment and industrial planning needed to probe the exceedingly complex molecular mechanisms of disease, heritage, and even life itself. It was this experience rather than the genome sequence itself (which in reality was not *the* human genome but rather *a* genome or a handful of genomes) that provided the dialectic and direction taken on by postgenomic research. At this defining moment, industrial biology becomes recognised as a primary vehicle for the continued implementation and expansion of *knowledge based economies* (Lisbon Strategy 2000).

While the word "knowledge-based economy" was popularized by Peter Drucker in the 1960s, its conceptual origins at least go back to the 1940s with Vannevar Bush and Joseph Schumpeter's imaginative thinking. In a rough sense, the concept of knowledge-based economies attempts to solve the problem of unlimited economic and industrial expansion, given that raw materials and land, and to some extent labour, are limited resources. In Rosa Luxemburg's analysis, capitalism "needs pre-capitalist societies as a market for its surplus value" (Lee 1971: 848). Hence, in its imperialist stage capitalism is intrinsically leading toward collapse when all *hinterland* has been assimilated.

What Luxemburg deemed a collapse, and for Europe proved to be a catastrophe, took on the meaning of victory for the strongest of the capitalist nations. With victory in sight, President Roosevelt asked Vannevar Bush what positive lessons could be learnt from scientific progress in wartime for the postwar era: "New frontiers of the mind are before us, and if they are pioneered with the same vision, boldness, and drive with which we have waged this war we can create a fuller and more fruitful employment and a fuller and more fruitful life" (Bush 1945). Bush' optimistic answer is well known: Basic science is the keystone on which war against disease could be waged, national security would be preserved, and equally important, national welfare, and prosperity would be increased:

> One of our hopes is that after the war there will be full employment. [...] To create more jobs we must make new and better and cheaper products. We want plenty of new, vigorous enterprises. But new products and processes are not born full-grown. They are founded on new principles and new conceptions which in turn result from basic scientific research. Basic scientific research is scientific capital (Bush 1945).

It has been the basic United States policy that the Government should foster the opening of new frontiers. It opened the seas to clipper ships and furnished land for pioneers. Although these frontiers have more or less disappeared, the frontier of science remains. "It is in keeping with the American tradition – one which has made the United States great – that new frontiers shall be made accessible for development by all American citizens" (Bush 1945). In other words, when all land has been encloed by fences, permanent economic and industrial growth can be achieved through the development of "new ideas and knowledge" leading to "new consumer products." It is in this sense that Bush sees science as the endless frontier. Though the frontier on land had come full circle, the world of ideas and knowledge could not be fenced in, not even by the shifting horizons of scientific world views. Just as science became the endless frontier of the economy, the economy opened up endless new spaces for action and entrepreneurship in "the staggering small worlds below", as Feynman (1960) later would call them.

Another solution to the problem of unlimited expansion, this time theoretical, was proposed only 1 year earlier, in 1943, by Joseph Schumpeter. According to Schumpeter's classic work, *Capitalism, Socialism and Democracy*, capitalism evolves through a Heraclitean process – "Creative Destruction" – which enables the capitalist structure to incessantly "mutate" into new forms, inevitably destroying old ones and rendering others obsolete (Schumpeter 2005: 83). The capitalist structure is thus never stationary, but evolves.

This has two implications for the asymmetrical relation we so far have exemplified between expansion and land. First, if the capitalist structure exhibits self-organizing structures that incessantly reinvent themselves, then they will continue to do so, even after all natural resources have been depleted and all land annexed, simply because the reinvention – what we may call the immanent creative propensity of the capitalist structure – opens up new markets, new spaces to be annexed, and new resources to be exploited. The industrial recycling of waste is an example that, more or less, drives this point home. Although this may be interpreted as matters of emphasis and biography, for Schumpeter it is not basic science that represents the endless frontier, as it was for Bush, but the capitalist structure.

Second, any evaluation of the capitalist structure that disparages it for its imperfections, caprice, and inefficiencies fails to see that these weaknesses, in fact, are simultaneously its strengths. The capitalist structure is never optimal; there has never been a golden age of unfettered commerce simply because the structure is open-ended, information is always incomplete, and change uncertain. Indeed, the uncertainty provides the opportunity for unpredictable profit and is as such a necessary condition for the provision of venture capital and the functioning of the stock market (Knight 1921). Hence, capitalism does not lead to an apocalyptic collapse, as in the historical determinism of Marx, but is incessantly falling apart while it reassembles its parts into ever new configurations of expansion.

In the following, we will critically assess Schumpeter's concept of "Creative Destruction" in some depth, not only because it is so powerfully and profoundly expressed by its author, but also even more so, because it forms an indispensable backdrop for understanding the theoretical underpinnings of the *Lisbon Strategy*, and the European ambition to become "the World's most innovative and competitive knowledge-based economy by the year 2010" (Lisbon Strategy 2000). In this subsection as well as in the next, our focal point will be on *normativity* in the theoretical framing of knowledge-based economies and industrial biology.

At first, Schumpeter dismisses what he considers to be aspects of the capitalist structure, but not its *essence*:

> Capitalism, then, is by nature a form or method of economic change and not only never is but never can be stationary. And this evolutionary character of the capitalist process is not merely due to the fact that economic life goes on in a social and natural environment which changes and by its change alters the data of economic action; this fact is important and these changes (wars, revolutions and so on) often condition industrial change, but they are not its prime movers. Nor is this evolutionary character due to a quasi-automatic increase in population and capital or to the vagaries of monetary systems, of which exactly the same thing holds true (Schumpeter 2005: 82).

In fact, according to Schumpeter's theory, there are either two *essences* of capitalism, or they amount to the same: (a) "The *essential* thing to grasp is that in dealing with capitalism we are dealing with an evolutionary process" (Schumpeter 2005: 82); (b) "This process of Creative Destruction is the *essential* fact about capitalism. It is what capitalism consists in and what every capitalist concern has got to live in" (Schumpeter 2005: 83; our italics). This essential feature or features, however, are elucidated through qualitatively different analogies: Sometimes "Creative Destruction" is explicated through biological analogies such as "evolution" and "mutation"; sometimes through mechanistic analogies such as "engine"; sometimes through vitalistic analogies such as "organic process"; and sometimes even by analogies of metaphysical origin, i.e. "prime movers":

> The fundamental impulse that sets and keeps the capitalist engine in motion comes from the new consumers, goods, the new methods of production or transportation, the new markets, the new forms of industrial organization that capitalist enterprise creates (Schumpeter 2005: 83).

"Impulse" has its modern origin in vitalistic theories of evolutionary creation (Bergson 1954), while "mutation" was invented by de Vries and Bateson to describe a perennial biological mechanism of evolutionary change at the turn of the nineteenth century (Canguilhem 1994a). "The opening up of new markets, foreign or domestic, and the organizational development from the craft shop and factory to such concerns as U.S. Steel illustrate the same process of industrial mutation – if I may use that biological term – that incessantly revolutionizes the economic structure *from within*, incessantly destroying the old one, incessantly creating a new one" (Schumpeter 2005: 83). However, these qualitative differences cannot be reduced to that between macroeconomic and microeconomic scales of description, enabling mechanistic explanation at the microlevel and vitalistic on the macrolevel of societies (Schumpeter 2005: 83–84). In his drive to describe the volatile vitality of the capitalist structure by biological analogies, Schumpeter seems to have forgotten that for the organism "many mutations are subpathological, and a fair number lethal, so the mutant is less viable than the original organism" (Canguilhem 1994a, b: 318). Are we then with Georges Canguilhem allowed to say that the capitalistic structure is a "monster," since "many mutations are 'monstrous'"?

> – but from the standpoint of life as a whole, what does "monstrous" mean? Many of today's life forms are nothing other than "normalized monsters", to borrow an expression from the French biologist Lois Roule. (Canguilhem 1994a, b: 318)

If life and societies were organized through identical norms and structures, then perhaps we could say that the capitalistic structure is a "normalized monster." However, contrary to societies, an organism has little or no choice regarding the unfolding of its ontogenesis, it must "be" and "become" in a continuum, that is, sustain one form of being, while new forms are being constructed (Rose 1997). For living organisms, organization is a *fact*, for society it is a *task* (Canguilhem 1991). Hence, if the ceaselessly mutating capital structure is a fact, it is because societies have set it as their task to organize their existence on the expansion of this particular structure. This is the task put forth by the Lisbon strategy.

Industrial biology and information technology have become primary catalysts for strengthening European implementation of knowledge-based economies and competition with other capitalist economies, especially the United States. With the inception of industrial biology, the normativity of the capitalist structure and the normativity of living organisms have found a common ground. The technoscientific *imperium*[4] over the molecular space of human life enables human ends to be inscribed into the teleology of life itself, as a function of "mutations" in the capitalist structure. When the capitalist structure begins to capitalise on the human organism through annexation and technoscientific command, may we conclude then that the capitalist structure may be "pathological", "monstrous" or even "lethal" for societies too?

[4] Latin "command".

Industrial Biology

How does one assert that science indeed is an endless frontier? Not all sciences serve equally well for production of consumer goods. First, the intermediate step of technology development is needed to turn knowledge into material products. However, not all forms of scientific knowledge are well suited for technological development. Second, it is not obvious that scientific pioneering can escape the spatial bounds eventually experienced by American pioneers as well as stargazing imperialists such as Cecil Rhodes. Perhaps what Horgan later would announce as "the end of science" could not be avoided? The solution in physics was to increase the energies, while in biology the molecular space of life was recast in terms of information – an apparently endless resource.

The differences between these solutions can be observed in the problem of "manipulating and controlling things on a small scale" (Feynman 1960). Feynman observed that the problem entails something thoroughly different in biology than in physics. Prophesying that physics would one day be able to write the entire bibliography of mankind on a cube of materials, or as he put it, on "the barest piece of dust," Feynman recognized that "biology is not simply writing information," because the information, according to Crick's Central Dogma, is already there. For biology, it is insufficient to merely study Nature's own "information system," and "it is *doing something* about it" (Feynman 1960). This program is to some degree contained in Vannevar Bush's answer to President Roosevelt, in so far as it is a program for constructing Nature in the image of our own needs. "To some degree" only, because when Feynman asks us to "consider the possibility that we too can make a thing very small which does what we want" (Feynman 1960), he also includes the construction of life forms in the image of our own desires.

The important thing here is the concept of information, because information is always relative to a *purpose*. While industrial biology entails both an expansion and contraction of scale, allowing the study of technoscientific molecular objects at an industrial level, the sequencing of these objects into information enables them to be defined in terms of anthropomorphic ends. Hence, in the life sciences and biotechnology, the concept of biological (genetic) *information* has opened up spaces beyond those defined by Euclidean geometry (Aristotle B 2 996 a; Canguilhem 1994a).

The role of industrial biology in the knowledge-based economy is at least twofold: On the one hand, to produce *new ideas and knowledge* leading to *new consumer products* endowed with biological information. On the other hand, its role is to advance permanent economic expansion, innovation, and industrial growth. The problem, however, is that contrary to life, industrial biology, as of yet, cannot produce its own information. In other words, it needs an immense input of biological information, both genotypic and phenotypic. What makes this problem special is that the "natural resource" needed as input is situated in the human body (cells, tissue, and DNA) and in the lives of citizens (health data, biographic information, genealogy) – persons with interests of their own and fundamental rights. What is the solution to these problems? To convince the public and oneself that the new framing of collecting, storing, and utilising information for research purposes – so called research biobanking – in fact, is nothing less than the latest progression toward improved health, cheaper health care, and personalized medicine.

The idea of collecting and storing information about a population's health, however, is much older than research biobanking in the age of industrial biology. As Michel Foucault has pointed out, the birth of health registries coincides with the birth of the clinic at the turn of the eighteenth century and the development of national strategies to counter epidemics threatening the Republic:

> [...] there were requests for a statistical supervision of health based on the registration of births and deaths (which would have to mention the disease from which the individual suffered, his mode of life, and the cause of his death, thus constituting a pathological record); [...] in fact, that a medical topography of each department should be drawn up, "with detailed observations concerning the region, housing, people, principal interests, dress, atmospheric constitution, produce of the ground, time of their perfect maturity and harvesting, and physical and moral education of the inhabitants of the area" (Foucault 1997: 31).

As the industrial mode of production disseminates from Britain into other European countries, America, and Asia, the work force's ability to produce surplus capital becomes targeted as a field of its own for expansionist policies. If it is labor that creates surplus capital (Locke 1884), then from investing in the health of the work force, one can predict that the surplus capital each individual worker can produce, on average, will increase, even if work hours are reduced. Lowered work hours will again increase consumption, as leisure is confined to well defined hours of the working day. When health registries are created as political instruments of the state to measure, survey, and modify the population's health to such ends, biopolitics is born:

> The mechanisms introduced by biopolitics include forecasts, statistical estimates, and overall measures. And their purpose is not to modify any given phenomenon as such, or to modify a given individual insofar as he is an individual, but, essentially, to intervene at the level of their generality. The mortality rate has to be modified or lowered; life expectancy has to be increased; the birth rate has to be stimulated (Foucault 2003: 246).

The up-scaling of biology realized at the turn of the twentieth century, however, transforms the social and moral space formerly occupied by biopolitics in a number of important and unforeseen ways:

1. Although biopolitics saw the health in light of progressive political stability and the productive propensity of a healthy and content population, health is increasingly being defined in terms of the individual's investment in his or her body.
2. Aiming for a more personal tailored medicine, mechanisms to intervene at the general level are being sought replaced by interventions at the cellular, subcellular, or molecular level (Kay 1992).
3. The collection and storage of tissue and health information in spatially confined collections and registries can no longer be justified solely as measurement and surveillance of the population's health, but are increasingly being connected together in research infrastructures to facilitate industrial development and economic exploitation.

4. Biological materials have either been appropriated without consent or framed as an altruistic act donation, while human tissue and bioinformation has become a natural resource that can be converted into fiscal value.
5. The information stored in traditional registries and repositories was rather limited; the new bioinfrastructures come with information technology that can increment this limit indefinitely.
6. While the progress in health being made until the industrialization of medicine after WW2 had mostly to do with clean water and sanitary conditions, biopolitics allowed this progress to be measured. Expansion of health became a good in itself. The contentious nature – and power – of the new technoscientific possibilities of industrial biology, however, has opened up for a political recognition that even though progress is a good in itself, it needs to be calibrated by ethics.

The Political Concept of Biocolonialism

Along with the machine gun, telegraph, and steam-propelled ships, the "locomotive clearly had a unique propensity for integrating and annexing territory, for monopolizing its resources, and for pre-empting the future of great stretches of country" (Robinson et al. 1991). Often being built from the coastline into the heartland of annexed territories, the railway was a centrepiece infrastructure needed for transport of raw materials and manufactured goods. An efficient infrastructure is vital for efficient transactions for a number of reasons. Its logistical efficiency can be measured as a function of: (a) the number of possible transactions per time unit, (b) the volume of possible transaction per time unit, and (c) the quality of transactions per time unit – such as security, safety, and punctuality. In this respect, an efficient infrastructure will add value to the goods and raw material that move along its trajectory, by speeding up the number, volume, and quality of transactions between markets, raw materials, investments, and industry. The "railroad imperialism" of the second half of the nineteenth century became a precondition for economic integration and industrial expansion, while the information highways of the twentieth century not only integrate the knowledge-based economy through an unfathomable network of binary transactions but also allow biobanks to be linked in vast bioinfrastructures. The bioinfrastructures of twentieth century biocolonialism do not link harbor and minerals, but integrate a range of biobanks and databases into rhizome-like nodes that, in principle, can be extended indefinitely.

The European Biobanking and Biomolecular Resources Research Infrastructure (BBMRI) is a pan-European *type 3* research biobank.[5] Its mission is fourfold:

> (1) "To prepare to construct a pan-European Biobanking and Biomolecular Resources Research Infrastructure (BBMRI), building on existing infrastructure, resources and technologies, specifically complemented with innovative components and properly embedded into European ethical, legal and societal frameworks"; (2) "To have a sustainable legal and financial conceptual framework for a pan-European BBMRI."; (3) "To increase scientific

[5] For this, see also chapter "In the Ruins of Babel: Should Biobank Regulations be Harmonized?".

> excellence and efficacy of European research in the biomedical sciences and discovery"; (4) "To expand and secure competitiveness of European research and industry in a global context" (Zatloukal et al. 2007: 2).

In this subsection, we will analyze the formation of the political concept of biocolonialism. Political concepts differ from, say, scientific concepts, in that their normativity is directed at organising and translating political problems into thoughts and actions that need means and justifications to reach their ends. At the heart of every concept is a problem, or clusters of problems, that spurred its formation in the first place. Once the concept becomes operational, this problem is often pushed aside, or tacitly forgotten. "Biocolonialism" first becomes operational as a political concept when industrial biology has reached a certain "critical mass". Although Alvin Weinberg, the man who coined the phrase "Big Science," had alluded to industrial biology as early as the 1960s (Weinberg 1999), this "critical mass" is first achieved with the completion of the human genome project:

> The sequencing of the human genome, completed at the dawn of the 21st century, allows researchers to integrate new data on genetic risk factors with demographic and lifestyle data collected via modern communication technologies. The technical prerequisites now exist in Europe for merging large volumes of molecular genetic data obtained by using new high throughput DNA analysis platforms with clinical, epidemiological and national health registry data (Zatloukal et al. 2007: 2).

Simultaneously as the human genome project transforms human biological material and its concomitant bioinformation into a "natural resource" that can be converted into monetary means of exchange (Annas 1999), a spring of policies pertaining to postgenomic research biobanking begins to surface in industrialized nations (Althingi 2000; Riksdagen 2002; Stortinget 2003). In these policies, the underlying problems pertaining to postgenomic biocolonialism are articulated. How does one facilitate: (1) mass transactions of biological material and bioinformation as well as (2) annexation of the molecular space of human life without transgressing the moral and political bounds of the institutions of the state? To translate and organize the former problem into policies, an automated, and sanitized, form of informed consent practice was installed to secure the "informed" compliance of "donors," enabling transactions of "natural resources" to be perceived as voluntary and altruistic donations. Translating and organizing the latter problem into policies, a distinction was struck between "human biological materials *as such*" and "human biological materials that have been made into *a product or process or a part of a product or process*." The distinction is here justified on grounds of innovation and intellectual property rights, which can be traced back to the Lockean doctrine of labor. According to this doctrine, it is labor that is the active denominator of value while Nature is passive, receiving, worthless (Locke 1884).

Although policies to facilitate mass transactions of human biological material became designed to "ensure that noncommercial donation and procurement of human tissue continues whilst allowing the development of medical advances by commercial activities based on these samples and the data attached to them" (European Commission 2004: 44), policies facilitating the annexation of life go as far back as the early 1980s. In the US Supreme Court decision in *Diamond v. Chakrabarty*, General Electric was allowed to hold a patent on biotechnologically engineered oil digesting bacteria. The case is interesting on several accounts, especially because the relevant line of demarcation between patentability is not drawn between inanimate objects and life, but between nature and invention, thus creating an economic incentive for legal biocolonial expansion (Kimbrell 1996).

We can now see how this line of demarcation reproduces an asymmetrical relation pervading the colonial mindset and rhetoric, of *pristine emptiness*. For imperialists, minerals and other natural resources could simply be taken from colonies, because these resources had no real value for natives lacking the technology and infrastructure needed to exploit them for themselves. Similarly, human biological materials are of no value to the "donor" because, in general, he or she does not possess the technology, training, and infrastructure needed to transform them into biotechnological "products and processes." Hence, human biological materials can be exploited without remuneration to the "donor" since it is far better that these resources are put to good use than being unused, or even squandered by the incompliant donor herself. At its most extreme, this mindset will therefore have us conclude that an incompliant donor is egoistic, if not a thief.

While bioprospectors may still move along the meridians and parallels of the Earth, their defining mode of movement is *inwards*, into the body of the individual. This space is defined not only in terms of Euclidean geometry, but also through a scientific conceptualisation of life as information (Canguilhem 1994a). What we are dealing with is the annexation and technological domination over physiological processes at a molecular and cellular scale – in the biocolonial conceptualisation of life as human invention and intellectual property. In this reconfigured biopolitical space, the individual's body is triangulated as gratis "natural resource," consumer of personalized health products and unlimited molecular space *in one*. One can ask oneself whether the purpose of the bioinfrastructures of industrial biology is really promulgating healthy individuals and populations, or whether their forthcoming results and applications will take medicalization to a new and yet unprecedented level. Will the knowledge-based economy, if catalyzed through the unlimited industrial expansion of biology, produce a "critical mass" of orthopharmaceutical needs for the *healthy sick*? By "healthy sick" we mean an individual whose diversity is identical with his pathological probabilities, so that he must consume health products to retain an optimal wellbeing – in a probabilistic sense of the word, because any genetic variation can potentially increment the risk for disease by exposing it to the "wrong" environment. Hence, by ever expanding the technoscientific power over vital and pathological phenomena, one may still be able to get scientific evidence for one of the most fundamental, and thus trivial, experiences of human life. As Canguilhem points out, in one sense it is perfectly normal to be sick (Canguilhem 1991: 125–149). Hardwiring an indefinite number of biobanks and health registries will undoubtedly reveal why it must be so, thus enabling human kind's first faltering steps into the age of genomic medicine.

Discussion: "Things Fall Apart"

> Turning and turning in the widening gyre,
> The falcon cannot hear the falconer;
> Things fall apart; the centre cannot hold;
> Mere anarchy is loosed upon the world
>
> W.B. Yeats, "The Second Coming" (1920)

Hegel rightly noted that "the owl of Minerva spreads its wings only with the falling of the dusk." It is too early to understand well our own time; the attempts of this chapter are in this sense as partial, fallible, and risky as any other.

Still, we will insist that attempts at understanding must be made during daytime, while there still might be time for something more than simply being pushed along, meliorating the most conspicuous "unethical" features of industrial biology with new and refined instruments of informed consent or techniques of patient dialogue and public participation, and in general strengthening what was called the *unpolitics of ethics* (Felt and Wynne, 2007).[6] Understanding was always important. In the current situation, however, there are two additional reasons for its urgency. First, there is an unmatched strength and the velocity of the driving forces of the technoscientific complex of industrial biology combined with globalised capitalism. Second and possibly more importantly, at stake in the development is our own personhood and nature, being under direct attack of the colonializing forces. In this sense, things fall apart as we may be falling apart. We shall conclude with a few paragraphs clarifying this conclusion:

> In contrast to the economic structure, the political structure cannot be expanded indefinitely, because it is not based upon the productivity of man, which is, indeed, unlimited. Of all forms of government and organizations of people, the nation-state is least suited for unlimited growth because the genuine consent at its base cannot be stretched indefinitely, and is only rarely, and with difficulty won from conquered peoples, since such conscience comes only from the conviction of the conquering nation that it is imposing a superior law upon barbarians (Arendt 1951: 126).

Colonialism and more generally expansion come at a cost: Something is being consumed; somebody is being displaced or dominated. Among the legitimizing myths of colonialism – as seen from the perspective of the colonial power – two are particularly powerful. The first is that of *otherness*, the morally relevant differences that could justify any action towards non-Europeans whatsoever, as they were not white, good Christians or even human. More fundamentally, and with deeper phenomenological implications, is the myth of *pristine emptiness*: that nothing there was to be seen and accordingly no harm was done. How could they have failed to note the presence of the others, humans, and non-humans? While already Maurice Merleau-Ponty pointed to the inherent presence of intentionality in our perception, stronger explanations will be needed in the future to explain why the blind ravaging of non-European societies could be followed by a no less wild and blind destruction of wilderness, habitats for the non-humans, and a number of crucial natural resources.

[6] For this, see also chapter "In the Ruins of Babel: Should Biobank Regulations be Harmonized?".

Since the 1970s, the destruction of the ecosystem has become visible as a problem and a paradox. As for the annexation of life, the problem is still only hardly visible, forcing us into unusual comparisons and unorthodox rhetoric in texts such as the present one. What could possibly be at stake? What possibly could be wrong with curing the cancer of the loved ones, and generating economic growth at the same time?

The aim of the chapter, however, is not to identify and pass judgements on what is *wrong*. The aim is to provide an understanding of the dynamics and the driving forces of the development of industrial biology. The point is not that anything is *wrong*. The point is that the development is *out of control,* but governed as if the outcome could be predicted and known to be good.

The main conclusion of our analysis is that the driving force for expansion is so strong that it overrides other values and relegates them to secondary positions, where they may serve as *means*. Hence, modern biomedicine and industrial biology become means in the political concept of biocolonialism. We say this not as an exaggeration: The successful introduction of new health consumer products – made *for* and *of* the people – is easily recognized as one of the measures taken by the European Union and other big economies to uphold economic growth. The value of life does not hold a primary position in a regime of expansionism: Humanitarian disasters do give rise to rescue operations, but do not enter the political agenda with the same force as a crisis in the financial markets.

Ethicists might argue that this is exactly why there is bioethics to keep the industrial biology and its inscription in a global economy within the limits of human welfare and human dignity. The problem, however, is twofold, if our analysis is correct. First, there is no balance of power between economy and ethics: Economy is stronger, practically as well as axiologically, and the processes of creative destruction also mould the ethics into what serves the market. Second, what is being colonized now is not an overseas territory, but our own bodies. As argued earlier, this is not simply about discovering a geometrical space but about annexation of biological *information*. Information stands in a relation to technological purposes in the same way as there will be a relation between the formal and final causes of artefacts. What appears to be happening in industrial biology is accordingly a production of health products that as a side effect, and without much attention or concern, mangles the *normativity of human life,* changing what humans are and how a human life is lived. Eugenics falling into disrepute in the ruins of Berlin, the war against cancer was a convenient paradigmatic case to put forward to avoid the moral and political challenging of such a project. Since then, the debates of medicalization, selective reproductive technologies, and the issue of human enhancement have entered the scene much to the embarrassment of medical expertise and biotechnological policies. We fear this is only the beginning: that there is no definite set of "ethical issues" or even a "slippery slope" to be avoided by a principled decision. What is there is an expansionist creative destruction guided by the appearance of opportunities to internalize new externalities into the economy. The ramifications of these internalizations will only be fully evident "in the falling of the dusk". By then it might be too late.

We will not try to give an answer to what should be done in the present situation. Indeed, our strategy has been to take one step back rather than once again climbing the barricades. Michel Serres pointed to the need for what he called "the Natural Contract" (Serres 1995): If one believes that there once was a natural state of mute violence that developed into civilized rules of war, and that such rules are preferable because they provide guidance for how wars can come to an end, then perhaps what is needed is a contract to regulate our violence on Nature. Environmental management, the Declaration of Rio, and the emergence of climate politics are perhaps inventions that enter as parts of such a contract. In optimistic moments, we dream of a Human Contract for industrial biology, whereby one may arrive at rules of war and possibly even ceasefire, if not peace, between the historical subject of industrial biology and its object, the humans. Nature, however, lying there, might be mute; still, it can be trusted to represent itself. As for the humans, we rapidly seem to assimilate into the logic of industrial biology, accepting and readjusting ourselves to the economic demands of consumption of health technologies. The old conceptions of things and humans falling apart, how could the Human Contract become anything other than unconditional surrender?

Acknowledgments The financial support of the project "Mapping the language of Research Biobanking" (Contract No: 159864/S10) of the Norwegian Research Council and the Norwegian Institute of Public Health and the project "Research biobanking and the ethics of transparent communication (Contract No. 182269)" of the Norwegian Research Council are greatly acknowledged. Besides, this work has been supported by GeneBanC, an EU-FP6 supported STREP contract number 036751.

References

Althingi (2000) Act on Biobanks. Iceland No 110/2000

Annas G (1999) Waste and longing – The legal status of placental-blood banking. New England Journal of Medicine 340:1521–1524

Arendt H (1951) The Origins of Totalitarianism. A Harcourt Brace & Company, New York

Bergson H (1954) Creative Evolution. Mitchell A (trans.) Macmillian & Co Ltd, London

Bush V (1945) Science – The Endless Frontier. A Report to the President by Vannevar Bush, Director of the Office of Scientific Research and Development, July 1945 United States Government Printing Office, Washington. Last accessed 14 January 2009 at URL http://www.nsf.gov/od/lpa/nsf50/vbush1945.htm

Canguilhem G (1991) The Normal and the Pathological. Fawcett CR & Cohen RS (trans.) Zone Books, New York

Canguilhem G (1994a) Etudes d'histoire et de philosophie des sciences concernant les vivants et la vie. Librairie Philosophique J. Vrin, Paris

Canguilhem G (1994b) A Vital Rationalist Selected Writings from Georges Canguilhem. Goldhammer A (trans.). Zone Books, New York

European Commission (2004) Ethical, legal and social aspects of genetic testing: research, development and clinical applications, Brussels

Felt U, Wynne B (2007) Taking European Knowledge Society Seriously. European Commission. Luxembourg, Office for Official Publications of the European Communities

Feynman RP (1960) There's plenty of room at the bottom. Engineering and Science magazine XXIII: 22–36
Foucault M (1997) The Birth of the Clinic. Routledge, London
Foucault M (2003) Society Must Be Defended: Lectures at the Collège de France, 1975–1976. Bertani M & Fontana A (eds.). Marcey D (trans.). Picador, New York
Hobson JA (2005) Imperialism: A study. Cosimo Classics, New York
Illich I (1975) Medical Nemesis: The Exploration of Health. Marion Boyars, London
Lee G (1971) Rosa Luxemburg and the impact of imperialism. The Economic Journal 81: 847–862
Lisbon Strategy (2000). Last accessed 9 January 2009 at URLs http://europa.eu/scadplus/glossary/lisbon_strategy_en.htm and http://ec.europa.eu/growthandjobs/index_en.htm
Locke J (1884) Two Treatises on Civil Government. George Routledge and Sons, London
Kay LE (1992) The Molecular Vision of Life. Caltech, The Rockefeller Foundation, and the Rise of the New Biology. Oxford University Press, New York
Kimbrell A (1996) Biocolonization. In: Mander J & Goldsmith E (Eds.) The Case Against the Global Economy. Sierra Club Books, San Fransisco
Knight F (1921) Risk, Uncertainty and Profit. Hart, Schaffner and Marx; Houghton Mifflin Co, Boston
Riksdagen (2002) The Swedish Act on Biobanks. Sweden SF 2002:297
Robinson RE et al. (Eds.) (1991) Railway Imperialism (Contributions in Comparative Colonial Studies). Greenwood Press, San Francisco
Rose S (1997) Lifelines. Penguin Books, London
Sachidanandam R et al. (2001) A map of human genome sequence variation containing 1.42 million single nucleotide polymorphisms. Nature 409:928–933
Schumpeter JA (2005) Capitalism, Socialism and Democracy. Routledge, New York
Serres M (1995) The Natural Contract. University of Michigan Press, Ann Arbor
Stortinget (2003) Act Relating to Biobanks. Norway No 12 21.2 2003
United Nations' General Assembly (1960) Resolution 1514 (XV)
Venter JC et al. (2001) The sequence of the human genome. Science 16:1304–1351
Weinberg A (1999) The birth of Big Biology. Nature 401:738
White House (2000) President Clinton announces the completion of the first survey of the entire human genome (Press release). Last accessed 9 December at URL http://www.ornl.gov/sci/techresources/Human_Genome/project/clinton1.shtml
Zatloukal et al. (2007) Information on the proposal for European Research Infrastructure European Bio-Banking and Biomolecular Resources'. BBMRI (Biobanking and Biomolecular Resources Research Infrastructure): Last accessed 8 December 2008 at URL http://www.bbmri.eu/

In the Ruins of Babel: Should Biobank Regulations be Harmonized?

Jan Reinert Karlsen, Jan Helge Solbakk, and Roger Strand

Abstract The current thrust toward greater integration and harmonization of transnational biobank regulations is analyzed with analogical reference to the Tower of Babel account in The Book of Genesis. Raising the question whether biobank regulations should be harmonized, we distinguish between three types of research biobanks: local, regional, and international biobanks. The question is critically addressed in terms of the social and ethical robustness of such regulations. Our conclusion is that current proposals for integrating and harmonizing research biobanking in an increasingly international and industrial context are not socially and ethically robust. Rather, they amplify the inherent problems arising from this challenging task.

Introduction

The need for harmonizing existing regulatory frameworks and practices seems to be a widely shared perception among different stakeholders within the field of research biobanking (BBMRI 2008; PHOEBE 2008; OECD 2008). Even among ELSA researchers in the field, "harmonization" has become a highly estimated petword (Maschke 2005; Knoppers 2007; Fleming 2007; Chalmers 2008). For example, when the editors of this book in 2003 drafted the research proposal "Mapping the language of Research-Biobanks and Health Registries – From traditional biobanking to research biobanking," a proposal that 5 years later resulted in the publication of the present book, the need for developing a common regulatory language for research biobanking was stated as one of the main objectives:

J.R. Karlsen (✉)
Section for Medical Ethics, Faculty of Medicine, University of Oslo, Oslo, Norway
e-mail: j.r.karlsen@medisin.uio.no

J.H. Solbakk, et al. (eds.), *The Ethics of Research Biobanking*,
DOI: 10.1007/978-0-387-93872-1_21,

> As stated in the Report, Biobanks for health,[1] 'Europe already has a comparative advantage in the field of biobanks' compared to the US and Japan, but "Many European countries are lacking legislation and regulations to collection and storage of human biological material and its use for research purposes".[2] This claim is confirmed by a recent empirical survey on biobanking of human genetic material and data, conducted in six EU countries.[3] According to this study there is a need for education on biobanking, for guidelines on the quality of collections, for *harmonising* the framework of consent forms as well as for addressing the issues of secondary use of samples, of gene ownership and of benefit sharing. *Thus, the need for harmonising existing regulatory frameworks and mapping areas currently suffering from the lack of a transparent regulatory framework, represents a challenge that not only requires cooperation between different research disciplines, but also cooperation across national boundaries* (our emphasis).

The aim of this chapter is to critically assess whether – and eventually to what extent – harmonization can still be perceived as a justifiable aim. More specifically, what we aim at is an analogical reading of the famous narrative about harmonization in Genesis 11, i.e., the narrative of the Tower of Babel, and through this reading to draw attention to some of the unacknowledged pitfalls involved in the attempt at building a uniform regulatory language for research biobanking.

The Tower of Babel is not just an allegory about the loss of a unifying language. It is also a narrative about the negative and positive implications of such a loss. Finally, it is a narrative about the potential of *hubris* embedded in the striving for a world in which nobody falls apart, i.e., a world in which everybody is linked to each other through "one language and a common speech."

When the descendants of Noah moved eastward and settled in Shinar, they were in the privileged possession of a global community speaking the same language. For this reason, says the Biblical author, they were living unified lives void of any form of confusion. Besides, the common speech functioned as a safeguard against dispersion of lives; everyone became an inhabitant of the global city. Nobody got lost. What this part of the narrative unveils to the reader is the unifying power of common speech: it binds people together; it builds and preserves a community. When we compare this situation of harmony and unity with the language situation of research biobanking, it may seem reasonable to say that the different stakeholders here live in dispersed communities in need of a language that could breach the gaps and do away with the confusions between them. Consequently, the quest and striving for a common regulatory language also seems to be a morally and legally justifiable endeavor.

It is in view of this widely shared perception that the second part of the narrative about the Tower of Babel may contain some fruitful clues for further deliberation. That brings us first to an earlier chapter of the Genesis, "Trust, Distrust and Co-Production: The Relationship between Research Biobanks and Donors" by

---

[1] Report and recommendations from an EU workshop held at Voksenåsen Hotell, Oslo, 28–31 January 2003, pp. 5–6.

[2] Report and recommendations from an EU workshop held at Voksenåsen Hotell, Oslo, 28–31 January 2003, p. 19.

[3] Hirtzlin et al. 2003.

Ducournau and Strand, where the author gives an account of God's Covenant with Noah and his descendants:

> Then God blessed Noah and his sons, saying to them, "Be fruitful and increase in number and fill the earth" (Genesis 9, 1).

This was their prime mission and responsibility, according to the Biblical author. However, instead of pursuing these noble goals, they turned their attention elsewhere: to the building of a tower to empower and glorify themselves. At first sight, this may seem to represent a completely irrelevant part of the narrative as the end result communicated is Jahve's punishment in terms of linguistic confusion. A second reading, however, may raise the critical question whether the striving for harmonization in research biobanking has taken on a role that stand at risk of turning the attention away from the primary responsibilities of research biobanking toward society, including what we later in the chapter shall call the production of socially and ethically robust knowledge and technology.

The Thrust for Harmonization: Industrial Biology

Earlier, we have introduced harmonization of biobank regulation as a language issue: the introduction of a shared set of concepts and vocabulary to ensure communication and understanding, and avoid confusion. The focus on language can be confirmed by considering the relevant ethics and policy literature. First, there are requests for shared language and harmonized regulations, as well as attempts to provide them (Maschke 2005; Knoppers 2007; Fleming 2007; BBMRI 2008; Chalmers 2008; PHOEBE 2008; OECD 2008). Second, the international ethics and policy literature can in itself be seen as implying harmonization through its universalist discourse. This is in part due to the nature of the type of expertise enrolled in these endeavours. In particular, medical research ethics expertise has typically centred around arguments based in universal principles rather than casuistry, while, say, social anthropologists have often paid more attention to the particularities of culture, time, place and *locale.*

This is by no means unique to the ethical and regulatory aspects of modern science and technology. Indeed, in contrast to the nineteenth century and the early decades of the twentieth century, science in general and biomedicine in particular have become increasingly monolingual, with English as the predominant scientific language. Why have French, German, Italian, and other language skills more or less disappeared as a relevant and prestigious component in higher education and university life in modern science?

Not entering into the difficult field of historical explanation in the history of science, which is neither simple on its own account nor a field to be studied in isolation from the rest of society, we would like to point to the changes in the nature, scale, and organization that science in general and biobank-driven research in particular have undergone over the last 100 years or so. Science in general has *grown.* What once was a curiosity-driven vocation for an elite has become a vast sector of prime importance to national and international economy. Science is now an integral part and an important locomotive force in modern economies – “the endless frontier,” as characteristically put by Vannevar Bush (1945).

Likewise, biobank research has changed. Let us first recall that the creation of repositories of human biological material for diagnostic, therapeutic, and research purposes – *biobanks* – are by no means a new phenomenon. Pathologists have collected and systematized human biological material for diagnostic and research purposes long before high-throughput sequencing techniques became part of the scientific repertoire. Although the construction of health registries and blood banks, to some degree, has been a matter of national health politics, one should not forget that many of these biobanks were originally the result of individual scientific and clinical efforts, often assembled according to what resources and circumstance allowed. With the development of medical science and public health services, the number and types of biobanks increased over the twentieth century, as evident in the myriad of small, medium, and large biobanks for therapeutic purposes existing today. Blood transfusion, in vitro fertilization techniques and diagnostics based on individual tissue samples all rely on the capacity to collect, preserve, and store human biological material.

Hence, what is new in the last one or two decades is neither the existence nor the quantity of collections of human biological material. The novelty lies in the increased *value* ascribed to these collections as biochemistry and molecular biology have advanced to a stage that allows their direct utility and integration into clinical science. It is sufficient to remind the reader of the arrival of recombinant DNA technologies in the 1970s, the consolidation of molecular genetics in the 1980s and 1990s, and the emergence of functional genomics, proteomics, and quantitative system biology approaches in early twenty-first century. We may say with Rommetveit (2007) that the massive interest in the biobanks and health registries that has emerged is due to new prospects of constructing new and powerful diagnostic and therapeutic techno-medical objects, based on molecular genetics.

In our view, there are two noteworthy features of the regulatory context of research biobanking. Both are related to the value that biobanks now take on. Somewhat simplified we may say that they relate to the *creation* and *distribution* of that value. What makes the biobanks so valuable in the genomic era? In part, the answer is that they allow so-called “brute force” methodologies in which large numbers of genes or genetic activities can be correlated with large numbers of phenotypic features through multivariate analysis. In a sense, this represents an advancement of science that escapes the old trade-off or even dilemma in biomedicine, already understood by Bernard (1865/1957): On the one hand, epidemiological approaches may provide representativity through large numbers of real patients, but tend to be too crude to cast light upon causal mechanisms. On the other hand, physiology, biochemistry, and molecular biology provide causal understanding, but at the cost of studying simple and idealized (sub)systems that often were little more than artefacts in the biomedical sense. The combination of biobanks and health registries apparently allows for the combination of molecular rigor with epidemiological comprehensiveness. This is how they may create value: they may give rigorous, representative medical knowledge of the population. Furthermore, the comprehensiveness also means that unique biological features will be found: The needle in the haystack, be it a variant form of a gene, enzyme, or cell tissue, does no longer represent a near-impossible task as it can be found by automated routine techniques.

Two conditions must be met, however, for these values to be created. First, the research needs access to extensive, well-characterized, and high quality collections of human biological material and health information: Size matters (Khourty 2004). This is why harmonization is needed – data must be pooled to achieve the required statistical sensitivity and power, and this is a *scientific* necessity:

> The protocols that are the most important to harmonise, and consequently warrant the greatest attention, are those pertaining to data management, genotyping, phenotyping and ethical-legal constructs. These are at the core of PHOEBE (PHOEBE 2008, Executive summary 1: 1).

Similarly, it is at the core of the HuGENet (2008) and its handbook, and more generally in modern biomedicine, as evidenced in the Cochrane collaboration.

Second, the research is capital-intensive in the sense that the necessary technical rigor and excellence is (also) a matter of investment. But it is also *attractive* to investors because of the expectations of new powerful medical technologies. In part, we see attempts at exploiting this situation for the benefit of national competitiveness. In part, however, these matters are already inscribed in a globalized reality both in terms of research and capital.

Accordingly, there is a focus on international collaboration to facilitate exchange and combination of data and material through the construction of the adequate infrastructures. In our view, *the biobanks as well as the attempts at harmonization of their regulations* across borders may both be seen as parts of the venture of providing such an infrastructure.

We are now ready to sum up what is at stake in the current context of biobanking. First, the biobanks are perceived to be of potentially large value in terms of novel medical technologies. This value is manifested both in terms of *health benefits* and in *monetary benefits*. The production of value is capital-intensive, but there is both public and private investment willingness on a globalized level due to the expectations of profit. Furthermore, the production of value hinges on the *size* of the biobanks, and the possibility of their combination and hi-tech utilization across national borders. From the research perspective, both biobanks and their utilization, as well as the regulatory solutions to facilitate their construction and utilization, is a matter of building adequate infrastructures for the research activities, namely investigations of molecular epidemiology, functional genomics, and the like. This is how the uniformity of language is related to the tower of Babel: to build the tallest tower, the workers must speak the same language, know the same methods of construction, and use the same materials. It is not a question first and foremost of understanding or confusion; it is a question of coordination and efficiency of industry.

Weaker Ethics for Stronger Technoscience

The question *whether* biobank regulations should be harmonized is clearly dependent on how the question "*What* is to be harmonized?" is answered. Conversely, we cannot infer from any answer to the latter question a negative answer to the former, because it is equally clear that there are many ways of regulating biobanks, some of which could be ethically and politically sound to harmonize, while others would not. Moreover, there are numerous organizational, logistical, and technological challenges pertaining to the harmonization and coordination of the biobanks themselves. A minimum criterion for a harmonized regulatory framework, however, should be *social and ethical robustness*. The concept of socially robust knowledge and technology was introduced and developed further by Nowotny (Nowotny 1999; Nowotny et al. 2001). It is a commonplace that knowledge claims to be knowledge proper must be valid within their scientific context (under ideal, controlled, or typical circumstances). Nowotny, however, pointed to the need for validity in other, atypical, social contexts, where complexity avails, and uses, interests, values, and consequences may differ greatly from those of the laboratory. Social robustness was hence introduced as a concept for discussing the unforeseen consequences of science and technology in their long-term contexts of *application* and *implication*. The normativity of the concept is not easily seen to include the ethical dimension, and this chapter will employ an expanded concept of *social and ethical robustness*. By ethical robustness, we mean that ethical guidelines and institutions are legitimate in their own right, and that this legitimacy will hold when assessed from a variety of value perceptions and social contexts, including those of affected parties outside the power centers.

Until now, the regulations adopted for research biobanking have to a considerable extent been cast in a language originally developed to address ethical challenges in clinical trials, medical research, and transplantation medicine (WMA 2009). The language we have in mind is particularly the language of informed consent. Adopting informed consent to a new context has enabled policy makers and NGOs to frame the ethical challenges of research biobanking within existing ethical principles and institutions – with some modifications – *as if* the nature of these challenges is the same for research biobanking as for clinical research involving human beings (OECD 2008; Council of Europe 1997). Hence, we must ask *whether* currently adopted informed consent regulations and practices provide sufficient social and ethical robustness for (a) the whole spectrum of biobanks that potentially could be linked together, (b) the transactional interface between participant and biobank, and (c) the protection of participants' interests and basic rights.

Our concern in these matters is that there is an *asymmetry* between the social and ethical robustness of informed consent, and the increased techno-scientific *power* that research biobanking attains when traversing the spectrum from local biobanks (type 1), through regional biobanks (type 2), to the integration of local and regional biobanks in global research bioinfrastructures (type 3) (Table 1). If this is the case, then integrating biobanks into such infrastructures will stand at risk of boosting the techno-scientific power beyond the ethical foundation, institutional framework,

Table 1 The table shows a typology of research biobanks based on their anticipated techno scientific power

Type of biobank	Localization	Examples	Organizational model
Type 1	Local	Hospital, Ph.D. project	Laboratory
Type 2	Regional, national	HUNT, Biobank for Health	Industrial sized laboratory
Type 3	International, global	BBMRI, GenomeEUtwin	Hubs, nodes

and control mechanisms supported by adopted regulations. A socially and ethically fragile regulation, if harmonized, will dramatically increment the stakes for persons and society, potentially short-circuiting the democratic and moral legitimacy of research biobanking.

The primary aims of harmonization are to (1) boost the technoscientific power of research biobanking as well as their expected and (2) industrial and monetary benefits. The establishment of a panEuropean type 3 biobank, the Biobanking and Biomolecular Resources Research Infrastructure (BBMRI), is to consolidate these two objectives before becoming fully operational in 2009:

> The [BBMRI] will coordinate and harmonise biomolecular resource infrastructure with the four complementary biobank formats, each with its own strengths, to render their combination much more powerful than each resource type individually. This will provide the best opportunity to define and correlate healthy, pre-clinical and clinical profiles, and will strongly boost the integrated study of biological and genetic disease mechanisms, improve the delineation of clinical phenotypes and establish biomarker spectra for disease prognosis and therapy monitoring. [...] Combining broad access and flexible data integration with securely protected access, using genome-wide approaches and broadband automation technology, this globally unmatched Research Infrastructure will speed up development of personalised medicine and prevention covering the needs of basic science as well as biotech and pharmaceutical industries. (BBMRI: 4–5).

The construction cost of the BBMRI is estimated to 170 M€ (Péro 2006). Taking the size of the investment into account along with the basic scientific and industrial outcomes expected, the system "will strongly boost political and scientific momentum to harmonize ethical, legal and quality standards across Europe" (BBMRI: 5). In other words, the ethical, legal, and quality standards of the system will be supplied *afterwards*. Although this stratagem may enhance the technoscientific power of the research infrastructure itself, it may come with a yet unrecognized "prize," because ethical values will have to be weighted against a scientific and industrial investment that has already been allowed to become indispensable for biomedical progress and economic growth. The question on whether to harmonize has thus become a question on how to harmonize – a technical question in which the scales are unevenly set.

An ethics that is refrained from asking whether an action should be pursued at all or refrains itself from asking such questions, thus limiting itself to questions on *how* any given action could become ethically permissible, has abandoned the ethical in favor of a sanitized and servile form of the political. From an axiological perspective on research ethics, however, the adopted stratagem may in the long run prove to be a setback for the social and ethical robustness of the BBMRI and similar endeavors, because the core value of medical research ethics, viz. "the interests and welfare of the human being shall prevail over the sole interest of society or science." (CoE 1997: § 2), is brought in too late in the process. If the so-called "primacy of the individual" is politically and institutionally allowed to be bypassed by scientific and industrial interests, then existing research ethics will neither be legitimate in its own right nor with regard to its universalistic claims. Must the core ethical value of research ethics then be taken as a value among other values, one that has to be weighted against other interests?

The Biobank of Babel: Should Biobank Regulations be Harmonized?

In the preceding section, we argued that global research biobank infrastructures may accumulate vast technoscientific power, the uses, and consequences of which to some extent still remain to be seen. Furthermore, we argued that the current concepts, practices, and institutions of medical research ethics cannot match this power. On the contrary, their force is weakened the farther away they are transposed from their original context: the doctor–patient relationship and the clinical intervention or experiment.

The question remains, however, to what extent our criticism of the adequacy of regulatory harmonization of biobanks applying the standard repertoire of medical research ethics is contingent to this particular technical solution, or if our argument may be extended into a more general critique. To approach this question, we have to enter into a discussion of the intended functions of harmonization.

We have mentioned earlier how the spokes-men and women of harmonization point to the objective of scientific progress and the increase in public welfare it can be assumed to entail. Their reason for harmonization is that it would allow more comprehensive population-level research, more efficient research, and more streamlined use of the data resources of the biobanks. According to PHOEBE, for example, the aim of biobank harmoniozation is to:

> ...promote, both now and in the future, the effective interchange of valid information and samples between a number of studies or biobanks, accepting that there may be important differences between those studies. Prospective harmonization is aimed at modifying study design and conduct, ahead of time, in order to render subsequent data and sample pooling more efficient and more straightforward. Retrospective harmonization is aimed at optimizing the pooling of data, samples and phenotypes that have already been collected – between studies with inevitably heterogeneous designs (PHOEBE 2008).

In other words, the objective of harmonization is to accelerate the development of this field of technoscience. It also appears that it is this intention the regulatory resources of the standard repertoire of medical research ethics, if successfully harmonized, could serve: they would eliminate, or at least minimize, the so-called "ethical obstacles," the apparent nuisance and impediment produced by individual and rights-based ethical principles and practices. We use the negatively laden expression "ethical obstacles" to evoke the technological optimist, utilitarian perspective from which these "obstacles" have appeared close to irrational due to the lack of visible risk in the act of donation of human biological material. From such a perspective, the long-term uncertainties of the social impacts of scientific and technological development have never been seen as a decisive problem for decision-making: it has been assumed that science and technology on the whole improves human life.

We are not arguing against the objective of scientific progress. What constitutes progress, however, is not exclusively to be assessed from within the value systems of scientific disciplines producing it. With the increasing power of technoscience, and with the increasing diversity of use and dispersion of technologies into society, the question must also be asked whether knowledge and technologies are introduced and applied in socially robust ways; indeed, if given technologies and forms of knowledge *are* socially robust (Nowotny et al. 2001). From the above considerations, there follow two general points, regarding the regulatory harmonization of ethics with the intended function to facilitate scientific progress. First, harmonization of ethical principles and practices comes at a potential *cost*: it implies a reduction of the multitude of values and concerns represented in legal and ethical discourse. Hence, it is conceivable that harmonization results in institutions and practices that are less rather than more socially and ethically robust. Second, as has been argued in the preceding sections, there are some aspects of the scientific development in this field, i.e., the construction of global infrastructures carrying personal information and possibly producing vast biomedical predictive power of the individual, which by themselves indicate a risk of decreased social and ethical robustness. Let us add that all this happens within a global capitalist *Realpolitik* of the so-called knowledge-based economy in which the objectives of scientific progress and public health and welfare are inscribed in expansive and competitive innovation policies. Is it then conceivable that attempts at harmonization could have *social and ethical robustness* rather than facilitation and efficiency as their objectives?

The answer, when considering regulatory and institutional harmonization in other sectors, appears to be affirmative. An interesting example in this respect is the international nuclear inspectorate function of the IAEA, the International Atomic Energy Agency (see http://www.iaea.org). IAEA has a threefold mandate: to promote science and technology, to promote safety and security, and to promote safeguards and verification.

European political and intellectual history gives ample opportunity to corroborate the utility in differentiation of mandates and balance of powers. In the case of IAEA, it would intuitively appear dangerous if their mandate was collapsed into the one of promoting science and technology. Indeed, it is reassuring that the inspectorate function is one that identifies with, and is identified with, a globally recognized power to *slow down science* when called for in a given installation or country. The inspectorate function is not solely to facilitate efficiency; within the nuclear sector, efficiency, and progress is not the only value.

In the biomedical sector, ethics was constituted as the power to balance the intrinsic values of the scientific endeavour, laying the rights of the patient-subject into the other scalepan. This is clearly seen in the Declaration of Helsinki. In the preceding section, we argued that this scalepan becomes lighter as biomedical science is lifted out of the individual laboratory and into the global infrastructures. Ethics thus risks to be transformed into a means of dedifferentiation. It is this phenomenon that was called "*the unpolitics of ethics*" by a European expert group (Felt and Wynne 2007).

In our view, it becomes urgent to pose the question: what is then required? From our analysis of the *function* of harmonization, we conclude that harmonization of biobank regulations also needs to address the function of promoting social and ethical robustness. Such robustness is a matter of long-term societal consequences of the science and technology, and it should not be assessed from within one value system only, such as the currently hegemonic ideologies of innovation-based capitalism and biological reductionism, assuming that citizens will be happier in a wealthier society, and healthier when provided more services of biomedical technology. In other words, the value perception of biobanks as a great scientific, societal as well as biopolitical asset needs to be viewed in the light of other, existing value perceptions, such as the perception of biobanks as institutions of a *dangerous* kind or the perception of biobanks as instutions of a *mysterious,* yet undetermined and poorly understood, nature. The first – and dominating – kind of perception leads to focus on the need for resolving disagreements about the real *value* of biobanks, while resolving disagreements about the *medical*, the *scientific* as well as the *commercial* and *societal* value of biobanks focuses attention on ways of coping with the different possible forms of danger that may arise from research biobanking, such as infraction upon privacy, forms of discrimination, and secret policing. Finally, the perception of biobanks as a mysterious kind of institution frames the moral landscape in a third possible direction, i.e., into a discourse about the *institutional nature* of biobanks (What is this thing called "biobanks"?), the *ontological* nature of a society with such institutions as well as the bioethical and legal nature of biobanks. Consequently, the question of social and ethical robustness needs to be approached by asking "what-if"-questions (Ravetz 1997) along a number of value dimensions and scales of implication of the science and technology in society: What if the political situation drastically changes in the countries with access to global biobank infrastructures? What if state racism reappears in Europe? What if the current affluence in Western/Northern countries disappears? In the words of Nowotny et al., the contexts of application and implication must be continuously addressed (Nowotny et al. 2001).

Following the European expert group (Felt and Wynne 2007), it is not apparent that this function, which really requires a supra-national, if not harmonized, approach to gain any force, should be delegated to the academic expertise or political institutions and practices labeled "ethics" and "ethical." It may be that "ethics" as a professionalized phenomenon is too entrenched in a certain social order to be able to deal with the questions of *goodness* of which social and ethical robustness eventually is a matter. Let that remain an open question; in any case it appears sound to ensure methodological interdisciplinarity and transversality when dealing with such a difficult issue as social and ethical robustness. Another useful comparison in this respect, closer to biomedicine, is the political and academic debates on nanotechnology. A key concept in these debates is that of *governance* of science and the ideal of more open, transparent, and inclusive decision-making and social dialogue at an early stage in research (Toumey 2006; Kearnes and Wynne 2007; Rogers-Hayden and Pidgeon 2007).[4] In the nanotechnology sector, 2008 saw the first comprehensive governmental attempt at harmonization for the sake of social and ethical robustness: The Code of conduct for responsible nanosciences and nanotechnologies research, published in February 2008 as a recommendation from the European Commission to its member states (European Commission 2008). Among its principles, it states that nanoresearchers do not only have duties in terms of achieving scientific excellence, innovation, but also sustainability, ethics, precaution, and – more innovatively – toward *inclusiveness* and *meaning*:

> N&N research activities should be comprehensible to the public. They should respect fundamental rights and be conducted in the interest of the well-being of individuals and society in their design, implementation, dissemination and use (EC 2008, principle 4.1).

The recommendations are an instance of "soft law," giving the member states the choice of how to implement them. Of course, the method and impact of implementation are far from obvious; rather, the recommendations may be seen as an opportunity for the creative design of novel instruments of governance. Again, we see how the value of efficiency and internally measured progress is intended to be balanced by other values.

Conclusion: Back to Babel

We have tried to show what we hold to be important problems in current attempts to harmonize research biobank regulations. Indeed, the problems appear to reside not only in the design of these attempts, but also in their very objectives, arising in a general political culture that needs more contemplation in contemporary medical research ethics.

We have contrasted these developments with examples from two other sectors: the nuclear inspectorate function of IAEA and the recently published European Code of conduct for nanoscience and technology. We believe that ethicists should articulate, delineate, and clarify underlying controversies and conflicts between the values at stake in the issue, rather than always trying to give them an operationalizable resolution, which typically anyway will assume more or less invisible but substantive political assumptions. Let us, therefore, return to the tale of the tower of Babel.

No tree grows into Heaven. Culture, civilization, science and technology can be seen as attempts at getting further than what we may achieve in mere coexistence

[4] For a review of the academic literature, see Kjølberg and Wickson 2007.

with Nature. In this sense, humans do build towers to reach further. In what resided Jahve's discontent? We cannot know. We can note, however, the apparent surprise when He faces the fact that it is *this* they do – they build the one tower, they do not spread out into the country, all stakes are put into the one *monolith*, literally: "Behold, the people is one, and they all have one language, *and this they begin to do*" (Genesis 11, 6).

The words of Jahve do not carry legitimate weight in contemporary bioethics and biopolitics. At least not *these* words. We did not introduce the tale of Babel to draw upon its possible legitimacy. Rather than the legitimacy, we have been interested in the reflection it sparks off, believing that the value of reflection, hesitation, and broad dialogue is what needs to be incorporated in the discourse on harmonization of biobank regulation. We hold that this is vital also for ethics. Stephen Toulmin once explained how medicine once saved the life of ethics. (Toulmin 1986). It now seems that the patient emancipates himself from his doctor; be it for his happy health or his natural death, lending space for other institutions and practices for the socially and ethically robust governance of biomedicine.

Acknowledgments The financial support of the project 'Mapping the language of Research Biobanking' (Contract No.: 159864/S10) of the Norwegian Research Council and the Norwegian Institute of Public Health and the project 'Research biobanking and the ethics of transparent communication (Contract No. 182269)' of the Norwegian Research Council are greatly acknowledged. Besides, this work has been supported by GeneBanC, an EU -FP6 supported STREP contract number 036751.

References

BBMRI (Biobanking and Biomolecular Resources Research Infrastructure) (2008). Last accessed 27 November 2008 at URL http://www.bbmri.eu/

An Introduction to the Study of Experimental Medicine. Dover Publications, New York

Bernard C (1865/1957) An Introduction to the Study of Experimental Medicine. Dover Publications, New York

Bush V (1945) Science – The Endless Frontier. A Report to the President by Vannevar Bush, Director of the Office of Scientific Research and Development, July 1945, United States Government Printing Office, Washington. Last accessed 28 November 2008 at URL http://www.nsf.gov/od/lpa/nsf50/vbush1945.htm

Chalmers DRC (2008) International Co-Operation Between Biobanks: The Case for Harmonisation of Guidelines and Governance. In: Human Biotechnology and Public Trust: Trends, Perceptions and Regulation. Centre for Law and Genetics, Hobartt, pp. 237–246

Council of Europe (1997) Convention for the Protection of Human Rights and Dignity of the Human Being with regard to the Application of Biology and Medicine: Convention on Human Rights and Biomedicine

European Commission (2008) Code of Conduct Commision recommendation of 07/02/2008 on a code of conduct for responsible nanosciences and nanotechnologies research, C(2008) 424. Brussels. Last accessed 28 November 2008 at URL http://ec.europa.eu/nanotechnology/pdf/nanocode-rec_pe0894c_en.pdf

Felt U, Wynne B (2007) Taking European Knowledge Society Seriously. European Commission. Luxembourg, Office for Official Publications of the European Communities

Fleming JM (2007) Biobanks: Professional, Donor and Public Perceptions of Tissue Banks and the Ethical and Legal Challenges of Consent, Linkage and the Disclosure of Research Results. PhD Thesis, Institute for Molecular Bioscience, The University of Queensland, Australia

HuGENet (2008) The HuGENet Review Handbook. Last accessed 28 November 2008 at URL http://www.hugenet.org.uk/resources/handbook.php

Kearnes M, Wynne B (2007) On nanotechnology and ambivalence: The politics of enthusiasm. NanoEthics 1:131–142

Khourty MJ (2004) The case for a global human genome epidemiology initiative. Nature Genetics 36:1027–1028

Kjølberg K, Wickson F (2007) Social and ethical interactions with nano: Mapping the early literature. Nanoethics 1:89–104

Knoppers BM (2007) Biobanking: International Norms. The Journal of Law, Medicine & Ethics 33:7–14

Maschke KJ (2005) Navigating an ethical patchwork – Human gene banks. Nature Biotechnology 23:539–545

PHOEBE (Promoting Harmonisation of Epidemiological Biobanks in Europe) (2008) Epidemiology and Biostatistics. Last accessed 28 November 2008 at URL http://www.phoebeeu.org/eway/default.aspx?pid=271&trg=Main_5685&Main_5685=5709:0:10,1847:1:0:0:::0:0

Nowotny H (1999) The need for socially robust knowledge. TA-Datenbank-Nachrichten 8:12–16

Nowotny H et al. (2001) Re-thinking science: knowledge and the public in an age of uncertainty. Polity Press, Cambridge

Ravetz JR (1997) The Science of 'What-If?' Futures 29:533–539

Rogers-Hayden T, Pidgeon N (2007) Moving engagement "upstream"? Nanotechnologies and the Royal Society and Royal Academy of Engineering's inquiry. Public understand. Science 16:345–364

Rommetveit K (2007) Biotechnology: Action and choice in second modernity. PhD dissertation. University of Bergen, Bergen

OECD (2008) Draft Guidelines for Human Biobanks and Genetic Research Databases. Last accessed at 28 November 2008 at URL http://www.oecd.org/document/12/0,3343,en_2649_34537_40302092_1_1_1_1,00.html

Toulmin S (1986) How Medicine Saved the Life Of Ethics. In: DeMarco JP, Fox RM (Eds.) New Direction in Ethics: The Challenge of Applied Ethics. Routledge & Kegan Paul, New York, pp. 265–281

Toumey C (2006) National discourses in democratizing nanotechnology. Quaderni 61:81–101

WMA (2006) Declaration of Helsinki. Ethical Principles for Medical Research Involving Human Subjects

Index

Lightning Source UK Ltd.
Milton Keynes UK
30 April 2010

153587UK00007B/14/P